Quantum Information and Computing

QP–PQ: Quantum Probability and White Noise Analysis

QP–PQ: Quantum Probability and White Noise Analysis

Vol. 19: Quantum Information and Computing
eds. L. Accardi, M. Ohya and N. Watanabe

Vol. 18: Quantum Probability and Infinite-Dimensional Analysis From Foundations to Applications
eds. M. Schürmann and U. Franz

Vol. 17: Fundamental Aspects of Quantum Physics
eds. *L. Accardi and S. Tasaki*

Vol. 16: Non-Commutativity, Infinite-Dimensionality, and Probability at the Crossroads
eds. *N. Obata, T. Matsui and A. Hora*

Vol. 15: Quantum Probability and Infinite-Dimensional Analysis
ed. *W. Freudenberg*

Vol. 14: Quantum Interacting Particle Systems
eds. *L. Accardi and F. Fagnola*

Vol. 13: Foundations of Probability and Physics
ed. *A. Khrennikov*

QP–PQ

Vol. 10: Quantum Probability Communications
eds. *R. L. Hudson and J. M. Lindsay*

Vol. 9: Quantum Probability and Related Topics
ed. *L. Accardi*

Vol. 8: Quantum Probability and Related Topics
ed. *L. Accardi*

Vol. 7: Quantum Probability and Related Topics
ed. *L. Accardi*

Vol. 6: Quantum Probability and Related Topics
ed. *L. Accardi*

QP–PQ
Quantum Probability and White Noise Analysis
Volume XIX

Quantum Information and Computing

Editors

L. Accardi
Università di Roma, Italy

M. Ohya and **N. Watanabe**
Tokyo University of Science, Japan

World Scientific

NEW JERSEY • LONDON • SINGAPORE • BEIJING • SHANGHAI • HONG KONG • TAIPEI • CHENNAI

Published by

World Scientific Publishing Co. Pte. Ltd.

5 Toh Tuck Link, Singapore 596224

USA office: 27 Warren Street, Suite 401-402, Hackensack, NJ 07601

UK office: 57 Shelton Street, Covent Garden, London WC2H 9HE

British Library Cataloguing-in-Publication Data
A catalogue record for this book is available from the British Library.

QP–PQ: Quantum Probability and White Noise Analysis – Vol. 19
QUANTUM INFORMATION AND COMPUTING

ISBN 981-256-614-7

Printed in Singapore by B & JO Enterprise

Preface

This volume constitutes the proceedings of the international conference on "Quantum Information 2003" held in Tokyo University of Science/Tokyo (Japan) from 1 to 3 November 2003, in International Institute for Advanced Studies/Kyoto (Japan) from 5 to 7 November 2003, in University of Waseda/Tokyo (Japan) from 29 to 31 October 2003, in Meijo University/Nagoya (Japan) from 8 to 9 November 2003. These meetings were organized by the Embassy of Italy in Japan in collaboration with Tokyo University of Science (Tokyo), University of Roma II, International Institute for Advanced Studies, Meijo University (Nagoya) and University of Waseda (Tokyo). We very much appreciate the financial support of the Embassy of Italy in Japan, the Frontier Research Center for Computational Sciences of the Tokyo University of Science, International Institute for Advanced Studies and the universities of Waseda and Meijo.

A study of quantum information has been developed extensively in both sides of the foundation and application in the various fields, but it becomes more and more important but difficult to catch the whole aspects of these fields. In order to understand the latest research of quantum information and to find a future orientation of quantum information, the International Conference entited "Quantum information: mathematical, physical engineering and industrial aspects" was organized by inviting researcheres from mainly Europe and Japan who are actively working on each side of the foundation and application of quantum information. The main purpose of this volume is to emphasize the multidisciplinary aspects in this new and very active field in which concrete technological realizations require the combined efforts of experimental and theoretical physicists, mathematicians and engineers.

Luigi Accardi
Masanori Ohya
Noboru Watanabe

Contents

The Embassy of Italy
in Japan

Frontier Research Center
for Computational Science

International Institute
for Advanced Studies

COHERENT QUANTUM CONTROL OF Λ-ATOMS THROUGH THE STOCHASTIC LIMIT

L. ACCARDI

Centro Vito Volterra, Università di Roma "Tor Vergata", 00133, Via Columbia 2, Roma, Italy
E-mail: accardi@volterra.mat.uniroma2.it

S. V. KOZYREV

Steklov Mathematical Institute, Russian Academy of Sciences, Gubkin St. 8, 119991, Moscow, Russia
E-mail: kozyrev@mi.ras.ru

A. N. PECHEN

Steklov Mathematical Institute, Russian Academy of Sciences, Gubkin St. 8, 119991, Moscow, Russia
E-mail: pechen@mi.ras.ru

We investigate, using the stochastic limit method, the coherent quantum control of a 3-level atom in Λ-configuration interacting with two laser fields. We prove that, in the generic situation, this interaction entangles the two lower energy levels of the atom into a single qubit, i.e. it drives at an exponentially fast rate the atom to a stationary state which is a coherent superposition of the two lower levels. By applying to the atom two laser fields with appropriately chosen intensities, one can create, in principle, any superposition of the two levels. Thus *relaxation* is not necessarily synonymous of *decoherence*.

1. Introduction

Preparation of atoms and molecules in a predefined state plays an important role in modern atomic and molecular physics, in particular in atom optics and quantum information. The difficulty of the problem consists in the fact that, in order to control a system, one has to interact with it, but interaction introduces dissipative effects and hence, at least generically, decoherence. In this note we prove, in specific but important and physically realizable example, that in some cases dissipation can generate coherence.

One of the ways to drive the system, atom or molecule, to the desired state is to exploit its interaction with laser pulses, i.e., to use the coherent laser control or laser-induced population transfer [1,2]. In this approach monochromatized and near-resonant with the atomic Bohr frequencies radiation fields are used to force the system to the final state. The laser coherent control techniques can be used in applications to laser cooling based on coherent population trapping [3], quantum computing with trapped ions [5,4], etc.

The stochastic limit method [6] was applied in [7] to study the phenomenon of coherent population trapping in an atom in lambda–configuration with a doubly degenerate ground state. It was found that the action of the field drives the atom to a 1–parameter family of stationary states so that the choice of one state in this family is uniquely determined by the initial state of the atom and by the initial state of the field.

In [7] the initial state of the field was chosen to be an arbitrary mean zero, gauge invariant Gaussian state, e.g. the Fock vacuum or an equilibrium state at any temperature. On the other hand, in the usual experiments on coherent population trapping, the atom is driven by two laser beams, i.e. coherent states, resonant with the atomic frequencies and, up to now, there is no evidence of this 1–parameter family of invariant states [1–9].

In the present paper we prove that, for a three-level non-degenerate atom, the field drives the atom to a pure state, which is a superposition of the two lowest energy levels and that, applying to the atom two laser fields with appropriately chosen intensities one obtains a single superposition. In the case of a degenerate atom we show that, if one starts from a coherent state, then the family of invariant states is destroyed and the field drives the atom to the dark state.

The results of the present paper, combined with those of [7] also suggest that the emergence of the dark state could be experimentally realized not only by tuning two lasers, but also in a generic equilibrium state, by preparing the state of the atom in the domain of attraction of the dark state.

The present result is obtained as a consequence of a general property of the stochastic limit, which we find for the first time in the paper, according to which the effect of replacing a mean zero, gauge invariant state of the field by a coherent one, amounts to the addition of a hamiltonian term to the master equation while the dissipative part of the generator remains the same [cf. (12)–(14)].

This general property is proved in sections 2–4 of the present paper for an arbitrary atom. In sections 5–8 we specialize to the case of 2- and 3-

level atoms, find for these cases the explicit form of the stationary atomic states [(17) and (23) for a two-level, (21) and (22) for a three-level atom], and prove the results mentioned in the present introduction. In Sect. 8 a two level atom with 3-times degenerate ground state is considered. In this case there is not a unique but a family of stationary states. The master equation we find in the 3-level case coincides with the optical Bloch equation considered by Arimondo in [1] in the absence of scattering [Eq. (2.7)] with the only difference that we obtain the equation and the explicit formula for the coefficients Ω_j [cf. equations (9) and (20) below] without assuming that the field is classical. Instead, we start from the microscopic quantum dynamics of the total atom+radiation system.

The approach of the present paper to study the coherent quantum control or laser induced population transfer is based on the derivation, using the stochastic limit method, of a quantum white noise equation approximating the total dynamics in the weak coupling regime (Eq. (6), for a rigorous treatment see [6]). The programme to exploit exponentially fast decoherence as a control tool to drive an atom into a pre-assigned state was formulated in [10,11], where the control parameter is the interaction Hamiltonian. In the approach of the present paper the control parameter is the state of the radiation, which can be easily controlled in experiments and, instead of decoherence, it uses relaxation, due to dissipative dynamics, to create quantum coherence between the atomic states.

2. An atom in a laser field

The dynamics of an atom interacting with radiation is determined on the microscopic level by the interaction Hamiltonian and by the reference state of radiation.

The Hamiltonian describing an atom interacting with radiation is given by the sum of the free and interaction terms

$$H_\lambda = H_{\text{free}} + H_{\text{int}} = H_{\text{A}} \otimes 1 + 1 \otimes H_{\text{R}} + \lambda H_{\text{int}},$$

where H_{A} and H_{R} are free Hamiltonians of the atom and radiation, H_{int} interaction Hamiltonian, and λ is the coupling constant. The free Hamiltonian of the atom has discrete spectrum

$$H_{\text{A}} = \sum_n \varepsilon_n P_n,$$

where ε_n is an eigenvalue and P_n is the corresponding projection. The free

Hamiltonian of the radiation is

$$H_{\mathrm{R}} = \int d\mathbf{k}\omega(\mathbf{k})a^+(\mathbf{k})a(\mathbf{k}), \qquad \omega(\mathbf{k}) = |\mathbf{k}|,$$

where the creation and annihilation operators $a^+(\mathbf{k})$, $a(\mathbf{k})$ describe creation and annihilation of photons with momentum $\mathbf{k}$. The interaction Hamiltonian has the form

$$H_{\mathrm{int}} = i(D \otimes a^+(g) - D^+ \otimes a(g)),$$

where the operators D and D^+ describe the transitions between the atomic levels, $a^+(g) = \int d\mathbf{k}g(\mathbf{k})a^+(\mathbf{k})$ is the smeared creation operator, and the formfactor $g(\mathbf{k})$ describes the coupling of the atom with the $\mathbf{k}$ mode of the field.

The state of the radiation, which corresponds to a long time laser pulse of frequency ω and intensity f is a coherent state, which is determined by the coherent vector

$$\tilde{\Psi}_\lambda = W\left(\lambda \int_{S/\lambda^2}^{T/\lambda^2} S_t e^{-it\omega} f dt\right)\Phi_0,$$

where Φ_0 is the vacuum, $W(\cdot)$ the Weyl operator, which is defined as $W(f) = \exp[i(a(f) + a^+(f))]$, $S_t = e^{it\omega(\mathbf{k})}$ the one photon free evolution, and $[S/\lambda^2, T/\lambda^2]$ the time interval in which the laser pulse is active. This vector $\tilde{\Psi}_\lambda$ is an eigenvector of the annihilation operator:

$$a(g)\tilde{\Psi}_\lambda = c_\lambda \tilde{\Psi}_\lambda.$$

The eigenvalue

$$c_\lambda = \lambda \int_{S/\lambda^2}^{T/\lambda^2} dt \int d\mathbf{k} g^*(\mathbf{k}) f(\mathbf{k}) e^{it(\omega(\mathbf{k})-\omega)}$$

is determined by the intensity of the laser f and the form factor g. The pulse starts at time S/λ^2 and ends at time T/λ^2. Since the coupling constant λ is small the duration of the pulse, being of order λ^{-2}, is large. The function $f(\mathbf{k})$ describes the amplitude of the input field with momentum $\mathbf{k}$.

The state, which corresponds to several laser pulses of frequencies $\omega_1, \ldots, \omega_n$, intensities $f_1, \ldots, f_n$, and acting during time intervals $[S_1, T_1], \ldots, [S_n, T_n]$, is determined by the coherent vector

$$\Psi_\lambda = W\left(\lambda \sum_{l=1}^{n} \int_{S_l/\lambda^2}^{T_l/\lambda^2} S_t e^{-it\omega_l} f_l dt\right)\Phi_0. \tag{1}$$

The initial state of the total system is supposed to be factorized

$$\omega = \omega_{\hat{\rho}} \otimes \omega_\lambda = \mathrm{Tr}\,(\hat{\rho}\,\cdot) \otimes \omega_\lambda,$$

where $\omega_{\hat{\rho}}$ is the state of the atom determined by a density matrix $\hat{\rho}$ and ω_λ is the state of the radiation determined by the coherent vector Ψ_λ: $\omega_\lambda = \langle \Psi_\lambda, \cdot \Psi_\lambda \rangle$.

The dynamics of the total system is determined by the evolution operator which in interaction picture is defined as

$$U_\lambda(t) = e^{itH_{\mathrm{free}}} e^{-itH_\lambda},$$

where the index denotes the dependence on the coupling constant. The evolution operator determines the Heisenberg evolution of any observable of the atom, say X, as

$$X_\lambda(t) = U_\lambda^+(t) X U_\lambda(t).$$

The reduced dynamics of the observable is obtained by taking average over the state of the radiation

$$\langle X_\lambda(t) \rangle := \langle \Psi_\lambda, U_\lambda^+(t) X U_\lambda(t) \Psi_\lambda \rangle$$

and, by duality, it can be equivalently described in terms of the reduced density matrix $\hat{\rho}_\lambda(t)$, defined by the following equality:

$$\begin{aligned} &\mathrm{Tr}_{\mathrm{A}}(\hat{\rho}_\lambda(t) X) = \mathrm{Tr}_{\mathrm{A}}(\hat{\rho} \langle X_\lambda(t) \rangle) \\ &= \mathrm{Tr}_{\mathrm{A}}\Big\{ \overline{\mathrm{Tr}}_{\mathrm{R}} \Big[U_\lambda(t) \Big(\hat{\rho} \otimes |\Psi_\lambda\rangle\langle\Psi_\lambda| \Big) U_\lambda^+(t) \Big] X \Big\}, \end{aligned}$$

where Tr_{A} denotes the trace over the atomic degrees of freedom and $\overline{\mathrm{Tr}}_{\mathrm{R}}$ the partial trace over the field degrees of freedom.

The first step of the present paper is to give a non fenomenological derivation, from the first principles, of the quantum master equation for the reduced density matrix of the atom using the stochastic limit method. This is done in the following two sections.

3. The stochastic limit

It is impossible to find, for a general coupling D, the explicit form of the dynamics for the reduced density matrix for realistic models, while the weakness of the interaction suggests some approximations. For small time the dynamics can be effectively studied by using the perturbation series. However, the n-th term of the series behaves like $(\lambda^2 t)^n$ and the approximation with the lowest order terms of the series becomes invalid for time which is large enough.

Hence the study of the long time dynamics requires another approach. Such an approach to study the dynamics on the long time scale ($\sim \lambda^{-2}$) is the stochastic limit method [6]. In the stochastic limit one considers the long time dynamics (on the time scale of order λ^{-2}) for a system with weak interaction. Mathematically this means that one takes the limit as the coupling constant goes to zero, $\lambda \to 0$, time goes to infinity, $t \to \infty$, but in such a way that the quantity $\lambda^2 t$ remains fixed. In this limit the dynamics of the total system is given by the solution of a quantum white noise or stochastic differential equation (which is a unitary adapted process). Since the interaction of an atom with radiation is weak one can apply the stochastic limit procedure to study the long time dynamics of this system.

On the long time scale the behavior of the exact reduced density matrix is approximated by the limiting density matrix

$$\hat{\rho}(t) = \lim_{\lambda \to 0} \hat{\rho}_\lambda(t/\lambda^2)$$

so that $\rho_\lambda(t) \simeq \rho(\lambda^2 t)$. The quantum master equation for the limiting density matrix in the case when the radiation is in a Gibbs state was derived and discussed in many papers (see [6] and references therein). The purpose of the present paper is to derive the quantum master equation for the limiting density matrix $\hat{\rho}(t)$ in the case the radiation is in a coherent state and to study this equation for particular, but important cases of two and three-level atoms.

The evolution operator after the time rescaling satisfies the equation

$$\frac{dU_\lambda(t/\lambda^2)}{dt} = \sum_{\omega \in B} \left(D_\omega \otimes a^+_{\lambda,\omega}(t) - D^+_\omega \otimes a_{\lambda,\omega}(t) \right) U_\lambda(t/\lambda^2),$$

where B is the set of all Bohr frequencies of the atom (spectrum of the free atomic Liouvillian $i[H_\mathrm{A}, \cdot]$, i.e. the set of $\omega = \varepsilon_n - \varepsilon_m$, where $\varepsilon_n, \varepsilon_m$ are eigenvalues of the free atomic Hamiltonian), for each $\omega \in B$

$$a_{\lambda,\omega}(t) = \frac{1}{\lambda} \int d\mathbf{k} g^*(\mathbf{k}) e^{-i(\omega(\mathbf{k})-\omega)t/\lambda^2} a(\mathbf{k}) \tag{2}$$

is the rescaled time dependent annihilation operator, and the operator

$$D_\omega = \sum_{n,m:\, \varepsilon_n - \varepsilon_m = \omega} P_m D P_n \tag{3}$$

describes transitions between atom levels with energies ε_n and ε_m with the energy difference $\varepsilon_n - \varepsilon_m = \omega$.

In the stochastic limit one first proves that the time rescaled creation and annihilation operators converge, as $\lambda \to 0$, to a quantum white noise (for details of the stochastic limit procedure see [6]):

$$\lim_{\lambda\to 0} a_{\lambda,\omega}(t) = b_\omega(t),$$

where the quantum white noise operators $b_\omega(t)$ are δ-correlated in time and satisfy the commutation relations

$$[b_\omega(t), b^+_{\omega'}(t')] = \delta(t'-t)\delta_{\omega,\omega'} 2\mathrm{Re}\,\gamma_\omega. \tag{4}$$

The complex number

$$\gamma_\omega = \int d\mathbf{k} \frac{|g(\mathbf{k})|^2}{i(\omega(\mathbf{k}) - \omega - i0)} \tag{5}$$

is the generalized susceptibility and its real part

$$\mathrm{Re}\,\gamma_\omega = \pi \int d\mathbf{k} |g(\mathbf{k})|^2 \delta(\omega(\mathbf{k}) - \omega)$$

gives the decay rate of the ω-transition (cf. [6], sect. 4.20). The Kronecker δ-symbol $\delta_{\omega,\omega'}$ in (4) indicates the mutual independence of the white noise operators for different Bohr frequencies.

This allows us to derive the equation for the limiting evolution operator

$$U_t := \lim_{\lambda\to 0} U_\lambda(t/\lambda^2).$$

That is the white noise Schrödinger equation

$$\frac{dU_t}{dt} = -iV(t)U_t \tag{6}$$

with the white noise Hamiltonian

$$V(t) = i \sum_{\omega\in B} \Big(D_\omega \otimes b^+_\omega(t) - D^+_\omega \otimes b_\omega(t) \Big).$$

Let X be any observable of the atom and $X_t = U^+_t X U_t \equiv \lim_{\lambda\to 0} X_\lambda(t/\lambda^2)$ its limiting time evolution. Using the stochastic golden rule of the stochastic limit (cf. [6], sect. 5.9), one gets the quantum Langevin equation for X_t, that is a normally ordered white noise differential equation:

$$\frac{dX_t}{dt} = U^+_t \Theta(X) U_t$$
$$+ \sum_\omega \Big(b^+_\omega(t) U^+_t L^+_\omega(X) U_t + U^+_t L_\omega(X) U_t b_\omega(t) \Big) \tag{7}$$

(such an equation is equivalent to a certain quantum stochastic differential equation). Here

$$\Theta(X) = \sum_{\omega}\left(2\mathrm{Re}\,\gamma_\omega D_\omega^+ X D_\omega - \gamma_\omega X D_\omega^+ D_\omega - \bar{\gamma}_\omega D_\omega^+ D_\omega X\right)$$

is the generator of a quantum Markov semigroup and, in the notation (3),

$$L_\omega^+(X) = [X, D_\omega], \qquad L_\omega(X) = [D_\omega^+, X].$$

The normally ordered form of the quantum Langevin equation, when the annihilation white noise operators are put on the right of the evolution operator and creation operators are put on the left, is very convenient for study the reduced dynamics in the case the radiation is in a coherent state. In order to get the reduced dynamics one just has to average equation (7) over the state of radiation.

The coherent state Ψ_λ of the radiation is approximated by the limiting coherent vector

$$\Psi = \lim_{\lambda\to 0} \Psi_\lambda.$$

This vector is an eigenvector of the quantum white noise annihilation operator:

$$b_{\omega_l}(t)\Psi = \chi_{[S_l,T_l]}(t) c_{\omega_l}\Psi. \tag{8}$$

Here $\chi_{[S_l,T_l]}(t)$ is the characteristic function of the interval $[S_l, T_l]$, determining the duration of the pulse. The complex number c_{ω_l} depends on the state of the radiation as follows:

$$c_{\omega_l} = 2\pi \int d\mathbf{k} g^*(\mathbf{k}) f_l(\mathbf{k}) \delta(\omega(\mathbf{k}) - \omega_l) \tag{9}$$

and determines the Rabi frequencies of the field.

In fact, the coherent vector Ψ_λ is an eigenvector of the time rescaled annihilation operator $a_{\lambda,\omega_l}(t)$ (defined by (2)) with eigenvalue

$$\sum_{m=1}^{n} \int_{S_m}^{T_m} d\tau \frac{1}{\lambda^2} \int d\mathbf{k} g^*(\mathbf{k}) f_l(\mathbf{k}) e^{i(\tau - t)(\omega(\mathbf{k}) - \omega_l)/\lambda^2}$$
$$\times e^{i\tau(\omega_l - \omega_m)/\lambda^2}.$$

From that, using the limit (see [6] for details)

$$\lim_{\lambda\to 0} \frac{1}{\lambda^2} e^{i(\tau - t)(\omega(\mathbf{k}) - \omega_l)/\lambda^2} e^{i\tau(\omega_l - \omega_m)/\lambda^2}$$
$$= 2\pi\delta(t - \tau)\delta_{\omega_l,\omega_m}\delta(\omega(\mathbf{k}) - \omega_l)$$

one obtains the limiting eigenvalue

$$2\pi \sum_{m=1}^{n} \int_{S_m}^{T_m} d\tau \delta(t-\tau) \int d\mathbf{k} g^*(\mathbf{k}) f_m(\mathbf{k}) \delta_{\omega_l,\omega_m} = \chi_{[S_l,T_l]}(t) c_{\omega_l}$$

We will consider the case $S_l = S$, $T_l = T$ for all l, which means that all lasers are applied during the same time interval $[S,T]$.

Let X be an observable of the atom and denote by

$$\langle X \rangle_t = \langle \Psi, X_t \Psi \rangle$$

the average, over the limiting state of the radiation, of its time evolution. Taking the average of both sides of the quantum Langevin equation (7) and using property (8) one gets the following equation for the averaged observable:

$$\frac{d\langle X \rangle_t}{dt} = \langle \Theta(X) \rangle_t + i\chi_{[S,T]}(t) \sum_{\omega} \langle c_\omega L_\omega(X) - c_\omega^* L_\omega^+(X) \rangle_t. \tag{10}$$

This master equation for the case with presence of external field is one of the main results of the present paper. In the next section we will rewrite it in the equivalent form as an equation for the reduced density matrix.

4. The quantum master equation

The dynamics of the atom can be described either in terms of observables, or equivalently, in terms of density matrices. The equation for the reduced density matrix can be obtained using the relation

$$\mathrm{Tr}(\hat\rho(t) X) = \mathrm{Tr}(\hat\rho(0) \langle X \rangle_t)$$

and equation (10). It is a quantum master equation of the form:

$$\frac{d\hat\rho(t)}{dt} = \mathcal{L}_t(\hat\rho(t)). \tag{11}$$

The time dependent generator of this equation is given by the sum of its dissipative and Hamiltonian parts:

$$\mathcal{L}_t(\hat\rho) = \mathcal{L}_{\text{diss}}(\hat\rho) - i[H_{\text{eff}}(t), \hat\rho]. \tag{12}$$

As already stated in the Introduction, the dissipative part is the same as in the case the reservoir in the vacuum state:

$$\mathcal{L}_{\text{diss}}(\hat\rho) = \sum_{\omega} \mathrm{Re}\,\gamma_\omega \Big(2D_\omega \hat\rho D_\omega^+ - \hat\rho D_\omega^+ D_\omega - D_\omega^+ D_\omega \hat\rho \Big). \tag{13}$$

The Hamiltonian part is determined by the effective time dependent Hamiltonian

$$H_{\text{eff}}(t) = \sum_{\omega} \Big(\text{Im}\gamma_\omega D_\omega^+ D_\omega + \chi_{[S,T]}(t)(c_\omega^* D_\omega + c_\omega D_\omega^+) \Big). \tag{14}$$

The first term in the brackets is the standard term which appears in the stochastic limit when the reservoir is in the vacuum state. It commutes with the free atomic Hamiltonian H_{A}. The other terms are new and their presence leads to the important consequence that the effective Hamiltonian at time $t \in [S,T]$ does not commute with the free atom Hamiltonian

$$[H_{\text{eff}}(t), H_{\text{A}}] \neq 0. \tag{15}$$

Therefore the states of the atom driven by a long time laser pulse, which are approximated by stationary states of the master equation obtained from (11) replacing in the effective Hamiltonian (14) $\chi_{[S,T]}(t)$ by identity, will not be diagonal with respect to H_{A}. Our goal is to investigate the structure of these states. In generic situation the interaction with the laser field drives the atom to a (dynamical) equilibrium state at an exponential rate. In this sense the interaction prepares a state of the atom. Thus our problem can be reformulated as follows: What kind of states can we prepare applying a laser pulse? For a long laser pulse these states are described by the time independent generator $\mathcal{L}$ obtained from (10), by setting $S \to -\infty$, $T \to +\infty$, i.e. the characteristic function $\chi_{[S,T]}$ in (10) replaced by the identity.

In the limit $S \to -\infty$, $T \to +\infty$ the effective Hamiltonian becomes time independent

$$H_{\text{eff}} = \sum_{\omega} \Big(\text{Im}\gamma_\omega D_\omega^+ D_\omega + c_\omega^* D_\omega + c_\omega D_\omega^+ \Big).$$

In this case the stationary state $\hat{\rho}_{\text{st}}$ is a solution of the following equation

$$\mathcal{L}(\hat{\rho}_{\text{st}}) \equiv \mathcal{L}_{\text{diss}}(\hat{\rho}_{\text{st}}) - i[H_{\text{eff}}, \hat{\rho}_{\text{st}}] = 0. \tag{16}$$

In the following sections we solve explicitly this equation for two and three-level atoms.

5. Stationary states for a two-level atom

Let us consider a two-level atom under the rotating wave approximation.

The free Hamiltonian of the atom is

$$H_{\text{A}} = \varepsilon_0 |0\rangle\langle 0| + \varepsilon_1 |1\rangle\langle 1|,$$

where ε_0 and ε_1 are the energies of the ground and the excited states, so that the Bohr frequency $\omega = \varepsilon_1 - \varepsilon_0 > 0$ is positive. The ground state is denoted by $|0\rangle$ and the excited state by $|1\rangle$. The transitions between atomic levels are described by the operators

$$D = |0\rangle\langle 1| \equiv \sigma_-, \qquad D^+ = |1\rangle\langle 0| \equiv \sigma_+.$$

The interaction Hamiltonian has the form

$$H_{\text{int}} = i(\sigma_- \otimes a^+(g) - \sigma_+ \otimes a(g)).$$

The first term in the interaction Hamiltonian describes the process of emission of a photon due to transition of the atom from the excited to the ground state. The second term describes the opposite process, that is the absorbtion of a photon by the atom and transition to the excited state.

The real part $\gamma = \operatorname{Re}\gamma_\omega$ of the susceptibility (5) determines the inverse lifetime of the atom without interaction with external field and the imaginary part $\Delta\omega = \operatorname{Im}\gamma_\omega$ determines energy shift (cf. [6], sect. 5.5). The dissipative part of the generator and the effective Hamiltonian for this system have the forms

$$\mathcal{L}_{\text{diss}}(\hat{\rho}) = \gamma(2\sigma_-\rho\sigma_+ - \rho\sigma_+\sigma_- - \sigma_+\sigma_-\rho),$$
$$H_{\text{eff}} = \Delta\omega|1\rangle\langle 1| + \Omega^*\sigma_- + \Omega\sigma_+,$$

where $\Omega \equiv c_\omega$ is the complex Rabi frequency of the laser field. Equation (16) for a stationary state $\hat{\rho}_{\text{st}}$ of the two-level atom equivalent to the following equations on the matrix elements $\rho_{mn} = \langle m|\hat{\rho}_{\text{st}}|n\rangle$:

$$\gamma\rho_{11} = \operatorname{Im}(\Omega\rho_{01})$$
$$\gamma_\omega\rho_{10} = i\Omega(\rho_{11} - \rho_{00}).$$

The general properties $\operatorname{tr}\hat{\rho} = 1$, $\hat{\rho}^+ = \hat{\rho}$ of the density matrix lead to the relations $\rho_{00} + \rho_{11} = 1$, $\rho_{01} = \rho_{10}^*$, $\rho_{11}^* = \rho_{11}$. Using these relations one finds that, with the notation $\alpha = -i\Omega/\gamma_\omega$, the stationary state is

$$\hat{\rho}_{\text{st}} \equiv \begin{pmatrix} \rho_{11} & \rho_{10} \\ \rho_{01} & \rho_{00} \end{pmatrix} = \frac{1}{1+2|\alpha|^2} \begin{pmatrix} |\alpha|^2 & \alpha \\ \alpha^* & 1+|\alpha|^2 \end{pmatrix} \tag{17}$$

This state is pure only if $\alpha = 0$. For very intense illumination ($|\Omega|/|\gamma_\omega| \gg 1$) the quantity $|\alpha|$ becomes very large and the atom becomes saturated with equal probabilities in the upper and lower levels, so that $\rho_{00} = \rho_{11} = 1/2$, $\rho_{01} = 0$. For very low illumination, the stationary state is the ground state.

6. Stationary states for a three-level atom

The free Hamiltonian of a three-level atom is

$$H_{\mathrm{A}} = \varepsilon_0|0\rangle\langle 0| + \varepsilon_1|1\rangle\langle 1| + \varepsilon_2|2\rangle\langle 2|,$$

where $|0\rangle$, $|1\rangle$ and $|2\rangle$ denote the ground state, the first and second excited states. For a non degenerate atom the energies satisfy the inequalities $\varepsilon_2 > \varepsilon_1 > \varepsilon_0$. There are three positive Bohr frequencies $\omega_1 = \varepsilon_1 - \varepsilon_0$, $\omega_2 = \varepsilon_2 - \varepsilon_1$, $\omega_3 = \varepsilon_2 - \varepsilon_0$, which correspond to three possible transitions. The transitions between the atomic levels due to interaction with radiation are described by the operator

$$D = d_1^*|0\rangle\langle 1| + d_2^*|1\rangle\langle 2| + d_3^*|0\rangle\langle 2|,$$

where d_j are some complex numbers. In particular, the contribution of the first term in the sum to the interaction Hamiltonian corresponds to emission of a photon by the atom and transition from the first excited level $|1\rangle$ to the ground state $|0\rangle$.

For the three-level atom the three possible transitions are described by the operators

$$D_1 := D_{\omega_1} = d_1^*|0\rangle\langle 1|, \quad D_2 := D_{\omega_2} = d_2^*|1\rangle\langle 2|, \\ D_3 := D_{\omega_3} = d_3^*|0\rangle\langle 2|. \tag{18}$$

Hence, with the notations

$$\gamma_j := \gamma_{\omega_j}|d_j|^2 = a_j + ib_j, \tag{19}$$

(a_j and b_j are the real and imaginary parts of γ_j) the dissipative part of the generator, in the generic case, can be written as

$$\mathcal{L}_{\mathrm{diss}}(\hat{\rho}) = 2\left(a_1\rho_{11} + a_3\rho_{22}\right)|0\rangle\langle 0| + 2a_2\rho_{22}|1\rangle\langle 1| \\ - a_1(\hat{\rho}|1\rangle\langle 1| + |1\rangle\langle 1|\hat{\rho}) - (a_2 + a_3)(\hat{\rho}|2\rangle\langle 2| + |2\rangle\langle 2|\hat{\rho}),$$

where $\rho_{mn} = \langle m|\hat{\rho}|n\rangle$ are the matrix elements of the density matrix. Introducing the Rabi frequencies of the laser field

$$\Omega_j := c_{\omega_j} d_j, \tag{20}$$

the effective Hamiltonian can be written as

$$H_{\mathrm{eff}} = b_1|1\rangle\langle 1| + (b_2 + b_3)|2\rangle\langle 2| \\ +\Omega_1|1\rangle\langle 0| + \Omega_2|2\rangle\langle 1| + \Omega_3|2\rangle\langle 0| \\ +\Omega_1^*|0\rangle\langle 1| + \Omega_2^*|1\rangle\langle 2| + \Omega_3^*|0\rangle\langle 2|.$$

The matrix elements of the stationary state for arbitrary Rabi frequencies Ω_j satisfy the following system of equations:

$$\begin{aligned}
&a_1\rho_{11} + a_3\rho_{22} = \mathrm{Im}\,(\Omega_1\rho_{01} + \Omega_3\rho_{02})\\
&(a_2 + a_3)\rho_{22} = \mathrm{Im}\,(\Omega_2\rho_{12} + \Omega_3\rho_{02})\\
&(\gamma_2^* + \gamma_3^*)\rho_{02} + i\Omega_3^*(\rho_{22} - \rho_{00}) - i\Omega_2^*\rho_{01} + i\Omega_1^*\rho_{12} = 0\\
&\gamma_1^*\rho_{01} + i\Omega_1^*(\rho_{11} - \rho_{00}) - i\Omega_2\rho_{02} + i\Omega_3^*\rho_{21} = 0\\
&(\gamma_1 + \gamma_2^* + \gamma_3^*)\rho_{12} + i\Omega_2^*(\rho_{22} - \rho_{11}) + i\Omega_1\rho_{02} = i\Omega_3^*\rho_{10}.
\end{aligned}$$

Let us consider the case when the radiation, describing laser pulse, is in a coherent state which drives only transitions from the ground state to the highest excited state, i.e., a laser field with frequency near ω_3. In this case $\Omega_1 = \Omega_2 = 0$, $\Omega_3 \equiv \Omega \neq 0$ is the only nonzero Rabi frequency and the system of equations for the matrix elements of the stationary state becomes simpler:

$$\begin{aligned}
0 &= a_1\rho_{11} + a_3\rho_{22} - \mathrm{Im}\,(\Omega\rho_{02})\\
0 &= (a_2 + a_3)\rho_{22} - \mathrm{Im}\,(\Omega\rho_{02})\\
0 &= (\gamma_2^* + \gamma_3^*)\rho_{02} + i\Omega^*(\rho_{22} - \rho_{00})\\
0 &= \gamma_1^*\rho_{01} + i\Omega^*\rho_{21}\\
0 &= (\gamma_1 + \gamma_2^* + \gamma_3^*)\rho_{12} - i\Omega^*\rho_{10}.
\end{aligned}$$

This system of equations, together with the conditions $\mathrm{tr}\,\hat{\rho} = 1$, $\hat{\rho}^+ = \hat{\rho}$, can be easily solved. With the notations

$$\alpha = \frac{-i\Omega}{\gamma_2 + \gamma_3}, \qquad r = \frac{a_2}{a_1}$$

the solution, which is the stationary state, has the form

$$\hat{\rho}_{\mathrm{st}} \equiv \begin{pmatrix} \rho_{22} & \rho_{21} & \rho_{20} \\ \rho_{12} & \rho_{11} & \rho_{10} \\ \rho_{02} & \rho_{01} & \rho_{00} \end{pmatrix}$$

$$= \frac{1}{1 + (2 + r)|\alpha|^2} \begin{pmatrix} |\alpha|^2 & 0 & \alpha \\ 0 & r|\alpha|^2 & 0 \\ \alpha^* & 0 & 1 + |\alpha|^2 \end{pmatrix}. \qquad (21)$$

The obtained density matrix $\hat{\rho}_{st}$ is the stationary state of the three-level atom in the laser field driving only transitions from the ground state to the highest energy state of the atom.

If the ratio r is very large, i.e. $a_2 \gg a_1$, then the stationary state describes the population inversion between the first two excited atomic levels. In this case the population of the first excited state of the atom is greater than that of the ground state. To find the physical conditions for this effect let us note that the quantity

$$a_1 = |d_1|^2 \int d\mathbf{k} |g(\mathbf{k})|^2 \delta(\omega(\mathbf{k}) - \omega_1)$$

determines the inverse lifetime of the atom at the level $|1\rangle$ in the free space. That is $\tau \sim 1/a_1$. Hence, if the lifetime τ of the atom in the first excited state is long enough, then we obtain the population inversion between the ground and the first excited level.

As an example, consider a three-level atom in Λ configuration, so that the transitions between the ground state $|0\rangle$ and the first excited state $|1\rangle$ are forbidden. This means that $d_1 = 0$. In this case $a_1 = 0$, the lifetime of the first excited level is infinite, and the steady state is $|1\rangle\langle 1|$, i.e., in this case all the population is concentrated on the first excited state.

7. Three-level lambda-atom

Let us now consider a three-level atom in Λ configuration in the general coherent field, that is both Rabi frequencies Ω_2 and Ω_3 (20) are different from zero.

The system of dynamical equations for matrix elements of the density matrix is (matrix elements are time dependent, $\rho_{nm} \equiv \rho_{nm}(t)$, the dot denotes time derivative)

$$\begin{aligned}
\dot{\rho}_{00} &= 2a_3\rho_{22} - 2\mathrm{Im}\,(\Omega_3\rho_{02}) \\
\dot{\rho}_{22} &= -2(a_2 + a_3)\rho_{22} + 2\mathrm{Im}\,(\Omega_2\rho_{12} + \Omega_3\rho_{02}) \\
\dot{\rho}_{02} &= -(\gamma_2^* + \gamma_3^*)\rho_{02} - i\Omega_3^*(\rho_{22} - \rho_{00}) + i\Omega_2^*\rho_{01} \\
\dot{\rho}_{01} &= i\Omega_2\rho_{02} - i\Omega_3^*\rho_{21} \\
\dot{\rho}_{12} &= -(\gamma_2^* + \gamma_3^*)\rho_{12} - i\Omega_2^*(\rho_{22} - \rho_{11}) + i\Omega_3^*\rho_{10}.
\end{aligned}$$

The stationary solution of this system of equations is

$$\hat{\rho}_{\rm st} \equiv \begin{pmatrix} \rho_{22} & \rho_{21} & \rho_{20} \\ \rho_{12} & \rho_{11} & \rho_{10} \\ \rho_{02} & \rho_{01} & \rho_{00} \end{pmatrix} = \frac{1}{|\Omega_2|^2 + |\Omega_3|^2} \begin{pmatrix} 0 & 0 & 0 \\ 0 & |\Omega_3|^2 & -\Omega_2^*\Omega_3 \\ 0 & -\Omega_2\Omega_3^* & |\Omega_2|^2 \end{pmatrix}.$$

An important property is that it is a pure state: $\hat{\rho}_{\rm st} = |\psi\rangle\langle\psi|$, where

$$|\psi\rangle = \frac{1}{\sqrt{|\Omega_2|^2 + |\Omega_3|^2}} \Big(\Omega_2|0\rangle - \Omega_3|1\rangle\Big). \tag{22}$$

In particular, if the ground state is doubly degenerate and if the coupling does not distinguish between the two ground states, then $\Omega_2 = \Omega_3$ and the unique stationary state is the dark state (cf. [1] for a discussion).

8. Two-level 3-times degenerate atom

In the present section a two-level atom with 3-times degenerate ground state is considered.

Denote the ground states as $|0_1\rangle$, $|0_2\rangle$, and $|0_3\rangle$. The free atomic Hamiltonian is $H_{\rm A} = \varepsilon_0 P_0 + \varepsilon_1 P_1$, where P_0 and $P_1 = |1\rangle\langle 1|$ are the projectors on the ground and the excited subspaces. Matrix D, which describes transitions between each of the ground states and the excited state, has the form $D = \sum_{l=1}^3 d_l^* |0_l\rangle\langle 1|$, where complex numbers d_l determine coupling between levels $|0_l\rangle$ and $|1\rangle$. There is only one positive Bohr frequency $\omega_0 = \varepsilon_1 - \varepsilon_0$ and $D_{\omega_0} = D$.

Denote $\gamma = \operatorname{Re}\gamma_{\omega_0}$, $\Delta = \operatorname{Im}\gamma_{\omega_0}$, $c = c_{\omega_0}$. The dissipative part of the generator and the effective Hamiltonian have the form

$$\mathcal{L}_{\rm diss}(\hat{\rho}) = \gamma(2D\hat{\rho}D^+ - \hat{\rho}D^+D - D^+D\hat{\rho}),$$

$$H_{\rm eff} = \Delta \sum_{l=1}^{3} |d_l^*|^2 P_1 + c^* D + cD^+.$$

In this case, instead of a unique stationary state for a two-level atom with a 2-times degenerate ground state, there is the family of the stationary states. It is parameterized by the values of $\rho_{22} = \langle 0_1|\hat{\rho}_{\rm st}|0_1\rangle$ and $\rho_{23} = \langle 0_1|\hat{\rho}_{\rm st}|0_2\rangle$ and has the form

$$\hat{\rho}_{\rm st} = \begin{pmatrix} 0 & 0 & 0 & 0 \\ 0 & \rho_{22} & \rho_{23} & \rho_{24} \\ 0 & \rho_{32} & \rho_{33} & \rho_{34} \\ 0 & \rho_{42} & \rho_{43} & \rho_{44} \end{pmatrix} \tag{23}$$

where the matrix elements $\rho_{mn} \equiv \langle 0_{m-1}|\hat{\rho}_{\rm st}|0_{n-1}\rangle$ of the stationary density matrix for $m, n = 2, 3, 4$ are given by

$$\rho_{33} = \frac{|d_3|^2 - \rho_{22}(|d_1|^2 + |d_3|^2) - 2\mathrm{Re}\,(d_1^* d_2 \rho_{32})}{|d_2|^2 + |d_3|^2}$$

$$\rho_{44} = \frac{|d_2|^2 + \rho_{22}(|d_1|^2 - |d_2|^2) + 2\mathrm{Re}\,(d_1^* d_2 \rho_{32})}{|d_2|^2 + |d_3|^2}$$

$$\rho_{42} = -\frac{d_1\rho_{22} + d_2\rho_{32}}{d_3}$$

$$\rho_{43} = -\frac{d_2\rho_{33} + d_1\rho_{23}}{d_3}$$

and $\rho_{nn} \geq 0$. This can be shown by calculations similar to that of the previous section. The investigation of the convergence of an arbitrary initial state to a state in this family will be done in a future paper.

In general, if the ground energy state of an atom is degenerate then in the absence of laser field the generator of the master equation for the reduced density matrix has a family of invariant states. It follows from (22) that, for a two-level atom with 2-times degenerate ground state, in the presence of a laser field the family of invariant states is destroyed and there is only one invariant state

$$|\psi\rangle = \frac{1}{\sqrt{|d_2|^2 + |d_3|^2}}\Big(d_2|0\rangle - d_3|1\rangle\Big).$$

The result of the present section shows that for a 3-times degenerate ground state the laser field only partially destroys the family of stationary states.

9. Conclusions

In the present work we derive, starting from exact microscopic dynamics and using the stochastic limit method, the quantum master equation (11) for an arbitrary atom driven by a general laser field. As an application of this equation, we study the cases of two and three-level atoms. The explicit form of the stationary states of such atoms are found. It is shown that for a three-level lambda-atom in two laser fields, which drive both allowed transitions, the stationary state of the atom is a pure state given by a superposition of the two lowest energy levels. Varying the intensities of the lasers, any such a superposition can be obtained. Hence the two lower levels of the atom behave like a single qubit and any state of this qubit can be obtained in a stable way and in exponentially small time.

This work is supported by the EU Network QP-Applications, Contract No. HPRN-CT-2002-00279. A.Pechen and S. Kozyrev are grateful to Luigi Accardi for kind hospitality in the Centro Vito Volterra. A. Pechen and S. Kozyrev are partially supported by CNR-NATO fellowship and the grant of Russian Foundation for Basic Research RFFI 02-01-01084. S. Kozyrev is also partially supported by the grants CRDF UM1-2421-KV-02 and a Grant for support of scientific schools NSh 1542.2003.1.

References

1. E. Arimondo, *Coherent population trapping in laser spectroscopy*, E. Wolf, Progress in Optics XXXV 1996 Elsevier Science B.V.
2. N.V. Vitanov, T. Halfmann, B.W. Shore, and K. Bergmann, Annu. Rev. Phys. Chem. **52**, 763 (2001).
3. A. Aspect, E. Arimondo, R. Kaiser, N. Vansteenkiste, and C. Cohen-Tannoudji, Phys. Rev. Lett **61**, 826 (1988).
4. J.I. Cirac and P. Zoller, Phys. Rev. Lett. **74**, 4091 (1995).
5. J.J. Garcia-Ripoll, P. Zoller, and J.I. Cirac, Phys. Rev. Lett. **91**, 157901-1 (2003).
6. L. Accardi, Y.G. Lu, and I.V. Volovich, *Quantum Theory and Its Stochastic Limit* (Springer, Berlin, 2002).
7. L. Accardi and S.V. Kozyrev, Discrete Dynamics in Nature and Society (2004);
   http://arxiv.org/abstracts/quant-ph/0303185.
8. M.O. Scully, S.Y. Zhu, and A. Gavrielides, Phys. Rev. Lett. **62**, 2813 (1989).
9. S.Y. Zhu and M.O. Scully, Phys. Rev. Lett. **76**, 388 (1996).
10. L. Accardi and K. Imafuku, *Control of quantum states by decoherence*, Preprint Volterra N. 542 (2003).
11. L. Accardi and M. Ohya, to appear in: Open Systems and Information Dynamics (2004).

RECENT ADVANCES IN QUANTUM WHITE NOISE CALCULUS

LUIGI ACCARDI
Centro Vito Volterra, Universitá di Roma TorVergata
Via di TorVergata, 00133 Roma, Italy
E-mail: volterra@volterra.mat.uniroma2.it

ANDREAS BOUKAS
American College of Greece
Aghia Paraskevi, Athens, Greece 15342
E-mail: andreasboukas@yahoo.com

1. Emergence of white noise equations from classical quantum mechanics

In the Schroedinger picture of Quantum Mechanics the initial state of a quantum system is described by a ket-vector $|\psi(0)>$ evolving under the influence of a Hamiltonian operator $H = H_0 + H_I$, where H_0 and H_I are the "free" and "interaction" parts, respectively, to a state given at time t by a ket-vector $|\psi(t)>$ satisfying the Schroedinger equation

$$i\,\hbar\, d|\psi(t)> = H\,|\psi(t)>\, dt \tag{1}$$

or in operator form

$$|\psi(t)> = V_t\,|\psi(0)> \tag{2}$$

where the unitary evolution operator V_t is given by

$$V_t = e^{-\frac{i}{\hbar}\, t\, H}. \tag{3}$$

In the Heisenberg picture of Quantum Mechanics, it is the observables X, i.e the self-adjoint operators on the system space that vary with time generating a flow

$$X_t = V_t^* \, X \, V_t \tag{4}$$

where

$$V_t = e^{-\frac{i}{\hbar}\, t\, H} \tag{5}$$

and

$$dX_t = \frac{i}{\hbar}\,[H, X_t]\,dt, \qquad X_0 = X \tag{6}$$

In Interaction Representation one has

$$|\psi_I(t) >= e^{\frac{i}{\hbar}\, t\, H_0}\, |\psi(t) >= e^{\frac{i}{\hbar}\, t\, H_0}\, e^{-\frac{i}{\hbar}\, t\, H}\, |\psi(0) >= U_t\, |\psi(0) > \tag{7}$$

where the propagator

$$U_t = e^{\frac{i}{\hbar}\, t\, H_0}\, e^{-\frac{i}{\hbar}\, t\, H} = e^{\frac{i}{\hbar}\, t\, H_0}\, V_t \tag{8}$$

satisfies

$$dU_t = -i\, H_I(t)\, U_t\, dt, \quad U_0 = 1 \tag{9}$$

where

$$H_I(t) = \frac{1}{\hbar}\, e^{\frac{i}{\hbar}\, t\, H_0}\, H_I\, e^{-\frac{i}{\hbar}\, t\, H_0} \tag{10}$$

In the Stochastic Limit approach, developed by L. Accardi, Y. G. Lu, and I. Volovich [17], starting with a Hamiltonian $H = H(\lambda)$, where $\lambda \in \mathbb{R}$ is small (e.g $H = H_0 + \lambda\, H_I$, λ: coupling constant), equation (9) takes the form

$$dU_t^{(\lambda)} = -i\, H_I^{(\lambda)}(t)\, U_t^{(\lambda)}\, dt, \;\; U_0^{(\lambda)} = 1 \tag{11}$$

where $t \to +\infty$ sescribes the long-term behavior, (scattering theory), and $\lambda \to 0$ describes the weak effects e.g weak coupling, low density (perturbation theory). The time rescaling $t \to \frac{t}{\lambda^2}$ puts things together by considering the long term cummulative effects of weak actions. The solution of the rescaled Hamiltonian equation (11)

$$dU^{(\lambda)}_{\frac{t}{\lambda^2}} = -\frac{i}{\lambda^2} H_I^{(\lambda)}(\frac{t}{\lambda^2}) U^{(\lambda)}_{\frac{t}{\lambda^2}} dt, \quad U_0^{(\lambda)} = 1 \tag{12}$$

converges, in the sense of matrix elements of collective vectors (ref. [17]), as $\lambda \to 0$ to a process U_t satisfying

$$dU_t = -i\, H_t\, U_t\, dt, \quad U_0 = 1 \tag{13}$$

where H_t is a singular Hamiltonian which can be written in terms of the Hida white noise functionals b_t and $b_t^\dagger$ as

$$\begin{aligned} H_t &= \int_0^t \left(C_0(s) + \sum_{n,k} C_{n,k}(s) b_t^{\dagger^n} b_t^k \right) ds \\ &= \int_0^t C_0(s)\, ds + \sum_{n,k} C_{n,k}(s)\, dB_k^n(s) \end{aligned} \tag{14}$$

where we have used the noise differentials notation

$$dB_k^n(t) = b_t^{\dagger^n} b_t^k\, dt \tag{15}$$

2. Quantum white noise unitary evolutions

The first order quantum stochastic differentials

$$dA_t^\dagger = b_t^\dagger\, dt, \;\; dA_t = b_t\, dt \;\; d\Lambda_t = b_t^\dagger\, b_t\, dt \tag{16}$$

were considered by R. L. Hudson and K. R. Parthasarathy (ref. [26]). The underlying Lie algebra is that of Heisenberg and Weyl. In this case, equation (13) is defined on the tensor product of the system Hilbert space and the Boson Fock space over $L^2([0,+\infty),\mathbb{C})$ and is of the form

$$\begin{aligned} dU_t &= -((iH + \frac{1}{2} L^* L)\, dt + L^* W\, dA_t - L\, dA_t^\dagger + (1-W)\, d\Lambda_t)\, U_t, \\ U_0 &= 1 \end{aligned} \tag{17}$$

where H, L, W are bounded system operators, with W unitary. The corresponding quantum flow $j_t(X) = U_t^*\, X\, U_t$ satisfies the quantum Langevin equation

$$
\begin{aligned}
dj_t(X) = & \; j_t(i[H,X] - \frac{1}{2}(L^*LX + XL^*L - 2L^*XL))\,dt \\
& + j_t([L^*,X]\,W)\,dA_t + j_t(W^*\,[X,L])\,dA_t^\dagger \\
& + j_t(W^*\,X\,W - X)\,d\Lambda_t, \\
j_0\,(X) = & \; X
\end{aligned} \tag{18}
$$

The square of white noise and/or the unitarity of the solutions of the associated quantum stochastic evolution equations have been considered by L. Accardi, A. Boukas, T. Hida, H.H. Kuo, Y.G. Lu, N. Obata, M. Skeide, and I. Volovich, in [2-7, 9, 12, 14-21]. The underlying Lie algebra is $sl(2;\mathbb{R})$. Here the stochastic differentials are

$$
dB_t^\dagger = b_t^{\dagger^2}\,dt, \;\; dB_t = b_t^2\,dt, \;\; dM_t = b_t^\dagger\,b_t\,dt \tag{19}
$$

which, as proved by L. Accardi, U. Frantz, and M. Skeide in [15], admit a Boson Fock space representation:

$$
dM_t = d\Lambda_t(\rho^+(M)) + dt \tag{20}
$$

$$
dB_t^+ = d\Lambda_t(\rho^+(B^+)) + dA_t^\dagger(e_0) \tag{21}
$$

$$
dB_t^- = d\Lambda_t(\rho^+(B^-)) + dA_t(e_0) \tag{22}
$$

where

$$
\rho^+(B^{+^n} M^k B^{-^l})\,e_m = \theta_{n,k,l,m}\,e_{n+m-l} \tag{23}
$$

is a representation of $sl(2;\mathbb{R})$, hence of its universal enveloping algebra $U(sl(2;\mathbb{R}))$, on $l_2(\mathbb{N})$. Here $n,k,l = 0,1,2,...$ and e_m, $m = 0,1,2,\cdots$ is any orthonormal basis of $l_2(\mathbb{N})$. Moreover, for each $n,k,l,m = 0,1,2,...$, $\theta_{n,k,l,m}$ is a scalar whose precise combinatorial definition can be found in [9].

The Itô table for $dB_t^\dagger$, dB_t, dM_t is not closed and to overcome this difficulty one must use Stochastic Calculus on the Fock Module generated by the stochastic differentials (ref. [9])

$$
d\Lambda_{n,k,l}(t) = d\Lambda_t(\rho^+(B^{+^n} M^k B^{-^l})) \tag{24}
$$

$$
dA_m(t) = dA_t(e_m) \tag{25}
$$

$$
dA_m^\dagger(t) = dA_t^\dagger(e_m) \tag{26}
$$

In this context, it was shown in [9] that, square of white noise unitary evolutions take the form

$$
\begin{aligned}
dU_t &= ((-\frac{1}{2}(D_-^*|D_-^*) + iH)\,dt + d\mathcal{A}_t(D_-) + d\mathcal{A}_t^\dagger(-r(W)D_-^*) \qquad (27)\\
&\quad + d\mathcal{L}_t(W-I))\,U_t \\
U_0 &= 1
\end{aligned}
$$

where $D_- = \sum_m D_{-,m} \otimes e_m$ and $W = \sum_{n,k,l} W_{n,k,l} \otimes \rho^+(B^{+^n} M^k B^{-^l})$, with $D_{-,m}$ bounded and $W_{n,k,l}$ unitary, and r, l are the right and left module actions respectively. The corresponding Langevin equation satisfied by the flow $j_t(X) = U_t^*\, X\, U_t$ is

$$
\begin{aligned}
dj_t(X) &= j_t(i\,[X,H] - \frac{1}{2}((D_-^*|D_-^*)\,X + X\,(D_-^*|D_-^*)) \qquad (28)\\
&\quad + (r(W)D_-^*|X\,r(W)D_-^*))\,dt \\
&\quad + j_t(d\mathcal{A}_t^\dagger(D_-^*\,X - r(W^*\,X)r(W)D_-^*)) \\
&\quad + j_t(d\mathcal{A}_t(X\,D_- - l(X\,W)l(W^*)D_-)) \\
&\quad + j_t(d\mathcal{L}_t(W^*\,X \circ W - X)), \\
j_0(X) &= X
\end{aligned}
$$

where the $\circ$-product is defined in [9].

3. Higher powers of white noise

The theory of quantum white noise has recently been extende by L. Accardi and A. Boukas in [13] to higher powes of the Hida white noise functionals. The so called "renormalized powers of white noise" integrators

$$
B_k^n(t) = \int_0^t b_s^{\dagger^n}\, b_s^k\, ds \qquad (29)
$$

corresponding to the stochastic differentials (15) were shown in [13] to satisfy the commutation relations

$$
\begin{aligned}
&[B_K^N(t), B_k^n(s)] \\
&= \epsilon_{K,0}\epsilon_{n,0} \sum_{L\geq 1} \binom{K}{L} n^{(L)}\, c^{L-1}\, B_{K+k-L}^{N+n-L}(t\wedge s) \\
&\quad -\epsilon_{k,0}\epsilon_{N,0} \sum_{l\geq 1} \binom{k}{l} N^{(l)}\, c^{l-1}\, B_{K+k-l}^{N+n-l}(t\wedge s)
\end{aligned} \tag{30}
$$

where $c > 0$ is the "renormalization constant" corresponding to the renormalization $\delta(t)^2 = c\,\delta(t)$ introduced by L. Accardi, Y.G. Lu and I. Volovich in [18]. Also, $\epsilon_{n,k} = 1 - \delta_{n,k}$ and $\delta_{n,k}$ is Kronecker's delta. It was recently shown by L. Accardi and A. Boukas in [13] that the above commutation relations admit a Fock space representation. This was done by proving the positive definiteness of the kernel

$$
\langle B_0^{K_N}(f_N)\ldots B_0^{K_1}(f_1)\Phi, B_0^{n_M}(g_M)\ldots B_0^{n_1}(g_1)\Phi\rangle \tag{31}
$$

where Φ denotes the vacuum state. A program is currently underway to study the classical probability distributions obtained as powers of the Hida white noise functionals and to classify the corresponding classes of orthogonal polynomials (ref. [1]).

4. Appplications to quantum stochastic control

The quadratic cost control problem of classical stochastic control theory was extended to the quantum stochastic framework by L. Accardi and A. Boukas in [8], [10],[11],[22]-[24].

In the case of first order white noise it was shown that if $U = \{U_t \,/\, t \geq 0\}$ is a stochastic process satisfying on a finite interval $[0,T]$ the quantum stochastic differential equation

$$
dU_t = (F\,U_t + u_t)\,dt + \Psi\,U_t\,dA_t + \Phi\,U_t\,dA_t^\dagger + Z\,U_t\,d\Lambda_t, \quad U_0 = 1 \tag{32}
$$

then the performance functional

$$
Q_{\xi,T}(u) = \int_0^T [< U_t\,\xi, X^2\,U_t\,\xi > + < u_t\,\xi, u_t\,\xi >]\,dt - < u_T\,\xi, U_T\,\xi > \tag{33}
$$

satisfies

$$\min Q_{\xi,T}(u) = <\xi, \Pi\xi> \tag{34}$$

where the minimum is taken over all processes of the form $u_t = -\Pi\, U_t$, ξ is an arbitrary vector in the exponential domain of the tensor product of the system Hilbert space and the Boson Fock space over $L^2([0,+\infty),\mathbb{C})$, and Π is the solution of the Algebraic Riccati Equation

$$\Pi F + F^*\Pi + \Phi^*\Pi\Phi - \Pi^2 + X^2 = 0 \tag{35}$$

with the additional conditions

$$\Pi\,\Psi + \Phi^*\,\Pi + \Phi^*\,\Pi\, Z = 0 \tag{36}$$
$$\Pi\, Z + Z^*\,\Pi + Z^*\,\Pi\, Z = 0 \tag{37}$$

Using this we have proved (ref. [10], [11]) that if X is a bounded self-adjoint system operator such that the pair $(i\,H,\, X)$ is stabilizable, then the quadratic performance functional

$$J_{\xi,T}(L,W) = \int_0^T [\,\|j_t(X)\,\xi\|^2 + \frac{1}{4}\|j_t(L^*L)\,\xi\|^2\,]\,dt + \frac{1}{2}\|j_T(L)\,\xi\|^2 \tag{38}$$

associated with the quantum stochastic flow $\{j_t(X) = U_t^*\,X\,U_t\,/\,t \geq 0\}$ satisfying

$$\begin{aligned} dj_t(X) &= j_t(i[H,X] - \frac{1}{2}(L^*LX + XL^*L - 2L^*XL))\,dt \qquad (39)\\ &\quad + j_t([L^*,X]\,W)\,dA_t + j_t(W^*\,[X,L])\,dA_t^\dagger + + j_t(W^*\,X\,W - X)\,d\Lambda_t, \\ j_0(X) &= X \end{aligned}$$

where $U = \{U_t\,/\,t \geq 0\}$ is the solution of

$$\begin{aligned} dU_t &= -((iH + \frac{1}{2}\,L^*L)\,dt + L^*\,W\,dA_t - L\,dA_t^\dagger + (1-W)\,d\Lambda_t)\,U_t, \qquad (40)\\ U_0 &= 1, \end{aligned}$$

is minimized by choosing

$$L = \sqrt{2}\,\Pi^{1/2}\,W_1 \tag{41}$$
$$W = W_2 \tag{42}$$

where Π is the solution of the Algebraic Riccati Equation

$$i\,[H,\Pi] + \Pi^2 + X^2 = 0 \tag{43}$$

and W_1, W_2 are bounded unitary system operators commuting with Π. Moreover

$$\min_{L,W} J_{\xi,T}(L,W) =< \xi, \Pi\,\xi > \tag{44}$$

In the case of quantum stochastic differential equations driven by the square of white noise processes, we have shown (ref. [11]) that if X is a bounded self-adjoint system operator such that the pair $(i\,H,\, X)$ is stabilizable then the performance functional

$$\begin{aligned} J_{\xi,T}(D_-,W) = &\int_0^T [\,\|j_t(X)\,\xi\|^2 + \frac{1}{4}\|j_t((D_-^*|D_-^*))\,\xi\|^2\,]\,dt \\ &+\frac{1}{2} < \xi, j_T((D_-^*|D_-^*))\,\xi > \end{aligned} \tag{45}$$

associated with the quantum flow $\{j_t(X) = U_t^*\,X\,U_t\,/\,t \geq 0\}$, where $U = \{U_t\,/\,t \geq 0\}$ is the solution of the quantum stochastic differential equation

$$\begin{aligned} dU_t &= ((-\frac{1}{2}\,(D_-^*|D_-^*) + iH)\,dt + d\mathcal{A}_t(D_-) \\ &\quad + d\mathcal{A}_t^\dagger(-r(W)D_-^*) + d\mathcal{L}_t(W-I)) \\ U_0 &= 1 \end{aligned} \tag{46}$$

is minimized by choosing

$$D_- = \sum_n D_{-,n} \otimes e_n \tag{47}$$

and

$$W = \sum_{\alpha,\beta,\gamma} W_{\alpha,\beta,\gamma} \otimes \rho^+({B^+}^\alpha M^\beta {B^-}^\gamma) \tag{48}$$

so that

$$\frac{1}{2}\,(D_-^*|D_-^*) = (\frac{1}{2}\sum_n D_{-,n}\,D_{-,n}^*) \otimes 1 = \Pi \tag{49}$$

and

$$[D_{-,n}, D_{-,m}] = [D_{-,n}, D^*_{-,m}] = [D_{-,n}, W_{\alpha,\beta,\gamma}] \\ = [D_{-,n}, W^*_{\alpha,\beta,\gamma}] = 0 \tag{50}$$

for all $n, m, \alpha, \beta, \gamma$, where Π is the positive self-adjoint solution of the Algebraic Riccati equation

$$i\,[H, \Pi] + \Pi^2 + X^2 = 0 \tag{51}$$

Moreover

$$\min_{D_-, W} J_{\xi,T}(D_-, W) =< \xi, \Pi\, \xi > \tag{52}$$

References

1. L. Accardi, *Meixner classes and the square of white noise*, Talk given at the: AMS special session "Analysis on Infinite Dimensional Spaces (in honor of L. Gross)" during the AMS-AMA Joint Mathematics Meetings in New Orleans, LA, January 10-13, 2001. AMS Contemporary Mathematics 317, Kuo H.–H., A. Sengupta (eds.) (2003) 1–13.
2. L. Accardi, A. Boukas, *Unitarity conditions for stochastic differential equations driven by nonlinear quantum noise*, Random Operators and Stochastic Equations, vol. 10, no.1, pp. 1-12 (2002).
3. L. Accardi, A. Boukas, *Stochastic evolutions driven by non-linear quantum noise*, Probability and Mathematical Statistics, vol.22.1 (2002).
4. L. Accardi, A. Boukas, *Stochastic evolutions driven by non-linear quantum noise II*, Russian Journal of Mathematical Physics, v.8, no.4, (2001).
5. L. Accardi, A. Boukas, *Square of white noise unitary evolutions on Boson Fock space*, International conference on stochastic analysis in honor of Paul Kree, Hammamet, Tunisie, October 22-27, 2001.
6. L. Accardi, A. Boukas, *Unitarity conditions for the renormalized square of white noise*, Trends in Contemporary Infinite Dimensional Analysis and Quantum Probability, Natural and Mathematical Sciences Series 3, 7-36, Italian School of East Asian Studies, Kyoto, Japan (200).
7. L. Accardi, A. Boukas, *The semi-martingale property of the square of white noise integrators*, Stochastic Partial Differential Equations and Applications, pp. 1-19, eds g. Da Prato and L. Tubaro, Marcel Dekker, Inc. (2002).
8. L. Accardi, A. Boukas, *Control of elementary quantum flows*, Proceedings of the 5th IFAC symposium on nonlinear control systems, July 4-6, 2001, St. Petersburg, Russia".

9. L. Accardi, A. Boukas, *The unitarity conditions for the square of white noise*, Dimensional Anal. Quantum Probab. Related Topics , Vol. 6, No. 2 (2003) 1-26.
10. L. Accardi, A. Boukas, *Quadratic control of quantum processes*, Russian Journal of Mathematical Physics (2002).
11. L. Accardi, A. Boukas, *Control of quantum Langevin equations*, Open Systems and Information Dynamics.9:1-15, 2002.
12. L. Accardi, A. Boukas, *Quantum stochastic Weyl operators*, IEEE Proceedings of the International Conference "Physics and Control" (PhysCon 2003), August 20-22, 2003, St. Petersburg, Russia.
13. L. Accardi, A. Boukas, *Higher powers of quantum white noise*, to appear (2003).
14. L. Accardi, A. Boukas, H.H. Kuo, *On the unitarity of stochastic evolutions driven by the square of white noise*, Infinite Dimensional Analysis, Quantum Probability, and Related Topics, vol. 4, no. 4, pp.1-10, (2001).
15. L. Accardi, U. Franz, M. Skeide, *Renormalized squares of white noise and non- gaussian noises as Levy processes on real Lie algebras*, Comm. Math. Phys. 228 (2002), no. 1, 123-150.
16. L. Accardi, T. Hida, H.H Kuo, *The Itô table of the square of white noise*, Infinite Dimensional Analysis, Quantum Probability, and Related Topics, 4, (2001) 267-275.
17. L. Accardi, Y.G Lu, I.V Volovich, *Quantum theory and its stochastic limit* , Springer 2002.
18. L. Accardi, Y.G Lu, I.V Volovich, *White noise approach to classical and quantum stochastic calculi*, Lecture Notes of the Volterra International School of the same title, Trento, Italy, 1999, Volterra Center preprint 375.
19. L. Accardi, Y.G Lu, I.V Volovich, *Non-linear extensions of classical and quantum stochastic calculus and essentially infinite dimensional analysis*, Volterra Center preprint no.268, 1996.
20. L. Accardi, N. Obata, *Towards a non-linear extension of stochastic calculus*, Publications of the Research Institute for Mathematical Sciences, Kyoto, RIMS Kokyuroku 957, Obata N. (ed.), (1996), 1-15.
21. L. Accardi, M. Skeide, *On the relation of the square of white noise and the finite difference algebra*, Infinite Dimensional Analysis, Quantum Probability, and Related Topics, 3, (2000), 185-189.
22. Boukas A.,*Linear Quantum Stochastic Control*, Quantum Probability and related topics, 105-111, QP -PQ IX, World Scientific Publishing, River Edge NJ,1994.
23. Boukas A.,*Application of Operator Stochastic Calculus to an Optimal Control problem*, Mat. Zametki 53, (1993), no5, 48-56 Russian).Translation in Math.Notes 53 (1993), No 5-6, 489 -494, MR 96a 81070.
24. Boukas A.,*Operator valued stochastic control in Fock space with applications to noise filtering and orbit tracking*, Journal of Probability and Mathematical Statistics, Vol .16, 1, 1994.
25. Boukas A.,*Stochastic Control of operator-valued processes in Boson Fock space*, Russian Journal of Mathematical Physics ,4 (1996) , no2, 139-150, MR 97j 81178
26. K. R. Parthasarathy, *An introduction to quantum stochastic calculus*, Birkhauser Boston Inc., 1992.

CONTROL OF QUANTUM STATES BY DECOHERENCE

L. ACCARDI AND K. IMAFUKU

Centro Vito Volterra, Università di Roma "Tor Vergata", 00133, Via Columbia 2, Roma, Italy

E-mail: accardi@volterra.mat.uniroma2.it

Homepage: http://volterra.mat.uniroma2.it

We propose a new technique to control quantum states by exploiting the decoherence due to the coupling with environment. With this technique we can get any target state, in a stable way and in an exponentially small time, as a stationary state of a semigroup (master equation) canonically derived from a microscopic Hamiltonian model. The stationary state of the system depends (i) on the interaction with the environment; (ii) on the initial state of the environment and it "inherits" some properties of it in a sense that will be explained in sec. 5 of the present paper. Moreover this is true not only for by thermal or vacuum environments but also for more general non-equilibrium environments. We prove that, by appropriately choosing this control parametes (interaction system–environment, initial environment state) one can drive the system to an arbitrary pre–assigned quantum state.

1. Introduction

Recent developments of technology greatly improved our ability to control individual quantum systems. This brings quantum technology beyond academic research to the level of concrete industrial programs [1,2]. For the requirements of any quantum technology the requirement of stability is essential: it is not only required that at time T the system is in a given quantum state, but also that it remains in this state sufficiently long time to allow the manipulations required by quantum computation. One possible way to achieve this goal is to exploit a general principle of the stochastic limit[3,4,5,6], namely: under explicit and easily realizable conditions, *the interaction of a quantum field with a discrete system (e.g. an N-level atom) drives the system to a stationary state which is uniquely determined by the initial state of the field and by the form of the interaction.*

Already now many manipulations on microscopic objects are achieved through their interaction with appropriate fields (for example, a creation of entangled state with ion trap (ions + phonon) or with cavity QED (atom + EM field) have been already reported [7,8,9,10].) The scenario we are proposing generalizes this approach by extending it to a large class of interactions and integrates it with the additional requirement of stability. In some sense, this scenario realizes the converse program of the stochastic limit: there one starts from a given interaction and a given state of the field and looks for the corresponding Langevin and master equations and their stationary states. Here one starts from a state of the system, say atom or lattice, and looks for an interaction and an initial state of the field such that the associated master equation will drive the system to the given state. In other words: in our approach the initial state of the field and the interaction Hamiltonian are seen as *control parameters*.

The advantage of the stochastic limit approach is that it gives a quite explicit description of the parameters which control the final state of the system. Therefore, if we are able to act on these parameters by suitably choosing the initial state of the field and the interaction, we could drive the system, in a stable way and in an exponentially small time, to a large class (in principle, any) of pre-assigned states.

In this paper, we consider the conditions on the microscopic systems which realize the above scenario. As shown in sec.2, it is easy to write down a general master equation which drives any quantum state to a given target state. Our interest in this paper is to derive this master equation from a microscopic Hamiltonian model. This allows to make clear the physical conditions to realize this control. From sec. 3 we show that as far as the system degrees of freedom are finite for any target states we can always find an appropriate microscopic model. We also discuss the concrete conditions on the initial state of the environment. For a given target state, the initial state of the environment can be chosen in a variety of ways whereas the condition for the interaction are more restrictive. This means that, in real experimental situations, one can choose an easily realizable initial state which depends on the individual case. In sec. 5, we show that the system has the "assimilation property", i.e. if the initial state of the environment is non-equilibrium (resp. equilibrium or vacuum) for the free environment Hamiltonian, then the system is driven to a non-equilibrium (resp. equilibrium or ground) state for the free system Hamiltonian. In the equilibrium case, this assimilation property is true for rather general models, but in non-equilibrium case, it appears as a consequence of the specific choice of the interaction. In sec. 6 we discuss purification as a special case of our general framework: any pure state can be obtained by a general assimilation property with a vacuum environment. As an example in sec. 7, the singlet state of two spin-1/2 particles.

2. A master equation driving to a pre–assigned state

Let us consider a master equation

$$\frac{d}{dt}\rho(t) = \mathcal{L}_*\rho(t) \tag{1}$$

with GKSL generator [12,13]

$$\mathcal{L}_*\rho(t) = -i[H,\rho(t)] - \gamma \sum_{j\neq k} p_j \left(\frac{1}{2}\left\{L_{jk}^\dagger L_{jk},\rho(t)\right\} - L_{jk}\rho_t L_{jk}^\dagger\right) \tag{2}$$

with H, p_j, L_{jk} given respectively by

$$H = \sum_j \epsilon_j |\mu_j\rangle\langle\mu_j|, \quad \sum_j p_j = 1, \quad (p_j \geq 0), \tag{3}$$

$$L_{jk} = |\mu_j\rangle\langle\mu_k|, \quad L_{jk}^\dagger = |\mu_k\rangle\langle\mu_j|. \tag{4}$$

Notice that

$$\sum_{jk} p_j L_{jk}^\dagger L_{jk} = 1, \quad \sum_{jk} p_j L_{jk}\rho L_{jk}^\dagger = \sum_j p_j |\mu_j\rangle\langle\mu_j|. \tag{5}$$

Therefore one can rewrite the master equation as

$$\begin{aligned}
\mathcal{L}_*\rho(t) &= -i[H,\rho(t)] - \gamma \sum_{j,k} p_j \left(\frac{1}{2}\left\{L_{jk}^\dagger L_{jk},\rho(t)\right\} - L_{jk}\rho_t L_{jk}^\dagger\right) \\
&\quad + \gamma \sum_j p_j \left(\frac{1}{2}\left\{L_{jj}^\dagger L_{jj},\rho(t)\right\} - L_{jj}\rho_t L_{jj}^\dagger\right) \\
&= -i[H,\rho(t)] - \gamma\ (\rho(t) - \mu) \\
&\quad + \gamma \sum_j p_j \left(\frac{1}{2}\left\{|\mu_j\rangle\langle\mu_j|,\rho(t)\right\} - \langle\mu_j|\rho(t)|\mu_j\rangle|\mu_j\rangle\langle\mu_j|\right),
\end{aligned} \tag{6}$$

where

$$\mu := \sum_j p_j |\mu_j\rangle\langle\mu_j|\ . \tag{7}$$

One can see that, denoting $\rho_{mn}(t) = \langle\mu_m|\rho(t)|\mu_n\rangle$, (2) becomes equivalent to

$$\frac{d}{dt}\rho_{mm}(t) = \gamma(p_m - \rho_{mm}(t)), \tag{8}$$

$$\frac{d}{dt}\rho_{mn}(t) = \left(i(\epsilon_n - \epsilon_m) - \gamma\left(1 - \frac{p_m + p_n}{2}\right)\right)\rho_{mn}(t). \tag{9}$$

Thus the state of the system is driven to the final state (7), i.e.

$$\rho(t) \to \mu, \quad \text{as } t \to \infty \tag{10}$$

Notice that the convergence in (10) is exponential and its speed of convergence is given by γ^{-1} which is thus interpreted as life time of the initial state $\rho(0)$. It means that if can we realize a physical system which is described by this master equation, we can control a system so that its state is driven to the target state μ. In the following section, we prove that the stochastic limit technique [3] allows to solve this problem, i.e. to construct microscopic models from which to derive the master equation (8), (9) and the explicit form of the life time γ.

Moreover these microscopic models are based on relatively simple, dipole type, interaction Hamiltonians. More precisely we prove that, given any state μ (cf. (7)) of a finite dimensional system, one can find an appropriate dipole type coupling of this system with a quantum field (concretely this means to specify the form of the operator D in the interaction (13)) such that the system will be asymptotically driven to the given state (in the sense of (10)) whatever its initial state is.

The explicit from of the operator D, as a function of the target state, will be given in section (5) below.

The practical implementation of the corresponding dipole interaction will of course depend on the system and on the target state and has to be discussed case by case. However just the possibility, in principle, of such a "control by decoherence" is quite a non trivial fact. We prove this possibility and we give an explicit and constructive description of the solutions.

3. A microscopic model

In the following sections, we show how to derive the master equation of sec.2 from a microscopic Hamiltonian model through the stochastic limit[3,4,5].

We consider the following Hamiltonian system:

$$H_{tot} = H_S + H_B + \lambda H_I, \quad H_0 = H_S + H_B \tag{11}$$

where, for the same μ_j as in (3), (7)

$$H_S = \sum_j E_j|\mu_j\rangle\langle\mu_j|, \quad H_B = \int dk\; \omega(k)a_k^\dagger a_k, \quad [a_k, a_{k'}^\dagger] = \delta(k - k') \tag{12}$$

$$H_I = \int dk \, \left(g(k) D a_k^\dagger + g^*(k) D^\dagger a_k \right) \tag{13}$$

and D is a system operator that will be specified in Section (5). For the initial state of the boson-field, we consider a general mean-zero gaussian state which is represented by the covariance matrix [3]

$$\begin{pmatrix} \langle a_k^\dagger a_{k'} \rangle & \langle a_k a_{k'} \rangle \\ \langle a_k^\dagger a_{k'}^\dagger \rangle & \langle a_k a_{k'}^\dagger \rangle \end{pmatrix} = \begin{pmatrix} \alpha_k^{(+)} \delta(k-k') & f_k \, \delta(k+k') \\ f_k^* \, \delta(k+k') & \alpha_k^{(-)} \delta(k-k') \end{pmatrix} \tag{14}$$

where

$$\alpha_k^{(+)} = |c_k|^2 n(k) + |s_k|^2 (n(k)+1) \tag{15}$$

$$\alpha_k^{(-)} = |c_k|^2 (n(k)+1) + |s_k|^2 n(k) \tag{16}$$

$$f_k = c_k s_{-k}^* n(k) + c_{-k} s_k^* (n(k)+1). \tag{17}$$

Notice that this is a squeezed state (i.e. $f_k \neq 0$) which can be obtained from the gauge invariant state of the $\tilde{a}_k \; \tilde{a}_k^\dagger$-field, with covariance

$$\begin{pmatrix} \langle \tilde{a}_k^\dagger \tilde{a}_{k'} \rangle & \langle \tilde{a}_k \tilde{a}_{k'} \rangle \\ \langle \tilde{a}_k^\dagger \tilde{a}_{k'}^\dagger \rangle & \langle \tilde{a}_k \tilde{a}_{k'}^\dagger \rangle \end{pmatrix} = \begin{pmatrix} n(k) & 0 \\ 0 & n(k)+1 \end{pmatrix} \delta(k-k') \tag{18}$$

by means of the Bogolyubov transformation

$$\begin{pmatrix} a_k \\ a_{-k}^\dagger \end{pmatrix} = \begin{pmatrix} c_k^* & s_k \\ s_{-k}^* & c_{-k} \end{pmatrix} \begin{pmatrix} \tilde{a}_k \\ \tilde{a}_{-k}^\dagger \end{pmatrix} \tag{19}$$

(cf. chap. 2 of [3] for details) where

$$|c_k|^2 - |s_k|^2 = 1. \tag{20}$$

4. The Master equation

In the following discussion, we assume that we can build up the system Hamiltonian H_S such that the genericity condition [5]

$$\omega_{ji} = E_j - E_i = \omega_{j'i'} \Leftrightarrow i = i', \; j = j'. \tag{21}$$

is satisfied. Under this condition, using the stochastic golden rule of [7] (Chap. 5), we can obtain the master equation for the reduced density matrix:

$$\frac{d}{dt}\rho_t = +i[\Delta, \rho_t] - \sum_{i \neq j} \Gamma_{ij} \left(\frac{1}{2} \{ L_{ij}^\dagger L_{ij}, \rho_t \} - L_{ij} \rho_t L_{ij}^\dagger \right) \tag{22}$$

where the L_{ij} are given by (29) and

$$\Delta = \sum_{ij} \Delta_{ij} |\mu_j\rangle\langle\mu_j| \tag{23}$$

$$\Delta_{ij} = \Delta^{(-)}_{c,ij} + \Delta^{(+)}_{s,ij} - \Delta^{(+)}_{c,ji} - \Delta^{(-)}_{s,ji}, \tag{24}$$

$$\Gamma_{ij} = \begin{cases} \Gamma^{(-)}_{c,ij} + \Gamma^{(+)}_{s,ij} \ (E_j > E_i) \\ \Gamma^{(+)}_{c,ji} + \Gamma^{(-)}_{s,ji} \ (E_i > E_j). \end{cases} \tag{25}$$

$$\Gamma^{(-)}_{c,ij} = 2\pi \int dk \ |g_{ij}(k) c_k|^2 \ (n(k)+1) \ \delta(\omega(k) - \omega_{ji}) \ \geq 0 \tag{26}$$

$$\Gamma^{(-)}_{s,ij} = 2\pi \int dk \ |g_{ij}(k) s_k|^2 \ (n(-k)+1) \ \delta(\omega(k) - \omega_{ji}) \ \geq 0 \tag{27}$$

$$\Gamma^{(+)}_{c,ij} = 2\pi \int dk \ |g_{ij}(k) c_k|^2 \ n(k) \ \delta(\omega(k) - \omega_{ji}) \ \geq 0 \tag{28}$$

$$\Gamma^{(+)}_{s,ij} = 2\pi \int dk \ |g_{ij}(k) s_k|^2 \ n(-k) \ \delta(\omega(k) - \omega_{ji}) \ \geq 0 \tag{29}$$

$$\omega_{ji} = E_j - E_i, \quad g_{ij}(k) = \langle\mu_i|D|\mu_j\rangle g(k) \, . \tag{30}$$

Notice that, because of the δ-function in (26)-(29), only those pairs ij with $\omega_{ji} = E_j - E_i > 0$ give a non-trivial contribution since $\omega(k) > 0$.

$$\Delta^{(-)}_{c,ij} = \text{P.P.} \int dk \ \frac{|g_{ij} c_k|^2 \ (n(k)+1)}{\omega(k) - \omega_{ji}} \tag{31}$$

$$\Delta^{(-)}_{s,ij} = \text{P.P.} \int dk \ \frac{|g_{ij} s_k|^2 \ (n(-k)+1)}{\omega(k) - \omega_{ji}} \tag{32}$$

$$\Delta^{(+)}_{c,ij} = \text{P.P.} \int dk \ \frac{|g_{ij} c_k|^2 \ n(k)}{\omega(k) - \omega_{ji}} \tag{33}$$

$$\Delta^{(+)}_{s,ij} = \text{P.P.} \int dk \ \frac{|g_{ij} s_k|^2 \ n(k)}{\omega(k) - \omega_{ji}} \tag{34}$$

Notice that the Γ_{ij} are strictly positive. Moreover if the condition

$$\Gamma_{ij} = \Gamma_i \tag{35}$$

holds, we can rewrite the master equation (22) as

$$\frac{d}{dt}\rho_t = +i[\Delta, \rho_t] - \Gamma \sum_{i \neq j} \tilde{p}_i \left(\frac{1}{2}\{L^\dagger_{ij} L_{ij}, \rho_t\} - L_{ij} \rho_t L^\dagger_{ij} \right) \tag{36}$$

where

$$\tilde{p}_i = \frac{\Gamma_i}{\Gamma}, \quad \Gamma = \sum_i \Gamma_i \ . \tag{37}$$

Comparing this master equation with (2), one can verify that the two equations coincide the Hamiltonian part in (36) commutes with the dissipative part whenever $H = -\Delta$, $\gamma = \Gamma$, $p_j = \tilde{p}_j$. Therefore to guarantee the convergence

$$\rho_t \to \sum_i p_i |\mu_i\rangle\langle\mu_i| \tag{38}$$

it is sufficient that condition (35) and the identities

$$\gamma = \Gamma \ ; \quad p_j = \tilde{p}_j \ . \tag{39}$$

Now, let us consider how to satisfy the condition (35) for a given family $\{\tilde{p}_j\}$. Introducing the notation

$$\mathcal{N}(k) = |s_k|^2(n(-k)+1) + |c_k|^2 n(k) \geq 0 \tag{40}$$

we can write

$$\Gamma_{ij} = \begin{cases} 2\pi|d_{ij}|^2 \int dk \ |g(k)|^2(\mathcal{N}(k)+1)\delta(\omega(k)-\omega_{ji}) \ (E_j > E_i) \\ 2\pi|d_{ji}|^2 \int dk \ |g(k)|^2\mathcal{N}(k)\delta(\omega(k)-\omega_{ij}) \qquad (E_i > E_j) \end{cases} \tag{41}$$

where

$$d_{ij} = \langle\mu_i|D|\mu_j\rangle. \tag{42}$$

We will only consider control environment states which satisfy the additional condition which is implied by the set of conditions:

$$n(k) = n_0(\omega(k)) \ ; \quad c_k = c_0(\omega(k)) \ ; \quad s_k = s_0((\omega(k)) \ ; \quad \omega(k) = \omega(-k) \tag{43}$$

$$\mathcal{N}(k) = N(\omega(k)) \tag{44}$$

under this condition (41) becomes

$$\Gamma_{ij} = \begin{cases} |d_{ij}|^2 Q_{ij} N_{ij} \qquad (E_j > E_i) \\ |d_{ji}|^2 Q_{ij}(N_{ij}+1) \ (E_i > E_j) \end{cases} \tag{45}$$

where

$$Q_{ij} = 2\pi \int dk \ |g(k)|^2 \delta(\omega(k) - |\omega_{ji}|) > 0 \tag{46}$$

and

$$N_{ij} = N(|\omega_{ji}|). \tag{47}$$

Our goal is to solve equations (45) in the unknowns d_{ij} (control interaction) and N_{ij} (control environment state) subject to the additional conditions (35, (37) and (39) which introduce the dependence on the target state. The Q_{ij}, given by (46), are additional control parameters, but in our case we will consider them as given. Introducing these conditions, (45) becomes equivalent to

$$\gamma p_i = |d_{ij}|^2 Q_{ij} N_{ij} \ ; \quad \text{for} \quad E_j > E_i \tag{48}$$

$$\gamma p_i = |d_{ji}|^2 Q_{ij} (N_{ij} + 1) \ ; \quad \text{for} \quad E_j < E_i \tag{49}$$

It is clear that, for any choice of the decay rate γ and of the target state (i.e. the p_i's) there exist a multiplicity of solutions depending on our choice of the d_{ij}, Q_{ij}, N_{ij}. This variety of choice will be very useful in the explicit construction of the control mechanism.

It should be emphasized that the freedom in the choice of the environment state is ample but not unlimited. For example if the target state is faithful (all $p_i > 0$) then equation (48) shows that the control state cannot be the Fock state (for which $N_{ij} = 0$ for all i, j). In the following sections we will show how to exploit this freedom in some concrete cases.

Let us summarize the properties which the Hamiltonian should have.

(1) H_S is discrete and is diagonal in the same basis which diagonalizes the target state μ.
(2) The spectrum of H_S is generic in the sense of (21).
(3) H_S must have a lowest eigenvalue and the eigenvalues E_i of H_S must be ordered from the lowest one while the $\tilde{p}_i$ are ordered from the largest one.
(4) The $d_{ij} = \langle \mu_i | D | \mu_j \rangle$ satisfy the equations (48), (49).

5. Assimilation of the state

In this section we prove the "assimilation lemma" (5.1) which shows that, for non degenerate target states, there is a universal relation between the control state of the environment and the target state of the field.

In the final part of this section we explain in what this relation is a natural extension of the fact that an equilibrium environment (at a given temperature for the free evolution) drives a system to an equilibrium state at the same temperature (for its own free evolution)

Lemma 5.1. *Suppose that:*
the coefficients Γ_{ij} of the master equation (22) satisfy condition (35), so that the master equation (22) takes the form (36), (37).
The probabilities $(\tilde{p}_i)$ satisfy the conditions

$$0 < \tilde{p}_i \neq \tilde{p}_j \quad ; \qquad \forall i \neq j \tag{50}$$

Then the $(\tilde{p}_j)$ are decreasing functions of the energies (E_j) and, if condition (43) is satisfied (i.e. the density of quanta in the environment state is a function of the energy density) then the following universal relation holds:

$$N_{ij} = N(E_i - E_j) = \frac{1}{\frac{\tilde{p}_j}{\tilde{p}_i} - 1} = \frac{1}{\frac{\Gamma_j}{\Gamma_i} - 1} \; ; \quad \forall E_j < E_i \tag{51}$$

Remark. The relation (51) is universal in the sense that it is independent of everything but the two energy level involved (E_i, E_j) and the pair correlations of the fied. The explicit formula for the quotient Γ_j/Γ_i,

$$\frac{\Gamma_j}{\Gamma_i} = \frac{1 + |\tanh(\omega_{ij})|^2 \dfrac{n(\omega_{ij})}{n(\omega_{ij})+1}}{|\tanh(\omega_{ij}|^2 + \dfrac{n(\omega_{ij})}{n(\omega_{ij})+1}} \,, \tag{52}$$

Proof. Solving (45) for N_{ij} with we

$$E_i > E_j \tag{53}$$

we get

$$N_{ij} = \frac{\Gamma_{ij}}{Q_{ij}|d_{ij}|^2}, \quad N_{ij} = \frac{\Gamma_{ji}}{Q_{ij}|d_{ij}|^2} - 1. \tag{54}$$

Therefore (54) is equivalent to

$$N_{ij} = \frac{\Gamma_{ij}}{Q_{ij}|d_{ij}|^2} = \frac{\Gamma_{ji}}{Q_{ij}|d_{ij}|^2} - 1$$

and, because of (35), (37) this is equivalent to

$$N_{ij} = \frac{\Gamma\tilde{p}_i}{Q_{ij}|d_{ij}|^2} = \frac{\Gamma\tilde{p}_j}{Q_{ij}|d_{ij}|^2} - 1\,. \tag{55}$$

From this equation we obtain

$$|d_{ij}|^2 = \frac{\Gamma}{Q_{ij}}\left(\tilde{p}_j - \tilde{p}_i\right). \tag{56}$$

Since Γ, $Q_{ij} \geq 0$, this and assumption (53) imply that $\tilde{p}_j$ is a decreasing function of E_j. Finally, using (50) and (55) we find

$$N_{ij} = \frac{\tilde{p}_i}{\tilde{p}_j - \tilde{p}_i} \tag{57}$$

which is (51). To prove universality notice that the expression of the quotient Γ_j/Γ_i in (51) is given by (35), (25), (26)–(29). Therefore assumption (43) implies that, for $E_i > E_j$

$$\Gamma_i = \Gamma_{ij} = \Gamma^{(+)}_{c,ji} + \Gamma^{(-)}_{s,ji} = 2\pi|c_0(|\omega_{ji}|)^2 n(|\omega_{ji}|)|d_{ij}|^2 Q_{ij} +$$

$$2\pi|s_0(|\omega_{ji}|)^2(n(|\omega_{ji}|)+1)|d_{ij}|^2 Q_{ij}$$

$$\Gamma_j = \Gamma_{ji} = \Gamma^{(-)}_{c,ij} + \Gamma^{(+)}_{s,ij} = 2\pi|c_0(|\omega_{ji}|)|^2(n(|\omega_{ji}|)+1)|d_{ij}|^2 Q_{ij}$$

$$+2\pi|s_0(|\omega_{ji}|^2)n(|\omega_{ji}|)|d_{ij}|^2 Q_{ij} \; .$$

Therefore, for $E_i > E_j$:

$$\frac{\Gamma_i}{\Gamma_i} = \frac{|c_0(\omega_{ij})|^2(n(\omega_{ij})+1) + |s_0(\omega_{ij})|^2 n(\omega_{ij})}{|c_0(\omega_{ij})|^2 n(\omega_{ij}) + |s_0(\omega_{ij})|^2(n(\omega_{ij})+1)}$$

$$= \frac{|c_0(\omega_{ij})|^2 + |s_0(\omega_{ij})|^2 \frac{n(\omega_{ij})}{n(\omega_{ij})+1}}{|c_0(\omega_{ij})|^2 \frac{n(\omega_{ij})}{n(\omega_{ij})+1} + |s_0(\omega_{ij})|^2} \; .$$

Let us consider two typical cases for the choice of c_k and n_k. Of course, it is always possible to consider other cases which are combination of the following two situations.

5.1. *Squeezed vacuum environment*

The simplest case corresponds to the squeezing of the vacuum state. In this case, since $n(k) = 0$, we get

$$\mathcal{N}(k) = |c_k|^2 - 1 = |s_k|^2 \tag{58}$$

Clearly, it is always possible to choose c_k so that $\mathcal{N}(k)$ satisfies (57). This means that we squeeze the a_k-field with a mode-dependent squeezing parameter.

5.2. *Non-squeezed but non-linear temperature state*

We can also think non-squeezed but non-linear temperature state. In this case, since $c_k = 1$ ($s_k = 0$) we get

$$\mathcal{N}(k) = n(k). \tag{59}$$

and again it is always possible to chose the density $n(k)$ so that $\mathcal{N}(k)$ satisfies (57). Comparing (58) and (59), one can see the similarity of the roles of squeezing and number density modification, which is the replacement of $|c_k|^2$ by $n(k)+1$ in (58).

Notice that, when the target state is a thermal equilibrium state for H_S, i.e.

$$\frac{\tilde{p}_j}{\tilde{p}_i} = e^{\beta(E_i - E_j)}, \tag{60}$$

from condition (57),

$$N_{ij} = \frac{1}{\frac{\tilde{p}_j}{\tilde{p}_i} - 1} = \frac{1}{e^{\beta(E_i - E_j)} - 1} = \frac{1}{e^{\beta\omega(k_{ij})} - 1}, \tag{61}$$

where $\omega(k_{ij}) = \omega_{ij} = E_i - E_j$. This tells us that a thermal state of the (free) environment at a given temperature drives the system to thermal state for its free Hamiltonian at the same temperature. In the general case, $n(k)$ is not the usual Gibbs factor but something else which can be described by a non-linear temperature $\beta(\omega(k))$ as

$$n(k) = \frac{1}{e^{\beta(\omega(k))\cdot\omega(k)} - 1}\,. \tag{62}$$

Because of the assimilation Lemma (5.1) this is equivalent to the following relation for the target state.

$$\frac{\tilde{p}_j}{\tilde{p}_i} = e^{\beta(\omega(k_{ij}))\cdot\omega(k_{ij})}. \tag{63}$$

So the nonlinear temperature function describing the stationary state of the system is the same one describing the control state of the environment. In this sense we say that the non-linear temperature environment drives the system to the same non-linear temperature state of the field. Notice that this nontrivial "assimilation" occurs as a consequence of an appropriate choice of the coupling to the environment, i.e. of the d_{ji} whereas in the linear case of a temperature environment (i.e. an usual equilibrium state) this is a quite general phenomenon[3,11], (See also [14].) If the d_{ji} don't satisfy condition (56), the non-linear temperature environment drives the system to a different stationary state. (See also [11].)

6. Purification

In this section, we discuss the case where the target state is pure,i.e.

$$\tilde{p}_0 = 1, \quad \tilde{p}_1 = \tilde{p}_2 = \cdots \tilde{p}_N = 0 \tag{64}$$

for $E_0 < E_1 < E_2 < \cdots < E_N$.

One can get such a pure state with the control procedure which we discussed in this paper. From conditions (57) and (56), we choose d_{i0} and N_{i0} as

$$|d_{i0}|^2 = \frac{\Gamma}{Q_{i0}}, \tag{65}$$

$$N_{i0} = 0. \tag{66}$$

As for the other d_{ij} and N_{ij}, we choose as

$$|d_{ij}|^2 N_{ij} Q_{ij} = 0. \tag{67}$$

Notice that (67) meets with d_{ij} chosen as (56), as far as N_{ij} and Q_{ij} are finite. Threfore the sufficient conditions for this purification is (56). Under these conditions, the state of the system converges to the lowest eigenstate $|\mu_0\rangle\langle\mu_0|$, obeying the master equation (36). This process can be understood as cooling the system by the vacuum. The energy which the initial state of the system has is released to the environment throught the coupling given by (65). Notice that, since $d_{ij} = 0$ for $j \neq 0$, N_{ij} doesn't affect this dynamics for $j \neq 0$. In other words, the initial energy is released by the direct transition from some energy level to the lowest eigenstate.

However, as discussed in the previous section, this cooling effect by the vacuum environment is more universal. As the thermal environment has assimilation property for a a large class of d_{ij}, the vacuum environment generally drives the system to the lowest eigenstate of the H_S. (In fact, the vacuum environment is the low temperature limit of the thermal state.) From this point of view, the specific choice for d_{ij} is not necessary for this cooling procedure. The necessary and sufficient condition for this cooling is the same with the condition for the assimilation in a linear temperature case (see previous section). Therefore $d_{ij} \neq 0$ for all $i \neq j$ is a sufficient condition for this thermal (vacuum) assimilation. Once one can prepare $\mathcal{N}(k) = 0$ (it means vacuum state of the environment), the state of the system is driven to the lowest energy eigen state through the proper but much more general interaction which satisfies the condition [14].

7. Example: Entangled state with Spin-Boson model

Let us illustrate our technique with one important example, that is, driving to the singlet state a system made of two 1/2-spin particles.

We consider the system Hamiltonian

$$H_S = \sum_{j=1}^{4} \epsilon_j |\phi_j\rangle\langle\phi_j| \tag{68}$$

where

$$|\phi_1\rangle = \frac{1}{\sqrt{2}}\Big(|+\rangle_A|-\rangle_B + |-\rangle_A|+\rangle_B\Big) \tag{69}$$

$$|\phi_2\rangle = \frac{1}{\sqrt{2}}\Big(|+\rangle_A|-\rangle_B - |-\rangle_A|+\rangle_B\Big) \tag{70}$$

$$|\phi_3\rangle = \frac{1}{\sqrt{2}}\Big(|+\rangle_A|+\rangle_B + |-\rangle_A|-\rangle_B\Big) \tag{71}$$

$$|\phi_4\rangle = \frac{1}{\sqrt{2}}\Big(|+\rangle_A|+\rangle_B - |-\rangle_A|-\rangle_B\Big). \tag{72}$$

$$\sigma_z^{(j)}|\pm\rangle_j = \pm|\pm\rangle_j, \quad j = A, B. \tag{73}$$

Notice that the Hamiltonian (68) can be rewritten as

$$H_S = \alpha\sigma_x^{(A)} \otimes \sigma_x^{(B)} + \beta\sigma_y^{(A)} \otimes \sigma_y^{(B)} + \gamma\sigma_z^{(A)} \otimes \sigma_z^{(B)} + \delta \tag{74}$$

i.e. like a standard Heisenberg model, where

$$\sigma_x^{(j)}|R\rangle_j = |R\rangle_j, \quad \sigma_x^{(j)}|L\rangle_j = -|L\rangle_j, \quad \sigma_y^{(j)}|U\rangle_j = |U\rangle_j, \quad \sigma_y^{(j)}|D\rangle_j = -|D\rangle_j$$

$$|R\rangle_j = \frac{1}{\sqrt{2}}\Big(|+\rangle_j + |-\rangle_j\Big), \quad |L\rangle_j = \frac{1}{\sqrt{2}}\Big(|+\rangle_j - |-\rangle_j\Big),$$

$$|U\rangle_j = \frac{1}{\sqrt{2}}\Big(|+\rangle_j - i|-\rangle_j\Big), \quad |D\rangle_j = \frac{1}{\sqrt{2}}\Big(|+\rangle_j + i|-\rangle_j\Big),$$

$$\alpha = \frac{+\epsilon_1 - \epsilon_2 + \epsilon_3 - \epsilon_4}{4}, \qquad \beta = \frac{+\epsilon_1 - \epsilon_2 - \epsilon_3 + \epsilon_4}{4} \tag{75}$$

$$\gamma = \frac{-\epsilon_1 - \epsilon_2 + \epsilon_3 + \epsilon_4}{4}, \qquad \delta = \frac{+\epsilon_1 + \epsilon_2 + \epsilon_3 + \epsilon_4}{4} \tag{76}$$

or

$$\epsilon_1 = +\alpha + \beta - \gamma + \delta, \qquad \epsilon_2 = -\alpha - \beta - \gamma + \delta \tag{77}$$

$$\epsilon_3 = +\alpha - \beta - \gamma + \delta, \qquad \epsilon_4 = -\alpha + \beta + \gamma + \delta. \tag{78}$$

Once one can control the parameters α, β, γ, (and δ), one can always prepare the Hamiltonian so that it becomes generic in the sense of (21).

$\omega_{ij} = \epsilon_i - \epsilon_j$	$i = 1$	$i = 2$	$i = 3$	$i = 4$
$j = 1$	0	$-2(\alpha + \beta)$	-2β	$-2(\alpha - \gamma)$
$j = 2$	$+2(\alpha + \beta)$	0	$+2\alpha$	$+2(\beta + \gamma)$
$j = 3$	$+2\beta$	-2α	0	$-2(\alpha - \beta - \gamma)$
$j = 4$	$+2(\alpha - \gamma)$	$-2(\beta + \gamma)$	$+2(\alpha - \beta - \gamma)$	0

We write the target state as $\mu = \sum_j p_j |\phi_j\rangle\langle\phi_j|$ with

$$p_1 \geq p_2 \geq p_3 \geq p_4. \tag{79}$$

The conditions for H_S we listed in the end of sec. 4 require to choose the parameter α, β, γ, δ so that

$$\epsilon_1 < \epsilon_2 < \epsilon_3 < \epsilon_4. \tag{80}$$

As for the interaction Hamiltonian with a bosonic environment, we take

$$\begin{aligned} H_I &= D \int dk \ g(k) a_k^\dagger + h.c. \\ &= \sum_{i \neq j} \int dk \ g_{ij}(k) |\phi_i\rangle\langle\phi_j| a_k^\dagger + h.c. \end{aligned} \tag{81}$$

where

$$D = \sum_{i \neq j} d_{ij} |\phi_i\rangle\langle\phi_j|, \quad g_{ij}(k) = \langle\phi_i|D|\phi_j\rangle g(k) = d_{ij} g(k). \tag{82}$$

The free Hamiltonian of the boson-field is

$$H_B = \int dk \ \omega(k) a_k^\dagger a_k, \quad [a_k, a_{k'}^\dagger] = \delta(k - k').$$

From (57), the proper d_{ij} for the control is obtained as

$$d_{ij} = e^{i\theta_{ij}} \left(\frac{\Gamma}{Q_{ij}}\right)^{1/2} \sqrt{p_i - p_j}, \quad \text{for} \quad i > j. \tag{83}$$

When $p_i \neq p_j$, the initial state of the field should satisfy

$$N_{ij} = \frac{p_i}{p_j - p_i} \tag{84}$$

with N_{ij} given by (47), (44), (40) and (18). Now let us consider the case when the target state is $|\phi_1\rangle\langle\phi_1|$. >From the discussion in the previous section, we find that

$$d_{1j} = e^{i\theta_{1j}} \left(\frac{\Gamma}{Q_{1j}} \right)^{1/2}, \quad j = 2, 3, 4, \qquad \text{and} \quad \text{other } d_{i<j} = 0 \tag{85}$$

is a sufficient condition. In addition, for simplicity, we assume that

$$Q_{1j} = Q(\text{i.e. it does not depend on } j). \tag{86}$$

In this case one possible choice of D is for example

$$D = \sqrt{\frac{\Gamma}{Q}} \sum_{j=2,3,4} |\phi_1\rangle\langle\phi_j|. \tag{87}$$

Notice the following condition holds:

$$\Gamma = \frac{1}{3} Q \langle\phi_1|DD^\dagger|\phi_1\rangle. \tag{88}$$

This tells us that the life time of the initial state Γ^{-1} depends on the choice of operator D. More precisely, the speed of the convergence is proportional to the norm of $D^\dagger|\phi_1\rangle$. In the general case without the additional condition (86), the Γ is constrained as

$$\frac{1}{3} Q^{(m)} \langle\phi_1|DD^\dagger|\phi_1\rangle \leq \Gamma \leq \frac{1}{3} Q^{(M)} \langle\phi_1|DD^\dagger|\phi_1\rangle \tag{89}$$

where

$$Q^{(m)} = \text{Min}_j Q_{1j}, \quad Q^{(M)} = \text{Max}_j Q_{1j}. \tag{90}$$

Acknowledgment

One of the authors (KI) is grateful to the Centro Vito Volterra for kind hospitality. This work is partially supported by JSPS for KI.

Appendix

Introducing a pair of independent creation and annihilation operators $\xi_1(k)$, $\xi_1^\dagger(k)$, $\xi_2(k)$, $\xi_2^\dagger(k)$ with commutation relations

$$[\xi_i(k), \xi_j^\dagger(k')] = \delta_{ij}\delta(k - k') \tag{91}$$

acting in a Fock space $\mathcal{H}_0$ with vacuum vector Φ_0

$$\xi_i(k)\Phi_0 = 0, \quad i = 1, 2, \tag{92}$$

the GNS representation of the a_k, $a_k^\dagger$-field with respect to the state (18) becomes

$$\begin{aligned}\tilde{a}_k &= \sqrt{n(k)+1}\xi_1(k) + \sqrt{n(k)}\xi_2^\dagger(k)\\ \tilde{a}_k^\dagger &= \sqrt{n(k)+1}\xi_1^\dagger(k) + \sqrt{n(k)}\xi_2(k).\end{aligned} \tag{93}$$

The Schrödinger equation in the interaction picture is therefore

$$\frac{d}{dt}U_t^{(\lambda)} = -i\lambda H_I(t)U_t^{(\lambda)} \tag{94}$$

where

$$\begin{aligned}H_I(t) &= e^{iH_0t}H_Ie^{-iH_0t}\\ &= \sum_{ij}\int dk\ g_{ij}^*(k)|\mu_j\rangle\langle\mu_i|a_ke^{-i(\omega(k)-\omega_{ji})t} + h.c.\\ &= \sum_{ij}\int dk\ g_{ij}^*(k)L_{ij}^\dagger\left\{c_k^*\left(\sqrt{n(k)+1}\xi_1(k) + \sqrt{n(k)}\xi_2^\dagger(k)\right)\right.\\ &\quad \left.+s_k\left(\sqrt{n(-k)+1}\xi_1^\dagger(-k) + \sqrt{n(-k)}\xi_2(-k)\right)\right\}e^{-i(\omega(k)-\omega_{ji})t}\\ &\quad +h.c.\\ &= \sum_{ij}L_{ij}^\dagger\left(\Xi_{c,ij}^{(-)}(t) + \Xi_{s,ij}^{(-)\dagger}(t) + \Xi_{c,ij}^{(+)\dagger}(t) + \Xi_{s,ij}^{(+)}(t)\right) + h.c.\end{aligned} \tag{95}$$

$$\Xi_{c,ij}^{(-)}(t) = \int dk\ g_{ij}^*(k)c_k^*\sqrt{n(k)+1}\ \xi_1(k)e^{-i(\omega(k)-\omega_{ji})t} \tag{96}$$

$$\Xi_{s,ij}^{(-)\dagger}(t) = \int dk\ g_{ij}^*(k)s_k\sqrt{n(-k)+1}\ \xi_1^\dagger(-k)e^{-i(\omega(k)-\omega_{ji})t} \tag{97}$$

$$\Xi_{c,ij}^{(+)\dagger}(t) = \int dk\ g_{ij}^*(k)c_k^*\sqrt{n(k)}\ \xi_2^\dagger(k)e^{-i(\omega(k)-\omega_{ji})t} \tag{98}$$

$$\Xi_{s,ij}^{(+)}(t) = \int dk\ g_{ij}^*(k)s_k\sqrt{n(-k)}\ \xi_2(-k)e^{-i(\omega(k)-\omega_{ji})t} \tag{99}$$

The stochastic limit describes the quantum dynamics in the regime of weak coupling ($\lambda \to 0$) and large times ($t \to t/\lambda^2$). Under rescaling the time $t \mapsto t/\lambda^2$, the Schrödinger equation becomes

$$\frac{d}{dt}U^{(\lambda)}(t/\lambda^2) = -i\frac{1}{\lambda}H_I(t/\lambda^2)U^{(\lambda)}(t/\lambda^2) \tag{100}$$

and the quantum field undergoes a frequency dependent rescaling

$$a_k \to \frac{1}{\lambda}e^{-i\frac{t}{\lambda^2}(\omega(k)-\omega)}a_k\,. \tag{101}$$

The main result of this theory is expressed by the stochastic golden rule [3,4,5] according to which in the limit $\lambda \to 0$:

(1) for each Bohr frequency ω, the rescaled field (101) becomes a quantum white noise (or master field) $b_\omega(t,k)$ satisfying the commutation relations

$$[b_\omega(t,k), b^\dagger_{\omega'}(t',k')] = \delta_{\omega\omega'} 2\pi\delta(t-t')\delta(\omega(k)-\omega)\delta(k-k') \quad (102)$$

(2) the solution $U^{(\lambda)}(t/\lambda^2)$ converges to the solution of the white noise equation

$$\frac{d}{dt}U(t) = -ih(t)U(t) \quad (103)$$

where the white noise Hamiltonian is given by

$$h(t) = \sum_{ij} L^\dagger_{ij}\left(b^{(-)}_{c,ij}(t) + b^{(-)\dagger}_{s,ij}(t) + b^{(+)\dagger}_{c,ij}(t) + b^{(+)}_{s,ij}(t)\right) + h.c. \quad (104)$$

and the master fields (white noises) are

$$b^{(\varepsilon)}_{x,ij}(t) = \lim_{\lambda\to 0}\frac{1}{\lambda}\Xi^{(\varepsilon)}_{x,ij}(t/\lambda^2), \qquad x = c, s, \quad \varepsilon = \mp. \quad (105)$$

In this case the commutation relations (102) become:

$$[b^{(\varepsilon)}_{x,ij}(t), b^{(\varepsilon)\,\dagger}_{y,i'j'}(t')] = \delta_{xy}\delta_{\varepsilon\varepsilon'}\delta_{\omega_{ji}\omega_{j'i'}}\Gamma^{(\varepsilon)}_{x,ij}\delta(t-t') \quad (106)$$

where, in addition, the commutation relation between U_t and these noise operators are obtained as

$$[b^{(-)}_{c,ij}(t), U_t] = \sum_{j'i'}\delta_{\omega_{ji}\omega_{j'i'}}(\Gamma^{(-)}_{c,ij} - i\Delta^{(-)}_{c,ij})L_{i'j'}U_t, \quad (107)$$

$$[b^{(-)}_{s,ij}(t), U_t] = \sum_{j'i'}\delta_{\omega_{ji}\omega_{j'i'}}(\Gamma^{(-)}_{s,ij} + i\Delta^{(-)}_{s,ij})L^\dagger_{i'j'}U_t, \quad (108)$$

$$[b^{(+)}_{c,ij}(t), U_t] = \sum_{j'i'}\delta_{\omega_{ji}\omega_{j'i'}}(\Gamma^{(+)}_{c,ij} + i\Delta^{(+)}_{c,ij})L^\dagger_{i'j'}U_t, \quad (109)$$

$$[b^{(+)}_{s,ij}(t), U_t] = \sum_{j'i'}\delta_{\omega_{ji}\omega_{j'i'}}(\Gamma^{(+)}_{s,ij} - i\Delta^{(+)}_{s,ij})L_{i'j'}U_t, \quad (110)$$

where

(3) From the step 1, by means of a standard procedure one can deduce the Langevin and the master equation (cf. [3]).

References

1. C.H.Bennet *et al*, Phys. Rev. Lett. **69**, 2881 (1992); **70**, 1895 (1993).
2. For example, *The Physics of Quantum Information*, edited by D. Bouwmeester, A. Ekert, and A. Zeilinger (Springer-Verlag, Heidelberg, 2000), and references therein.
3. L. Accardi, Y. G. Lu, and I. V. Volovich, *Quantum Theory and Its Stochastic Limit* Springer-Verlag (2002)
4. L. Accardi, A. Frigerio, and Y. G. Lu, Commun. Math. Phys. **131**, 537 (1990); L. Accardi, J. Gough, and Y. G. Lu, Rep. Math. Phys. **36**, 155 (1995); L.Accardi, S.V.Kozyrev, I.V.Volovich, Phys. Rev. A **57** (1997); Phys. Lett. A **260**, 31 (1999). G. Kimura, K. Yuasa, and K. Imafuku, Phys. Rev. A **63** (2001), 022103; Phys. Rev. Lett. **89** (2002), 140403.
5. L. Accardi, F. Fagnola (eds.), *Quantum interacting particle systems*, Lecture Note of Levico school, September 2000, Volterra Preprint N.431. in *Quantum interaction particles* L. Accardi, F. Fagnola (eds.) World Scientific (2002).
6. L. Accardi, A. Boukas, Control of quantum Langevin equations, Open Systems and Information Dynamics (2003); Control of elementary quantum flows, Proceedings of the 2–d Meijo Winter School, Quantum Information and Complexity, T. Hida (ed.), World Scientific (2003)
7. J.I. Cirac and P. Zoller, Phys. B **64**, 623 (1997)
8. C. Monroe, D.M. Meekhof, B.E. King, W.M. Itano, and D.J. Wineland, Phys. Rev. Lett. **75**, 4714 (1995).
9. P. Domokos, J.M. Raimond, M. Brune, and S. Haroche, Phys. Rev. A **52**, 3554 (1995).
10. A. Beige, D. Braun, B. Tregenna, and P.L. Knight, Phys. Rev. Lett. **85**, 1762 (2000).
11. L. Accardi, K. Imafuku, "Dynamical detailed balance and local KMS condition for non-equilibrium states" quant-ph/0209088.
12. Gorini V., Kossakowski A., Sudarshan E.C.G., J. Math. Phys **17**, 821-825 (1976).
13. Lindblad G. Comm. Math. Phys. **48**,119-130 (1976).
14. The necessary and sufficient condition of the interaction for the assimilation by the linear temperature environment (i.e. usual thermalization by equilibrium heat bath) is the generator in R.H.S of (22) generates full algebra on the system space [5].

LOGICAL OPERATIONS REALIZED ON THE ISING CHAIN OF N QUBITS

MASANARI ASANO, NORIHIRO TATEDA AND CHIKARA ISHII

Department of Physics, Faculty of Science, Tokyo University of Science, 1-3 Kagurazaka, Shinjuku-ku, Tokyo 162-8601, Japan
E-mail: asano@is.noda.tus.ac.jp

We propose *multi-qubit logical gates*, which implement logical operations on N qubits realized physically with the use of local manipulations of qubits before and after the one-time evolution of Ising chain (an array of N qubits connected with the Ising interactions), avoiding complicated tuning of the interactions between qubits. General rules of the action of our multi-qubit logical gate are derived by decomposing the whole process into a product of two-qubit logical operations. We construct the multi-qubit logical gate, which expressed as a group of controlled-NOT gates.

A quantum computer [1,2,3] performs parallel computations by operating a unitary transformation on the quantum states of N qubits. The explicit form of unitary operator depends on the problem to be solved, and the construction of such a unitary operator is one of the crucial tasks in the quantum computation (QC). Barenco *et al.* [4] showed that any unitary operations can be expressed as compositions of single-qubit and two-qubit Controlled-Not (CNOT) gates. In principle, QC is accomplished by performing these gates piece by piece.

In NMR quantum computers [5,6,7,8] that succeeded in realizing experimental quantum computations on a few qubits, the operation corresponding to a single-qubit gate is regarded as a manipulation of nuclear spin by applying of high-frequency pulses, and the CNOT gate is physically realized by effective tuning of two-qubit Ising interaction with an appropriate pulse train. Recently, Ladd *et al.* [9] proposed solid-state NMR implementation of quantum computation, in which a chain of ^{29}Si plays the role of qubit-system. The fabrication of this structure is feasible, and the quantum computations including a lot of qubits is expected. However, when size of the problem grows, an exponential or polynomial number of CNOT gates are needed to construct a quantum circuit. It is impossible to perform all these gates in the coherence time of a physical system. Furthermore, for realization of each of CNOT gates, complicated pulse trains will be needed to localize a relevant pair of spins from other qubits connected with multi-particle interactions, and it results in a huge loss of computation time.

To avoid these difficulties, we are going to propose a new scheme of QC, making effective use of the time evolution of the Ising chain consisting of multiple qubits. Here, the Ising chain is assumed as a simple model expressing the physical system, in which neighboring qubits are connected in the Ising interactions. We consider a series of simple multi-qubit gates with well-defined logical operations, which are realized with use of proper local manipulations at the beginning and end of the time evolution of N-qubit system. It is a great advantage that in the above physical processes, complicated tuning of the interactions is unneeded. Hereafter, we call these gates the *multi-qubit logical gate* . The final goal of our QC is to express any useful algorithm in terms of the above multi-qubit logical gates, just as the traditional sequential QC makes use of the single and CNOT gates. Needless to say, this is achieved only when we know; what kinds of logical operation our multi-qubit gates can implement? The present article is addressed just to this question.

The Ising spin chain is a well-known physical system, which is composed of an array of N spins playing the roles of qubits, coupled via nearest-neighbor Ising-interaction. The dynamics of this system is described by a Hamiltonian,

$$H = J\sum_{i=1}^{N} \sigma_i^z \sigma_{i+1}^z, \tag{1}$$

where σ_i^z stands for the Pauli matrix, and J being the Ising interaction constant. Let us choose the eigenstates $|0\rangle$, $|1\rangle$ of the i-th qubit σ_i^z as our computational basis. Then, the above Hamiltonian is diagonal in N qubits basis $|00\cdots 0\rangle$, $|00\cdots 1\rangle$, $\cdots$, $|11\cdots 1\rangle$. It is noted that the time evolution operator of the Ising chain $U_{Ic}(t) = \mathrm{e}^{-\mathrm{i}Ht}$ in the time interval t can be decomposed as follows,

$$\begin{aligned} U_{Ic}(t) &= \mathrm{e}^{-\mathrm{i}Ht} \\ &= \prod_{i=1}^{N} \mathrm{e}^{-\mathrm{i}(J\sigma_i^z\sigma_{i+1}^z)t} \\ &= \prod_{i=1}^{N} T_{i,i+1}^z(t). \end{aligned} \tag{2}$$

In this equation, $T^z_{i,i+1}(t) = \mathrm{e}^{-\mathrm{i}(J\sigma^z_i\sigma^z_{i+1})t}$ denotes a two-qubit unitary operator acting on the i-th and $(i+1)$-th qubits and $[T^z_{i,i+1}, T^z_{j,j+1}] = 0$.

Our multi-qubit logical gate is realized by the physical process of successive local manipulations acted on qubits at the time $t = 0$ and $t > 0$ and time evolution of the Ising system between these manipulations. By noticing that the Ising interactions need not be controlled, we can express this process in the form

$$U(t) = u_1 \otimes u_2 \otimes \cdots \otimes u_N U_{Ic} v_1 \otimes v_2 \otimes \cdots \otimes v_N, \tag{3}$$

where $u_i = u_i(t)$, $v_i = v_i(0)$ are local unitary operators acting on the i-th qubit at the time t and $t = 0$, expressing the effect of local manipulation due to external high-frequency pulses applied at each time. Here, by noticing that $[v_i, T^z_{i+1,i+2}] = 0$ and $[u_i, T^z_{i-2,i-1}] = 0$, we rewrite Eq. (3) in such a way

$$\begin{aligned} U(t) &= (u_1 \otimes u_2 T^z_{1,2} v_1 \otimes w_2)(w_2^\dagger \otimes u_3 T^z_{2,3} v_2 \otimes w_3) \\ &\quad \cdots \times (w_{N-1}^\dagger \otimes u_N T^z_{N-1,N} v_{N-1} \otimes v_N) \\ &= V_{1,2} V_{2,3} V_{3,4} \cdots V_{N-2,N-1} V_{N-1,N}. \end{aligned} \tag{4}$$

The w_i are local unitary operators, whose form can be chosen arbitrarily, since they do not have to be performed through the actual operations in Eq. (3). Here, we have introduced the two-qubit operator, $V_{i,i+1} = V_{i,i+1}(t)$ in Eq. (4) which consists of T^z and four local operators

$$V_{i,i+1} = w_i^\dagger \otimes u_{i+1} T^z_{i,i+1} v_i \otimes w_{i+1}, \tag{5}$$

where $w_1^\dagger = u_1$ and $w_N = u_N$ are assumed and the operators $V_{i-1,i}$ and $V_{i,i+1}$ are not commutable in general: $[V_{i-1,i}, V_{i,i+1}] \neq 0$. Thus, rewriting the form of unitary operation of Eq. (3) to the one of Eq. (4) leads us to know that N-qubit logical operations of our multi-qubit logical gate are specified if all $V_{i,i+1}$ in Eq. (4) are well-defined as two-qubit logical gates.

To derive the logical operation in Eq. (4) satisfying the above condition, we first consider the following periodic forms of the local operator u_i and v_i $(i = 1 \cdots N)$ in Eq. (5).

$$\begin{aligned} &u_1 = u_A, \quad u_{2n} = u_B, \quad u_{2n+1} = v_A^\dagger, \\ &v_N = v_B, \quad v_{2n} = u_B^\dagger, \quad v_{2n-1} = v_A, \\ &w_{2n-1} = u_A^\dagger, \quad w_{2n} = v_B, \\ &n = 1 \cdots N/2. \end{aligned} \tag{6}$$

Then, all two-qubit unitary operators $V_{i,i+1}$ can be expressed as

$$\begin{aligned} V_{2n-1,2n} &= u_A \otimes u_B T^z v_A \otimes v_B \equiv V, \\ V_{2n,2n+1} &= S(v_A^\dagger \otimes v_B^\dagger T^z u_A^\dagger \otimes u_B^\dagger) S \\ &\equiv SV'S, \end{aligned} \tag{7}$$

where, the S stands for the swap operator which exchanges the states of the $2n$-th and $(2n+1)$-th qubits, and except for global phases, the V' is equal to the transpose of V in the magic basis [10].

$$V' = e^{-2\mathrm{i}\delta} V^T. \tag{8}$$

This relation implies that both V and V' are well-defined logical gates, once we can construct either of which as a well-defined logical gate by choosing the local operators $u_A \otimes u_B$ and $v_A \otimes v_B$ appropriately.

Next, let us choose the local operators such as $u_A \otimes u_B = \mathrm{I} \otimes R_y(\frac{\pi}{2})$, $v_A \otimes v_B = R_z(-\frac{\pi}{2}) \otimes R_z(\frac{\pi}{2}) R_y(-\frac{\pi}{2})$ $(R_i(\theta) = e^{-\mathrm{i}\theta\sigma_i}/2)$ and the evolution time for $Jt = \frac{\pi}{4}$. Then the V turns out to the well known CNOT gate,

$$V = \mathrm{I}_i \otimes \sigma_{i+1}^{x_i} \tag{9}$$

where $\sigma = \begin{pmatrix} 0 & 1 \\ 1 & 0 \end{pmatrix}$ and I is a unit matrix. Moreover, the relation between V and V' in Eq. (8) leads us to know that the gate $SV'S$ in Eq. (7) also represents the CNOT gate such as,

$$(SV'S)_{j,j+1} = \sigma_j^{\bar{x}_{j+1}} \otimes \mathrm{I}_{j+1} \ (\bar{x}_{j+1} = 1 - x_{j+1}), \tag{10}$$

except for global phases. Consequently, the operation in Eq. (4) can be described in the following form.

$$\begin{aligned} U &= \mathrm{I} \otimes \sigma^{x_1 \oplus \bar{x}_3} \otimes \mathrm{I} \otimes \sigma^{x_3 \oplus \bar{x}_5} \otimes \\ &\cdots \otimes \mathrm{I} \otimes \sigma^{x_{N-3} \oplus \bar{x}_{N-1}} \otimes \mathrm{I} \otimes \sigma^{x_{N-1}} \\ &= I_{odd} \otimes \left[\left(\prod_{k=1}^{n-1} \sigma_{2k}^{x_{2k-1} \oplus \bar{x}_{2k+1}} \right) \sigma_N^{x_{N-1}} \right]_{even} \\ &\equiv I_{odd} \otimes \Lambda_{even}, \end{aligned} \tag{11}$$

where $x_i = 0, 1$ denotes computational basis of the i-th qubit, and the system size $N = 2n$ is assumed. The operator Λ_{even} acting on the states of qubits on the even sites consist of controlled operator $\sigma_{2k}^{x_{2k-1} \oplus \bar{x}_{2k+1}}$ and $\sigma_N^{x_{N-1}}$ which invert the state of each bit on the even sites when $x_{2k-1} \oplus$

$x_{2k+1} = 1$ and $x_{N-1} = 1$ ($a \oplus b$ represents exclusive-OR of a and b). By this U, a computational basis of N qubits denoted by

$$\left|x_1, x_2, \cdots, x_{N-1}, x_N\right\rangle \equiv \left|o_1, o_2, \cdots, o_{n-1}, o_n\right\rangle_{odd} \otimes \left|e_1, e_2, \cdots, e_{n-1}, e_n\right\rangle_{even} \quad (12)$$

is transformed as follows:

$$U\left|x_1, x_2, \cdots, x_{N-1}, x_N\right\rangle = \left|o_1, o_2, \cdots, o_{n-1}, o_n\right\rangle_{odd} \otimes \left|o_1 \oplus \bar{o}_2 \oplus e_1, o_2 \oplus \bar{o}_3 \oplus e_2, \cdots, o_{n-1} \oplus \bar{o}_n \oplus e_{n-1}, o_n \oplus e_n\right\rangle_{even}. \quad (13)$$

The above equation shows that the quantum state of every odd (even) site affects the results of transformation on the state of every even (odd)-numbered site. For example, in the case of $N = 4$, the controlled operator Λ_{even} has the variable forms depending on the four states of odd-numbered site (see Table 1). Here, let us note that the combination of three two-qubit CNOT gates constructs a swap gate. Correspondingly, our multi-qubit logical gate implementing operation U of Eq. (11) is used as a elementary gate for construction of a simple operation which swaps about states of even numbered qubits and those of odd numbered qubits. This topic will be discussed in detail in another paper.

o_1 o_2	Λ_{even}
0 0	$\sigma \otimes \mathrm{I}$
0 1	$\mathrm{I} \otimes \sigma$
1 0	$\mathrm{I} \otimes \mathrm{I}$
1 1	$\sigma \otimes \sigma$

Our approach will be applicable to other multi-qubit gates, realized on any physical systems, whose time evolution operator can be decomposed into two-qubit operators. How can we perform useful quantum algorithms by the use of combination of these N-qubits logical gates? The theoretical frameworks to answer the above question would be needed in future.

References

1. M. Nielsen and I. Chuang, *Quantum Computation and Quantum Information* (Cambridge University Press, Cambridge, UK, 2000).
2. D. DiVincenzo, Science **270**, 255 (1995).

3. A. Ekert and R. Jozsa, Rev. Mod. Phys. **68**, 733 (1996).
4. A. Barenco, C. H. Bennett, R. Cleve, D. P. DiVincenzo, N. Margolus, P. Shor, T. Sleator, J. A. Smolin, and H. Weinfurter, Phys. Rev. A **52**, 3457 (1995).
5. N. A. Gershenfeld and I. Chuang, Science 275, 350 (1997).
6. D. Cory, A. Fahmy, and T. Havel, Proc. Natl. Acad. Sci. U.S.A. **94**, 1634 (1997).
7. E. Knill et al., Phys. Rev. A **57**, 3348 (1998).
8. I. Chuang and N. A. Gershenfeld, M. Kubinec, and D. Leung, Proc. R. Soc. London, Ser. A **454**, 447 (1998)
9. T.D. Ladd, J.R. Goldman, F.Yamaguchi, and Y. Yamamoto, Phys. Rev. Lett. **89**, 017901 (2002).
10. S. Hill and W. K. Wootters, Phys. Rev. Lett. **78**, 5022 (1997)

JOINT EXTENSION OF STATES OF FERMION SUBSYSTEMS

HUZIHIRO ARAKI

Research Institute for Mathematical Sciences
Kyoto University, Sakyoku, Kyoto 606-8502, Japan
E-mail: araki@kurims.kyoto-u.ac.jp

The problem of existence and uniqueness of a state of a joint system, when its restrictions to subsystems are specified, is studied for a Fermion system, where a novel feature as compared with Boson or spin systems is non-commutativity between algebras of subsystems located at disjoint regions. The case of two subsystems with one of the subsystem states pure is completely analyzed.

1. Introduction

For Fermion systems such as electrons, the property of a state and its restrictions to subsystems is different from the case of tensor product systems such as photons and spin lattice systems due to mutual non-commutativity of Fermion subalgebras located at disjoint regions. For example, it is known that the Schmidt decomposition of a state in terms of subsystem states and some entropy inequalities, which hold for tensor product system, fail for Fermion systems although they remain to be valid for even states of Fermion systems [4,5].

In this review, the question of the existence and uniqueness of states of the total system, when its restrictions to subsystems localized at disjoint regions are specified, will be discussed for Fermion systems, emphasizing the difference with tensor product systems.

First, we consider a "product state extension" of given states of subsystems. While the product state extension always exists for tensor product systems, it exists for Fermion systems if and only if all of subsystem states with only one possible exception are even [1,6].

Next we will describe the situation for the case of two subsystems with one of the subsystem states being pure. In most cases we have the unique extension if the extension exists, but there are some cases where the extensions are not unique with somewhat involved extensions [1].

Lastly, we will mention some examples of multiple extensions [1] when both subsystem states are not pure.

2. Notation for Fermion Systems

We briefly introduce minimal notation. The joint system is described by a unital C*-algebra $\mathcal{A}$ and subsystems are C*-subalgebras of $\mathcal{A}$ denoted by $\mathcal{A}_i$, either i=1,2 for two subsystems or for an arbitrary number of the indices in Sec.3. In any case, $\cup_i \mathcal{A}_i$ generates $\mathcal{A}$ as a C*-algebra, and $\mathcal{A}_i \cap \mathcal{A}_j$ is a trivial subalgebra $\mathbb{C}\mathbf{1}$ for $i \neq j$.

There is an automorphism Θ of $\mathcal{A}$ such that

(1) Θ^2 =identity map,
(2) $\Theta(\mathcal{A}_i) = \mathcal{A}_i$ (the invariance as a set)

It is the Fermion even - odd indicator, multiplying each Fermion creation and annihilation operators with -1, while keeping any Boson or spin operators invariant.

Any $A \in \mathcal{A}$ is split as a sum of even and odd parts:

$$A = A_+ + A_-, \quad A_\pm \equiv \frac{1}{2}(A \pm \Theta(A)),$$
$$A_\pm \in \mathcal{A}_\pm \equiv \{A \in \mathcal{A} \mid \Theta(A) = \pm A\}.$$

By the above property (2) for Θ, we have also the splitting of the subalgebras:

$$\mathcal{A}_i = \mathcal{A}_{i+} - \mathcal{A}_{i-}, \quad \mathcal{A}_{i\pm} = \mathcal{A}_i \cap \mathcal{A}_\pm.$$

The crucial assumption is the following graded commutation relations. Let $A_{i\pm} \in \mathcal{A}_{i\pm}$. Then for $i \neq j$,

$$[A_{1+}, A_{2+}]_- = [A_{1+}, A_{2-}]_- = [A_{1-}, A_{2+}]_- = 0,$$
$$[A_{1-}, A_{2-}]_+ = 0,$$

where $[A, B]_\pm \equiv AB \pm BA$.

3. Product Extension

We consider subalgebras $\mathcal{A}_i$ of $\mathcal{A}$ as described in the preceding section. A state φ of $\mathcal{A}$ is called a product state of states φ_i of $\mathcal{A}_i$ if

$$\varphi(A_{i_1}, \cdots, A_{i_n}) = \prod_{k=1}^{n} \varphi_{i_k}(A_{i_k})$$

for any finite number of indices $i_1, \cdots, i_n$ and $A_{i_k} \in \mathcal{A}_{i_k}$ $(k = 1, \cdots, n)$. A state ψ of an algebra $\mathcal{A}_0$ is even if

$$\psi(\Theta(A)) = A$$

for all $A \in \mathcal{A}_0$. The following theorems hold [1,6].

Theorem 3.1. *(1) A product state of states φ_i of subalgebras $\mathcal{A}_i$ exists if and only if all states φ_i except at most one are even.*
(2) Suppose that all φ_i are pure. If there exists a joint extension φ of $\{\varphi_i\}$ (namely, a state φ of $\mathcal{A}$ such that its restriction to $\mathcal{A}_i$ is φ_i for every i), then it must be uniquely the product state extension of (1) and it is pure.

This result shows that there is a restriction on pure states of subsystems that all but one of them should be even in order to have any joint extension, in contrast to the situation for the tensor product systems such as Boson and spin systems. The above result determines all possible joint extensions (which turn out to be uniquely the product state extension) for the case of pure states of subsystems.

The "if" part of the first part of Theorem 3.1(1) is also given in Theorem 11.2 of [2], which plays an important role in that paper.

Theorem 3.2. *Let φ be the product state extension of states φ_i which are even except for φ_1.*
(1) φ_1 is pure if φ is pure.
(2) Assume that the GNS representations π_{φ_1} and $\pi_{\varphi_1\Theta}$ of states φ_1 and $\varphi_1\Theta$ $(\varphi_1\Theta(A) = \varphi_1(\Theta(A)))$ are not disjoint. Then φ is pure if and only if all φ_i are pure.

The assumption about π_{φ_1} and $\pi_{\varphi_1\Theta}$ in (2) is automatically satisfied if φ_i is even. Also, if the algebra $\mathcal{A}_1$ is of a finite dimension, the same assumption is automatic.

In the case not covered by Theorem 3.2, the following result gives a complete analysis if we lump together all subsystems except $i = 1$ as the subsystem 2.

Theorem 3.3. *Let φ be the product state extension of states φ_1 and φ_2 where φ_2 is even and the GNS representations π_{φ_1} and $\pi_{\varphi_1\Theta}$ are disjoint.*
(1) φ is pure if and only if φ_1 and the restriction φ_{2+} of φ_2 to A_{2+} are both pure.
(2) Assume that φ is pure. φ_2 is not pure if and only if

$$\varphi_2 = \frac{1}{2}(\widehat{\varphi}_2 + \widehat{\varphi}_2\Theta)$$

where $\widehat{\varphi}_2$ is pure and the GNS representations $\pi_{\widehat{\varphi}_2}$ and $\pi_{\widehat{\varphi}_2\Theta}$ are disjoint.

4. A Pair of Pure and General States for Two Subsystems

First we mention a result for two general states, which establish a background for the case where one of the two states is pure.

Theorem 4.1. *Let φ_1 and φ_2 be states of two subsystems. Assume that the GNS representations π_{φ_1} and $\pi_{\varphi_1\Theta}$ are disjoint. Then a joint extension of φ_1 and φ_2 exists if and only if φ_2 is even.*

Again we see from this Theorem that any joint extension does not exist for some type of a pair of states for two subsystems of Fermion systems, a situation which does not occur for tensor product systems.

In order to describe detailed results when one of the two subsystem states is pure, we have to introduce some quantities which indicate relations between states φ and $\varphi\Theta$.

Definition 4.1. (1) Consider any representation π of a C*-algebra $\mathcal{A}$ in a Hilbert space $\mathcal{H}$, containing unit vectors Φ and Ψ, which give rise to two given states φ and ψ:

$$\varphi(A) = (\Phi, \pi(A)\Phi), \quad \psi(A) = (\Psi, \pi(A)\Psi), \quad A \in \mathcal{A}.$$

(The inner product is linear in the second vector according to the physics convention.) The transition probability between φ and ψ is defined by

$$P(\varphi, \psi) = \sup |(\Phi, \Psi)|^2$$

where the supremum is taken over all $\mathcal{H}$, π, Φ and Ψ as described above.
(2) $p(\varphi) = P(\varphi, \varphi\Theta)^{\frac{1}{2}}$.

If φ is pure, then $\varphi\Theta$ is also pure and the GNS representations π_φ and $\pi_{\varphi\Theta}$ are irreducible. In this case, there are two alternatives:
(α) They are mutually disjoint and $p(\varphi) = 0$.
(β) They are unitarily equivalent and

$$p(\varphi) = |(\Omega_\varphi, \Omega_{\varphi\Theta})|$$

where Ω_φ and $\Omega_{\varphi\Theta}$ are vectors in the GNS space $\mathcal{H}_\varphi$, which represent the states φ and $\varphi\Theta$, respectively and are unique up to the multiplication of phase factors (i.e., complex numbers of modulus 1).

The quantity $p(\varphi)$ has the range $[0, 1]$. Even in the case (β) above, $p(\varphi)$ can take the value 0. On the other hand, $p(\varphi) = 1$ if and only if $\varphi = \varphi\Theta$, namely φ is even.

Definition 4.2. (1) For two states φ and ψ, define

$$\lambda(\varphi, \psi) = \sup\{\lambda \in \mathbb{R} \mid \varphi - \lambda\psi \geq 0\}.$$

(2) For a state φ, define

$$\lambda(\varphi) = \lambda(\varphi, \varphi\Theta).$$

The range of λ is also $[0, 1]$. If $\lambda(\varphi) = 1$, then $\varphi = \varphi\Theta$. In this definition, $\varphi - \lambda\psi \geq 0$ means $\varphi(A^*A) - \lambda\psi(A^*A) \geq 0$ for any element A of the algebra.

The following result gives a complete answer for a joint extension of states φ_1 and φ_2 of two subsystems when one of the subsystem states is pure.

Theorem 4.2. *Let φ_1 and φ_2 be states of two subsystems. Assume that φ_1 is pure.*
(1) A joint extension φ of φ_1 and φ_2 exists if and only if

$$\lambda(\varphi_2) \geq \frac{1 - p(\varphi_1)}{1 + p(\varphi_1)}.$$

(2) If the inequality in (1) is satisfied and $p(\varphi_1) \neq 0$, then a joint extension φ is unique and satisfies

$$\varphi(A_1 A_2) = \varphi_1(A_1)\varphi_2(A_{2+}) + \frac{1}{p(\varphi_1)} f(A_1)\varphi_2(A_{2-}),$$
$$f(A_1) = \left(\Omega_{\varphi_1}, \pi_{\varphi_1}(A_1) u_1 \Omega_{\varphi_1}\right)$$

for $A_1 \in \mathcal{A}_1$, $A_2 = A_{2+} + A_{2-}$, $A_{2\pm} \in \mathcal{A}_{2\pm}$ and the Θ-implementing unitary u_1 on the GNS representations $(\pi_{\varphi_1}, \mathcal{H}_{\varphi_1})$ of the state φ_1 which satisfies $\left(\Omega_{\varphi_1}, u_1\Omega_{\varphi_1}\right) > 0$ (these representations uniquely characterizing u_1) where Ω_{φ_1} is the GNS vector for φ_1.
(3) If $p(\varphi_1) = 0$, the inequality in (1) is equivalent to $\varphi_2 = \varphi_2\Theta$, and at least the product state extension of Theorem 1 exists.
(4) Assume that $p(\varphi_1) = 0$ and φ_2 is even. There exists a joint extension of φ_1 and φ_2 other than the unique product state extension if and only if φ_1 and φ_2 satisfy the following two conditions:
(4-1) π_{φ_1} and $\pi_{\varphi_1\Theta}$ are unitarily equivalent.
(4-2) There exists a state $\widetilde{\varphi}_2$ of $\mathcal{A}_2$ such that $\widetilde{\varphi}_2 \neq \widetilde{\varphi}_2\Theta$ and

$$\varphi_2 = \frac{1}{2}\left(\widetilde{\varphi}_2 + \widetilde{\varphi}_2\Theta\right).$$

(5) If all the conditions in (4) (including $p(\varphi_1) = 0$ and φ_2 is even) are satisfied, then, for every $\widetilde{\varphi}_2$, there exists a joint extension φ of φ_1 and φ_2 satisfying

$$\varphi(A_1 A_2) = \varphi_1(A_1)\varphi_2(A_{2+}) + \left(\Omega_{\varphi_1}, \pi_{\varphi_1}(A_1) u_1 \Omega_{\varphi_1}\right) \widetilde{\varphi}_2(A_{2-}).$$

Such extensions together with the product state extension exhaust all joint extensions of φ_1 and φ_2.

5. Examples

When states of subsystems are mixtures, there are many joint extensions of given subsystem states even for Fermion systems. We give two examples in the case of two subsystems $\mathcal{A}_1$ and $\mathcal{A}_2$, which can be full matrix algebras (i.e., of finite dimensions).

Example 5.1. Let ρ be an invertible density matrix of a state of the total system $\mathcal{A} = \mathcal{A}_{12}$ satisfying $\rho \geq \lambda \mathbf{1}$ for some $\lambda > 0$. Let $x = x^* \in \mathcal{A}_{1-}$ and $y = y^* \in \mathcal{A}_{2-}$ satisfying $\|x\| \, \|y\| \leq \lambda$. Let

$$\varphi_1(A_1) = Tr(\rho A_1), \quad A_1 \in \mathcal{A}_1,$$
$$\varphi_2(A_2) = Tr(\rho A_2), \quad A_2 \in \mathcal{A}_2.$$

Then

$$\varphi_{\rho'}(A) = Tr(\rho' A), \quad \rho' = \rho + ixy$$

is a state of $\mathcal{A} = \mathcal{A}_{12}$ and a joint extension of φ_1 and φ_2 for any choice of x and y (satisfying the above stated condition).

This construction gives a continuously many extensions of φ_1 and φ_2.

Example 5.2. Let φ and ψ be states of $\mathcal{A}_1$ and $\mathcal{A}_2$ with decompositions

$$\varphi = \lambda\varphi_1 + (1-\lambda)\varphi_2, \quad \psi = \mu\psi_1 + (1-\mu)\psi_2$$

where $0 < \lambda, \mu < 1$, φ_1 and φ_2 are even but ψ_1 and ψ_2 are arbitrary. Let $\varphi_i\psi_j$ be the product state extension of φ_i and ψ_j. Then

$$\begin{aligned}\chi = &(\lambda\mu + \kappa)\varphi_1\psi_1 + (\lambda(1-\mu) - \kappa)\varphi_1\psi_2 \\ &+ ((1-\lambda)\mu - \kappa)\varphi_2\psi_1 + ((1-\lambda)(1-\mu) + \kappa)\varphi_2\psi_2\end{aligned}$$

is a one parameter family of joint extensions of φ and ψ with the parameter κ satisfying

$$-\min(\lambda\mu, (1-\lambda)(1-\mu)) \leq \kappa \leq \min((1-\lambda)\mu, \lambda(1-\mu)).$$

References

1. Huzihiro Araki and Hajime Moriya, Joint extension of states of subsystems for a CAR systems, Commun. Math. Phys., 237 (2003), 105-122.
2. Huzihiro Araki and Hajime Moriya, Equilibrium statistical mechanics of Fermion lattice systems, Rev. Math. Phys. 15 (2003), 93-198.
3. Huzihiro Araki, Conditional expectation relative to a product state and the corresponding standard potentials, Commun. Math. Phys., 246 (2004), 113-132.
4. Hajime Moriya, Some aspects of quantum entanglement for CAR systems, Lett. Math. Phys., 60 (2002), 109-121.
5. Hajime Moriya, Validity and failure of some entropy inequalities for CAR systems, submitted to J. Math. Phys.
6. R. T. Powers, Representations of the canonical anticommutation relations, Princeton University Thesis, 1967.

QUANTUM FILTERING AND OPTIMAL FEEDBACK CONTROL OF A GAUSSIAN QUANTUM FREE PARTICLE

S. C. EDWARDS, V. P. BELAVKIN

Mathematics Department, University of Nottingham University Park, Nottingham NG7 2RD, United Kingdom

E-mail: Viacheslav.Belavkin@nottingham.ac.uk

Methods from quantum filtering theory and classical control theory are combined to give an analytic solution to the problem of optimal control of the state of a Gaussian quantum free particle. The solutions to the filtering problem show an asymptotic localization of the continuously observed free particle and we also observe a duality with the solutions of optimal control.

1. Introduction

Quantum control is growing into one of the most active exponents of quantum information. Its importance is now being realised with applications in quantum state preparation, stability theory, quantum error correction and substantial applications in quantum computation [27,3,25]. As such, there is a vast array of literature on the subject of quantum control encompassing a range of different methods and objectives. As with all experimental techniques, efficiency is an important factor so optimal control is a particularly rewarding subject for research. The optimality of a control strategy is judged by the way in which the objectives are achieved (e.g. we may require fast results, or more commonly, a cost-effective way of producing the results). There are two types of dynamical control - *open loop* (or blind) control where the controls are predetermined at the start of the experiment and *closed loop* (or feedback) control where controls can be chosen throughout the experiment and thus is preferable for stochastic dynamics. Previous work on the theory of optimal open loop control includes variational techniques on open (dissipative) [42] and closed [33,37] qubit systems. However, these approaches can only seek locally optimal solutions which can often be improved further with measurement and feedback [19], since an open quantum system inevitably loses information to its surrounding environment.

With technological advances now allowing the possibility of continuous monitoring and rapid manipulations of quantum systems [4,24], it is important that we understand the relevance of feedback control in the framework of optimal quantum control. The first breakthrough in quantum feedback control was the discovery that methods from classical control could be transferred into the quantum domain when applied to sufficient coordinates of the quantum system [8]. I.e. controllable parameters which describe the dynamics of the quantum system sufficiently for the purposes of feedback control. An obvious candidate for a set of sufficient coordinates is the components of the normalized posterior density matrix. These are used to calculate posterior probability distributions for all system observables which can then be used to formulate the required control objectives. However this becomes impractical when the dimension n of the Hilbert space is large as the number of components required to calculate an arbitrary posterior expectation is $n^2 - 1$. The simplest example of feedback control arises when considering linear systems with Gaussian dynamics, since sufficient coordinates are given in this case by the posterior means and covariances of the controllable quantities. This was first considered for the quantum oscillator in [6] and subsequent general solutions in discrete [23] and continuous time [21] appeared.

In this paper we start from a quantum stochastic model of an open system to derive the quantum Langevin equations for the Heisenberg dynamics of the position and momentum operators. Section 3 contains a derivation of the filtering equation which describes the dynamics of the posterior density matrix under a continuous non-demolition position measurement. This is then use to obtain the sufficient coordinates of the free quantum particle which are given by the posterior mean and covariances of position and momentum, restating earlier results ([13], [20]). Section 4 describes the application of dynamic programming in the quantum setting and optimal solutions for the controlled free quantum particle are derived in duality with the results from Section 3. We summarise with a discussion on the results and methods with particular reference to the observed duality between quantum filtering and control for this simple example.

2. The Model

Let $\mathcal{B}_s = \mathcal{B}(\mathcal{H}_s)$ be the algebra of bounded operators on the Hilbert space $\mathcal{H}_s$ of the quantum system, i.e. the von Neumann algebra generated by the initial observables the system. We model the measurement channel (bath) in the field by the symmetric Fock space $\mathcal{F}$ over the Hilbert space $L^2(\mathbb{R}_+ \to \mathcal{G})$ of square integrable functions from $[0, \infty)$ into the Hilbert space $\mathcal{G} = L^2(\mathbb{R})$ of the bath and denote by $\mathcal{W}$ the noise algebra of bounded operators on $\mathcal{F}$. From the properties of the symmetric Fock space, we can factorize the noise algebra

$$\mathcal{W} = \mathcal{W}_{[0,t)} \otimes \mathcal{W}_{[t,\infty)}$$

for arbitrary $t > 0$ where $\mathcal{W}_{[a,b)} = \mathcal{B}(\mathcal{F}_{[a,b)})$ and $\mathcal{F}_{[a,b)}$ is the symmetric Fock space over $L^2([a,b) \to \mathcal{G})$ for $0 < a < b$. In the weak coupling limit [1] (short bath memory), the unitary dynamics $U_t \mathcal{H}_s \otimes \mathcal{F}_{[0,t)} \to \mathcal{H}_s \otimes \mathcal{F}_{[0,t)}$ of this closed composite Markovian system can be described in terms of a Hudson-Parthasarathy quantum stochastic differential equation [28,55]

$$dU_t = \left\{ \left(-\frac{i}{\hbar} H - \frac{1}{2} L^* L \right) dt + L dA_t^* - L^* dA_t \right\} U_t \tag{1}$$

whose unique solution gives the Heisenberg dynamics or *flow* of an operator $X \in \mathcal{B}_s$

$$\begin{aligned} dX_t &= dj_t(X) = d\left(U_t^*(X \otimes I) U_t\right) \\ &= \mathcal{L}_t[X_t]dt + [X_t, L_t]dA_t^* - [X_t, L_t^*]dA_t \end{aligned} \tag{2}$$

If we take the expectation of (2) with respect to the vacuum state of the field we obtain the evolved Lindblad generator [31]

$$\mathcal{L}_t[X_t] = i[H_t, X_t] + L_t^* X_t L_t - \frac{1}{2}\{L_t^* L_t, X_t\} \tag{3}$$

for the semigroup of completely positive maps describing the dissipative evolution in the Markovian limit, where $L_t = j_t(L)$ is the time evolution of the coupling operator L between the system and the bath and $H_t = j_t(H)$ is the time evolved free Hamiltonian.

Note that we are viewing $\mathcal{B}_s$ and $\mathcal{W}$ as subalgebras of $\mathcal{B}_s \otimes \mathcal{W}$ by dropping the notation of tensoring with identities. The unitary operators U_t describe evolutions of the closed compostite system under the free Hamiltonian of the open quantum system and the interaction Hamiltonian between the system and the bath and satisfy the cocycle identity $U_{t+s} = S_{-s} U_t S_s U_s$, where $S_s \mathcal{F}_{[t_1,t_2)} \to \mathcal{F}_{[t_1+s,t_2+s)}$ is the second quantization of the left shift on $L^2(\mathbb{R}_+ \to \mathcal{G})$.

2.1. *Langevin equations for the controlled quantum particle*

We view the free particle as an open quantum system, weakly coupled to an electromagnetic field in which we perform an indirect homodyne measurement of its position. Let us denote the expectations of position and momentum by $\bar{q} = \mathbb{E}_\psi[q] = \psi^* q \psi$ and $\bar{p} = \mathbb{E}_\psi[p]$ respectively for the initial vector state $\psi \in \mathcal{H}_s$ of the particle. We also denote the initial symmetric covariances $\sigma_p = \mathbb{E}_\psi[p^2] - \bar{p}^2, \sigma_q = \mathbb{E}_\psi[q^2] - \bar{q}^2$ and $\sigma_{pq} = \frac{1}{2}\mathbb{E}_\psi[pq + qp] - \bar{p}\bar{q}$. We introduce control to the system by placing the particle in a linear potential which gives an effective Hamiltonian

$$H_t = \frac{p_t^2}{2m} - 2\sqrt{\mu} u_t q_t \tag{4}$$

where μ is the coupling strength to this potential, and u_t describes the control force applied to the particle at time t.

An indirect measurement of the particle's position can be obtained by a diffusive measurement of the evolved field quadrature $Y_t = U_t^*(A_t + A_t^*)U_t$ in the bath which is coupled to the system via the operator $L = \sqrt{\lambda} q$, where λ is the measurement accuracy. This can be seen in the Heisenberg picture by considering the flow of the input operators p and q and the evolution of the output operators Y_t

$$dq_t = \frac{1}{m} p_t dt \qquad dp_t = 2\sqrt{\mu} u_t dt + dV_t \tag{5}$$

$$dY_t = 2\sqrt{\lambda} q_t dt + dW_t \tag{6}$$

where $V_t = i\hbar\sqrt{\lambda}(A_t - A_t^*)$, $W_t = A_t + A_t^*$ are non-commuting field quadratures whose time derivatives describe the perturbation from the channel given by quantum white noises of intensity $\hbar^2\lambda$ and 1 respectively. So we see that the output is a noisy signal of the stochastic input operators which points towards an application of filtering.

3. Quantum Filtering

We parameterize the Gaussian quantum free particle by its posterior mean values of position and momentum. The dynamics of such conditional expectations are described by the quantum filtering or Belavkin equation [9], [12],[17]. Classically, filtering equations are used when we need to estimate the value of dynamical variables about which we have incomplete knowledge. For example, the Kalman-Bucy filter [29],[30] gives a continuous least-squares estimator for a Gaussian random variable with linear dynamics when we only have access to a correlated, noisy signal. Since closed composite systems exhibit fundamentally linear dynamics disturbed by Wiener quantum noise processes (c.f. (5),(6)), one would expect filtering theory to play an important role in quantum measurement. Belavkin was the first to realise this [6,7,9] and constructed the quantum filtering equation which describes the evolution of the density matrix conditioned on the classical non-demolition output of a noisy quantum channel and thus provides the required conditional expectation for system operators.

For the case where we initially have a vacuum field state δ_0 and pure system state, the quantum filtering equations of type A (linear, unnormalized) and type B (nonlinear, normalized) are derived below. More general quantum filtering equations are derived in [12,17] using martingale techniques.

Lemma 3.1. *The dynamics of the unnormalized posterior density matrix* ϱ_t *is described by*

$$d\varrho_t = \mathcal{L}^*[\varrho_t]dt + \frac{1}{\sqrt{\gamma}}\left(L\varrho_t + \varrho_t L^*\right) dW_t \tag{7}$$

where $\mathcal{L}^*[\varrho_t] = -i[H, \varrho_t] + L\varrho_t L^* - \frac{1}{2}\{L^*L, \varrho_t\}$ *is the dual to the Lindblad generator for dissipative Heisenberg dynamics.*

Proof. For an initial pure state $\varrho_0 = (\psi \otimes \delta_0)(\psi \otimes \delta_0)^*$ we denote its *a posteriori* state by $\varrho_t(\omega_t) = \hat{\psi}_t(\omega_t)\hat{\psi}_t^*(\omega_t)$ where $\hat{\psi}_t(\omega_t)$ is the reduced Schrödinger evolution $\psi_t = U_t(\psi \otimes \delta_0)$ conditioned on the measurement results ω_t over the interval $[0, t]$. Using $U_t(I \otimes dA_t) = (I \otimes dA_t)U_t$ (since $U_t \in \mathcal{B}_s \otimes \mathcal{W}_{[0,t)}$ and $dA_t \in \mathcal{W}_{[t,t+dt)}$) and also $dA_t\delta_0 = 0$, (1) gives

$$\begin{aligned} d\psi_t &= \left(\left(-\tfrac{i}{\hbar}H - \tfrac{1}{2}L^*L\right)dt + LdA_t^*\right)U_t(\psi \otimes \delta_0) \\ &= \left(\left(-\tfrac{i}{\hbar}H - \tfrac{1}{2}L^*L\right)dt + L(dA_t + dA_t^*)\right)U_t(\psi \otimes \delta_0) \end{aligned}$$

Now $A_t + A_t^* = W_t$ is a classical output process which can be viewed as a random variable $W_t \Sigma_t \to \mathbb{C}$ on the space of measurement results $\omega_t \in \Sigma_t$ over the interval $[0, t]$, so tracing over the field gives the reduced conditional dynamics

$$d\hat{\psi}_t(\omega_t) = \left(\left(-\frac{i}{\hbar}H - \frac{1}{2}L^*L\right)dt + LdW_t(\omega_t)\right)\hat{\psi}_t(\omega_t)$$

We now omit ω_t and view ϱ_t and $\hat{\psi}_t$ as random variables taking values in $\mathcal{T}[\mathcal{H}_s]$ and $\mathcal{H}_s$ respectively where $\mathcal{T}[\mathcal{H}_s]$ is the space of trace class operators on $\mathcal{H}_s$. So the Itô rule gives

$$\begin{aligned} d\varrho_t &= d\hat{\psi}_t\hat{\psi}_t^* + \hat{\psi}_t d\hat{\psi}_t^* + d\hat{\psi}_t d\hat{\psi}_t^* \\ &= \mathcal{L}^*[\sigma_t]dt + (L\varrho_t + \varrho_t L^*)\,dW_t \end{aligned}$$

□

Lemma 3.2. *The expectation $\rho_t^{\omega_t}(X) = tr(\varrho_t(\omega_t)X)/tr(\varrho_t(\omega_t))$ of an operator X conditioned on results ω_t up to time t has the following stochastic dynamics*

$$\begin{aligned} d\rho_t(X) = \rho_t(\mathcal{L}[X])dt + (\rho_t(L^*X + XL) - \rho_t(L+L^*)\rho_t(X)) \\ \times(dW_t - \rho_t(L+L^*)dt) \end{aligned} \tag{8}$$

where again we omit the argument ω_t and view ρ_t as a density matrix valued random variable.

Proof. Using the classical Itô rule on the definition of $\rho_t(X)$ we get

$$d\rho_t(X) = \frac{tr(d\varrho_t X)}{tr(\varrho_t)} + tr(\varrho_t X)d\frac{1}{tr(\varrho_t)} + tr(d\varrho_t X)d\frac{1}{tr(\varrho_t)}$$

where

$$d\frac{1}{tr(\varrho_t)} = -\frac{tr(d\varrho_t)}{tr(\varrho_t)^2} + \frac{tr(d\varrho_t)^2}{tr(\varrho_t)^3}$$

and (7) gives $tr(d\varrho_t) = tr((L+L^*)\varrho_t)dW_t$. The result (8) is then obtained using the cyclic property of the trace and the Itô multiplication $dW_t^2 = dt$. □

3.1. *Quantum Filtering for the Free Particle*

Using the quantum filtering equations, we can construct the expectation for the position and variance of the free particle conditional on the measurement data which we observe. The dynamics of these expectations $\hat{q}_t = \rho_t(q), \hat{p}_t = \rho_t(p)$ are obtained from (8) using the canonical commutation relation (CCR)

$$[q,p] = qp - pq = i\hbar\mathbf{1}$$

and give

$$d\hat{q}_t = \frac{1}{m}\hat{p}_t dt + 2\sqrt{\lambda}\sigma_{q,t}d\tilde{W}_t \tag{9}$$

$$d\hat{p}_t = 2\sqrt{\mu}u_t dt + 2\sqrt{\lambda}\sigma_{pq,t}d\tilde{W}_t \tag{10}$$

where $d\tilde{W}_t = dW_t - 2\sqrt{\lambda}\hat{q}_t dt$ is a Wiener increment called the *innovation process* which incorporates the measurement results and $\sigma_{pq,t} = \frac{1}{2}\rho_t(pq+qp) - \hat{p}_t\hat{q}_t$ is the symmetric conditional covariance. For Gaussian initial states, we have a linear quantum Kalman-Bucy type filter where the covariances satisfy the coupled Ricatti equations

$$\begin{aligned} \frac{d\sigma_{q,t}}{dt} &= \frac{2}{m}\sigma_{pq,t} - 4\lambda\sigma_{t,q}^2 \\ \frac{d\sigma_{pq,t}}{dt} &= \frac{1}{m}\sigma_{p.t} - 4\lambda\sigma_{q,t}\sigma_{pq,t} \\ \frac{d\sigma_{p,t}}{dt} &= \lambda\hbar^2 - 4\lambda\sigma_{pq,t}^2 \end{aligned} \tag{11}$$

with initial conditions

$$\sigma_{q,0} = \sigma_q, \quad \sigma_{pq,0} = \sigma_{pq}, \quad \sigma_{p,0} = \sigma_p.$$

Note that due to the Gaussian nature of the system, these equations are deterministic and so do not depend explicitly on the measurement results. These equations have an analytic solution [13], although for simplicity we will only consider stationary solutions here which give the asymptotic behaviour of the posterior covariances

$$\sigma_{q,t} \to \sqrt{\hbar/4m\lambda}, \qquad \sigma_{pq,t} \to \hbar/2, \qquad \sigma_{p,t} \to \hbar\sqrt{m\lambda\hbar} \tag{12}$$

and indicates a finite limit of dispersion. In the case where the measurement results are ignored (averaged over), the Ricatti equations become linear and the dispersions tend to infinity like t^3, which is faster than the t^2 dispersion for the case of a closed quantum system as one would expect from the dissipation caused by the measurement apparatus.

4. Control

As we have previously discussed, the problem of quantum control is similar to a classical control problem where we have imperfect state information. So optimality of the control procedure first requires an optimal estimation of the sufficient parameters (given by the filtering equation). Next we select control strategies which are optimal in the sense that the required objectives are obtained under certain restrictions. These objectives and restrictions are encoded into a *cost function*, in the sense that a minimum cost is achieved by an optimal strategy. In our case we aim to control the free particle into a desired position and momentum at a finite time T whilst constraining the amplitude of the controlling force for energy considerations. So a suitable cost function for these conditions is

$$J = \mathbb{E}\left\{\omega_p\hat{q}_T^2 + 2\omega_{pq}\hat{p}_T\hat{q}_T + \omega_q\hat{p}_T^2 + \int_0^T (u_t^2 + \alpha\hat{q}_t^2)dt\right\}. \tag{13}$$

The first terms defines the target state $\hat{p}_T = \hat{q}_T = 0$ (i.e. the objective) and the integral represents our desire to keep the chosen controls u_t as small in amplitude as possible whilst maintaining a trajectory towards the target state (i.e. the restrictions). The parameters $\omega_p, \omega_{pq}, \omega_q, \alpha > 0$ allow us to define the importance of each of the objectives and restrictions, i.e. if it is vital the we keep physical cost of control low, then all parameters should be small. Likewise, we could also insist that the final position of the particle is more important than its final momentum. Another reason for this choice of cost function will become apparent later when we discuss the duality between these parameters and the covariances used in filtering for this example. Since we are considering feedback control, each u_t is a function $u_t = f_t(\hat{p}_t, \hat{q}_t)$ of the expected state of the system which is conditional on the measurement outcomes received up to time t. So the expectation $\mathbb{E}\{\bullet\}$ is taken over all possible trajectories (measurement outcomes) for the duration of the experiment. The goal of optimal feedback control is to choose a suitable function f_t at each time t which minimizes the expected cost (13). Due to the implicit dependance of (13) on the conditional dynamics of the position and momentum, finding such an optimal feedback strategy is highly non-trivial. However, the problem of quantum control has now been reduced to classical control of the conditional expectations given by quantum filtering. This form of quantum feedback control via classical control of sufficient quantum coordinates was first suggested by Belavkin [8]. There is a wealth of literature on classical control [43,45] and a solution to this classical optimization problem was first given by Bellman [44]. This optimization algorithm is called *dynamic programming* and breaks (13) up into smaller time intervals representing the *cost-to-go* from time t

$$J(t, \hat{p}_t, \hat{q}_t) = \mathbb{E}_t \left\{ \omega_p \hat{q}_T^2 + 2\omega_{pq}\hat{p}_T\hat{q}_T + \omega_q \hat{p}_T^2 + \int_t^T (u_s^2 + \alpha \hat{q}_s^2) ds \right\}. \tag{14}$$

which is minimized recursively, working backwards from the terminal cost. The expectation is now taken over all possible trajectories for the time interval $[t, T]$ and we denote the minimization of (14) by $J^*(t, \hat{p}_t, \hat{q}_t)$.

Lemma 4.1. *Principle of Optimality*
If we have a control strategy $u_0^T = \{u_s 0 < s < T\}$ which minimizes (13), then its restriction u_t^T to the interval $[t, T]$ is also optimal for the cost-to-go (14).

Lemma 4.2. *The optimal control strategy for the quantum free particle is*

$$u_t = -2\sqrt{\mu}(\omega_{q,t}\hat{p}_t + \omega_{pq,t}\hat{q}_t) \tag{15}$$

where

$$\begin{aligned}
-\frac{d\omega_{q,t}}{dt} &= \frac{2}{m}\omega_{pq,t} - 4\mu\omega_{t,q}^2 \\
-\frac{d\omega_{pq,t}}{dt} &= \frac{1}{m}\omega_{p.t} - 4\mu\omega_{q,t}\omega_{pq,t} \\
-\frac{d\omega_{p,t}}{dt} &= \alpha - 4\mu\omega_{pq,t}^2
\end{aligned} \tag{16}$$

with terminal conditions

$$\omega_{q,T} = \omega_q, \quad \omega_{pq,T} = \omega_{pq}, \quad \omega_{p,T} = \omega_p.$$

Proof. The principle of optimality allows us to define a recursive minimum cost-to-go in infinitesimal time steps dt

$$J^*(t, \hat{p}_t, \hat{q}_t) = \mathbb{E}_t \left\{ \min_{u_t} \left[(u_t^2 + \alpha \hat{q}_t^2) dt + J^*(t + dt, \hat{p}_{t+dt}, \hat{q}_{t+dt}) \right] \right\} \tag{17}$$

We assume that $J^* = J^*(t, \hat{p}_t, \hat{q}_t)$ is suitably differentiable, i.e. that the partial derivatives $\partial_a J^*$ and second derivatives $\partial_{a,b}^2$ exist for $a, b \in \{t, p, q\}$. Then the Taylor expansion of the right hand side up to second order (sufficient for stochastic dynamics) gives

$$\begin{aligned}
J^*(t + dt, \hat{p}_{t+dt}, \hat{q}_{t+dt}) = J^* &+ \partial_t J^* dt + \partial_p J^* d\hat{p}_t + \partial_q J^* d\hat{q}_t \\
&+ \frac{1}{2}\partial_{p,p}^2 J^* d\hat{p}_t^2 + \frac{1}{2}\partial_{q,q}^2 J^* d\hat{q}_t^2 + \partial_{p,q}^2 J^* d\hat{p}_t d\hat{q}_t
\end{aligned}$$

So from (9), (10) and (17) and the quantum Itö multiplications, we get the Hamilton-Jacobi-Bellman (HJB) equation

$$\begin{aligned}
-\partial_t J^* = &\min_{u_t} \left\{ u_t^2 + 2\sqrt{\mu} u_t \partial_p J^* \right\} + \alpha \hat{q}_t^2 + \frac{1}{m}\hat{p}_t \partial_q J^* \\
&+ 2\lambda \sigma_{pq,t}^2 \partial_{p,p}^2 J^* + 2\lambda \sigma_{q,t}^2 \partial_{q,q}^2 J^* + 4\lambda \sigma_q \sigma_{pq} \partial_{p,q}^2 J^*
\end{aligned}$$

The minimum is easily calculated by completing the squares

$$u_t^2 + 2\sqrt{\mu} u_t \partial_p J^* = (u_t + \sqrt{\mu}\partial_p J^*)^2 - \mu(\partial_p J^*)^2$$

which is minimized by choosing $u_t = -\sqrt{\mu}\partial_p J^*$ where J^* now satisfies

$$\begin{aligned}
-\partial_t J^* = &\alpha \hat{q}_t^2 + \frac{1}{m}\hat{p}_t \partial_q J^* + 2\lambda \sigma_{pq,t}^2 \partial_{p,p}^2 J^* + 2\lambda \sigma_{q,t}^2 \partial_{q,q}^2 J^* \\
&+ 4\lambda \sigma_q \sigma_{pq} \partial_{p,q}^2 J^* - \mu(\partial_p J^*)^2
\end{aligned}$$

Using the ansatz $J^* = \omega_{p,t}\hat{q}_t^2 + 2\omega_{pq,t}\hat{p}_t\hat{q}_t + \omega_{q,t}\hat{p}_t^2 + \omega_t$ gives the control law (15) and comparing coefficients of $\hat{p}_t^2, \hat{q}_t^2$ and $\hat{p}_t\hat{q}_t$ gives the required Ricatti equations from the HJB equation. □

We obtain an equation for ω_t from the ansatz by looking at the constant terms in the HJB equation, which gives $-d\omega_t = 4\lambda(\omega_{q,t}\sigma_{pq,t}^2 + \omega_{p,t}\sigma_{q,t}^2 +$

$2\omega_{pq,t}\sigma_{q,t}\sigma_{pq,t})dt$ so that the minimum cost obtained for the whole experiment is given by

$$J^*(0,\bar{p},\bar{q}) = \omega_{q,0}\bar{p}^2 + 2\omega_{pq,0}\bar{p}\bar{q} + \omega_{p,0}\bar{q}^2 \qquad (18)$$
$$-4\lambda \int_0^T (\omega_{q,t}\sigma^2_{pq,t} + \omega_{p,t}\sigma^2_{q,t} + 2\omega_{pq,t}\varsigma_{q,t}\varsigma_{pq,t})dt$$

As with the filtering Ricatti equation, we can also seek stationery solutions to the control equations

$$\omega_{q,t} \rightarrow \frac{1}{2}\sqrt[4]{\frac{\alpha}{m^2\mu^3}}, \qquad \omega_{pq,t} \rightarrow \frac{1}{2}\sqrt{\frac{\alpha}{\mu}}, \qquad \omega_{p,t} \rightarrow \sqrt[4]{\frac{m^2\alpha^3}{\mu}} \qquad (19)$$

and so we can give an approximation to the total cost by assuming a steady state for the whole experiment and inserting the stationery solutions (12), (19) into (18).

5. Discussion

The example of the Gaussian free quantum particle is important since it exhibits an exact solution and emphasizes the similarities between the two ingredients of quantum feedback control, namely quantum filtering and optimal control. The Ricatti equations for filtering (11) and control (16) are in complete duality. This indicates the similarity between the roles of the covariances in filtering and the control parameters in optimal control. The duality can be understood when we examine the nature of each of the methods used. Both methods involve the minimization of a quadratic function for linear, Gaussian systems, (i.e. the least squares error for filtering and the quadratic cost for control). Whilst the results are not particularly novel here, we feel that the strength of this paper lies in the demonstration of the methods used and hope that they will fuel further research in this growing field.

References

1. L. Accardi and A. Frigerio and Y.G. Lu, The weak coupling limit as a quantum functional central limit, Commun. Math. Phys., 131, 537-570, 1990.
2. C. Ahn and H. M. Wiseman and G. J. Milburn, Quantum error correction for continuously detected errors, Phys. Rev. A, 67, 052310, 2003.
3. C. Ahn and A. C. Doherty and A. J. Landahl, Continuous quantum error correction via quantum feedback control, Phys. Rev. A, 65, 042301, 2002.
4. M. A. Armen and J. K. Au and J. K. Stockton and A. C. Doherty and H. Mabuchi, Adaptive homodyne measurement of optical phase, Phys. Rev. Lett., 89, 133602, 2002.

5. A. Barchielli and V. P. Belavkin, Measurements continuous in time and a posteriori states in quantum mechanics, J. Phys. A, 24, 1495-1514, 1991.
6. V. P. Belavkin, Optimal measurement and control in quantum dynamical systems, preprint 411, Institute of Physics, Nicolaus Copernicus University, Torun, 1979.
7. V. P. Belavkin, Quantum filtering of Markov signals with white quantum noise, Radiotechnika i Electronika, 25, 1445-1453, 1980 (English translation in Quantum communications and measurement, Belavkin, V.P., Hiroto, O., Hudson, R.L., eds, Plenum Press, 1994, 381-392).
8. V. P. Belavkin, On the Theory of Controlling Observable Quantum Systems, Automatica and Remote Control, 44, 2, 178–188, 1983.
9. V. P. Belavkin, Nondemolition stochastic calculus in Fock space and nonlinear filtering and control in quantum systems, Proceedings XXIV Karpacz winter school, Stochastic methods in mathematics and physics, Editor; R. Guelerak and W. Karwowski, World Scientific, Singapore, 310-324, 1988.
10. V. P. Belavkin, A new wave equation for a continuous nondemolition measurement, Physics Letters A, 140, 355-358, 1989.
11. V. P. Belavkin, Quantum Continual Measurements and a Posteriori Collapse on CCR, Commun. Math. Phys., 146, 611-635, 1992.
12. V. P. Belavkin, Quantum stochastic calculus and quantum nonlinear filtering, Journal of Multivariate Analysis, 42, 171-201, 1992.
13. V. P. Belavkin and P. Staszewski, Nondemolition observation of a free quantum particle, Phys. Rev. A, 45, No.3, 1347-1356, 1992.
14. V. P. Belavkin, Nondemolition Principle of Quantum Measurement Theory, Foundation of Physics, 24, No.5, 685-714, 1994.
15. V. P. Belavkin, Measurement, Filtering and Control in Quantum Open Dynamical Systems, Rep on Math Phys, 43, No.3, 405–425, 1999.
16. V. P. Belavkin, Quantum Causality, Stochastics, Trajectories and Information, Report on Progress in Physics, 65, 353-420, 2002.
17. L. M. Bouten and M. I. Gută and H. Maassen, Stochastic Schrödinger equations, J. Phys. A, 37, 3189-3209, 2004.
18. L. M. Bouten, Squeezing enhanced control, preprint, University of Nijmegen, 2004.
19. L. M. Bouten and S. C. Edwards and V. P. Belavkin, Bellman equations for optimal feedback control of qubit states, preprint, University of Nottingham, 2004.
20. A. C. Doherty and K. Jacobs, Feedback-Control of Quantum Systems Using Continuous State-Estimation, Phys. Rev. A, 60, 2700-2710, 1999.
21. A. C. Doherty and S. Habib and K. Jacobs and H. Mabuchi and S. M. Tan, Quantum Feedback Control and Classical Control Theory, Phys. Rev. A, 62, 012105, 2000.
22. A. C. Doherty and K. Jacobs and G. Jungman, Information, Disturbance and Hamiltonian Quantum Feedback Control, Phys. Rev. A, 63, 062306, 2001.
23. S. C. Edwards and V. P. Belavkin, On the Duality of Quantum Filtering and Optimal Feedback Control in Quantum Open Linear Dynamical Systems, Proc. IEEE Conf. Physics and Control, St. Petersburg, 768-772, 2003.

24. J. M. Geremia and J. K. Stockton and A. C. Doherty and H. Mabuchi, Quantum Kalman filtering and the Heisenberg limit in atomic magnetometry, Phys. Rev. Lett., 91, 250801, 2003.
25. M. Gregoratti and R. F. Werner, On quantum error-correction by classical feedback in discrete time, arXiv quant-ph/0403092, 2004.
26. I. Grigorenko and M. E. Garcia and K. H. Bennemann, Theory for the Optimal Control of Time-Averaged Quantities in Open Quantum Systems, Phys. Rev. Lett, 89, 233003, 2002.
27. R. van Handel and J. K. Stockton and H. Mabuchi, Feedback control of quantum state reduction, arXiv quant-ph/0402136, Caltech, 2004.
28. R. L. Hudson and K. R. Parthasarathy, Quantum Ito's formula and stochastic evolutions, Commun. Math. Phys., 93, 301-323, 1984.
29. R. E. Kalman, A New Approach to Linear Filtering and Prediction Problems, J. Basic Eng, 82, 34-45, 1960.
30. R. E. Kalman and R. S. Bucy, New Results in Linear Filtering and Prediction Theory, J. Basic Eng., 95-108, 1961.
31. G. Lindblad, On the generators of quantum dynamical semigroups, Commun. Math. Phys., 48, 119-130, 1976.
32. S. Lloyd, Quantum Controllers for Quantum Systems, quant-ph/9703042, 1997.
33. J.P. Palao and R. Kosloff, Quantum computing by an optimal control algorithm for unitary transformations, Phys. Rev. Lett, 89, 188301, 2002.
34. M. Sarovar and C. Ahn and K. Jacobs and G. J. Milburn, Practical scheme for error control using feedback, Phys. Rev. A, 69, 052324, 2004.
35. A. I. Solomon and S. G. Schirmer, Limitations on Quantum Control, International Journal of Modern Physics B, 2107-2112, 2002.
36. M. Takesaki, Conditional expectations in von Neumann algebras, J. Funct. Anal., 9, 306-321, 1971
37. C.M. Tesch and R.de Vivie-Riedle, Quantum computation with vibrationally excited molecules, Phys. Rev. Lett, 89, 157901, 2002
38. J. J. Ting, Alternative Method for Quantum Feedback-Control, quant-ph/0304089.
39. H. M. Wiseman and G. J. Milburn, Quantum theory of field-quadrature measurements, Phys. Rev. A, 47, 642-662, 1993
40. H. M. Wiseman and G. J. Milburn, Quantum-theory of optical feedback via homodyne detection, Phys. Rev. Lett., 70, 548-551, 1993.
41. H. M. Wiseman, Quantum theory of continuous feedback, Phys. Rev. A, 49, 2133-2150, 1994
42. R. Xu and Y. Yan and Y. Ohtsuki and Y. Fujimura and H. Rabitz, Optimal control of quantum non-Markovian dissipation Reduced Liouville-space theory, J. Chem. Phys, 120, 6600-6608, 2004.
43. K. J. Astrom, Introduction to Stochastic Control Theory, Academic Press, London, 1970.
44. R. Bellman, Dynamic Programming, Princeton University Press, Princeton, 1957.
45. D. P. Bertsekas, Dynamic Programming and Optimal Control, Athena Sci-

entific, Massachusetts, 1995.
46. V. B. Braginsky and F. Ya Khalili, Quantum Measurement, Cambridge University Press, 1992.
47. O. Bratteli and D. W. Robinson, Operator algebras and quantum statistical mechanics 2, Springer-Verlag, Berlin Heidelberg, 1981.
48. H. P. Breuer and F. Petruccione, The Thoery of Open Quantum Systems, Oxford University Press, 2002.
49. H. J. Carmichael, An Open Systems Approach to Quantum Optics, Springer-Verlag, Berlin Heidelberg New-York, 1993.
50. W. Feller, An Introduction to Probability Theory and its Applications, John Wiley and Sons, New York, 1950.
51. C. Gardiner and P. Zoller, Quantum Noise, Springer, Berlin, 2000.
52. G. R. Grimmett and D. R. Stirkazer, Probability and Random Proccess, Clarendon Press, Oxford, 1992.
53. S. P. Gudder, Quantum Probability, Academic Press, London, 1988.
54. M. A. Nielsen and I. L. Chuang, Quantum Computation and Quantum Information, Cambridge University Press, Cambridge, 2000.
55. K. R. Parthasarathy, An Introduction to Quantum Stochastic Calculus, Birkhäuser, Basel, 1992.
56. R. L. Stratonovich, Conditional Markov processes and their application to the theory of optimal control, American Elsevier Publishing Company, Inc, New-York, 1968.
57. J. von Neumann, Mathematische Grundlagen der Quantenmechanik, Springer, Berlin, 1932.

ON EXISTENCE OF QUANTUM ZENO DYNAMICS

PAVEL EXNER[A,B]
a) Department of Theoretical Physics, Nuclear Physics Institute,

Academy of Sciences, 25068 Řež, Czech Republic
b) Doppler Institute, Czech Technical University, Břehová 7,

11519 Prague, Czech Republic
E-mail: exner@ujf.cas.cz

TAKASHI ICHINOSE[C]
c) Department of Mathematics, Faculty of Science, Kanazawa

University, Kanazawa 920-1192, Japan
E-mail: ichinose@kenroku.kanazawa-u.ac.jp

We prove a product formula which involves the unitary group generated by a semibounded self-adjoint operator and an orthogonal projection P on a separable Hilbert space $\mathcal{H}$. The convergence is demonstrated in the space $L^2_{\mathrm{loc}}(\mathbb{R};\mathcal{H})$, which gives a partial answer to the question about existence of the limit describing quantum Zeno dynamics in the subspace $\mathrm{Ran}\, P$, the range of P. A stronger result with the convergence in $\mathcal{H}$ is proved in the case of a finite-dimensional P.

1. Introduction and the result

Recent years witnessed a new wave of interest to quantum Zeno effect. It is not a new topic, of course, recall that the fact that the decay of an unstable system can be slowed down, or even fully stopped in the ideal case, by frequently repeated measurements checking whether the system is still undecayed is known at least since the work of Beskow and Nilsson [2]. It became popular, however, only a decade later when Misra and Sudarshan [12] invented the name derived from the well-known Zeno aporia about a flying arrow. The new interest is motivated, in particular, by the fact that nowadays the effect becomes experimentally accessible. Mathematics of the effect is not simple. The first rigorous proof of the fact that continuous observation prevents decay can be found in [9]. This result as well as that of [12] leaves open two important questions, namely the existence of Zeno dynamics in the unstable-system state space, which will be in the following the subspace specified by the range $\operatorname{Ran} P$ of an orthogonal projection P, and the form of its effective Hamiltonian. The second problem is of particular interest in the case when $\dim \operatorname{Ran} P > 1$; a partial solution was given in [5] where it was shown that the results of Chernoff [3,4] allow one to determine the generator of the Zeno time evolution naturally through the appropriate quadratic form. A new interest to the problem was inspired by a paper by Facchi et al. [7] in which an important special case was studied, namely the situation when the presence of a Schrödinger particle in a domain of $\Omega \subset \mathbb{R}^d$ is repeatedly ascertained with the measurement frequency tending to infinity. Using the method of stationary phase the authors demonstrated on a heuristic level that the Zeno dynamics describes in this case the free particle confined to Ω, with the Dirichlet condition at the boundary of the domain. The aim of the present note is to analyze the problem rigorously in a general setting; we review here results derived in our recent work [6]. We combine a (modified) method of [11] with the results [3,4]. This allows us to show that if the natural effective Hamiltonian mentioned above is densely defined — which is a nontrivial assumption — then the Zeno dynamics exists and the said operator is its generator in a topology which includes an averaging over the time variable – cf. Theorem 1.1 for the exact statement. From the mathematical point of view our conclusion cannot be regarded as fully satisfactory, because the natural topology of the problem is given by the norm of the Hilbert space. We demonstrate, however, such a convergence in $\mathcal{H}$ for the particular case when the projections involved are finite-dimensional – cf. Theorem 1.2. On the other hand, let us stress that from the physical point of view the result given in Theorem 1.1 is quite plausible because any real measurement is always burdened with errors – see Remark 1.1 below. To formulate our results, let H be a nonnegative self-adjoint operator in a *separable* Hilbert space $\mathcal{H}$, and P an orthogonal projection. The nonnegativity assumption is made for convenience; our main result extends easily to any self-adjoint operator H bounded from below as well as one bounded from above, i.e. to each semi-bounded self-adjoint operator in $\mathcal{H}$. Suppose that the operator $H_P = (H^{1/2}P)^*(H^{1/2}P)$ is densely defined. This amounts to saying that the quadratic form $u \mapsto \|H^{1/2}Pu\|^2$ with form domain $D[H^{1/2}P]$ is densely defined, so that H_P is self-adjoint. It is obviously defined and acts nontrivially in a closed subspace $\operatorname{Ran} P$ of $\mathcal{H}$, while in the orthogonal complement $(\operatorname{Ran} P)^\perp$ it acts as zero. We denote by $L^2_{\text{loc}}(\mathbb{R};\mathcal{H}) = L^2_{\text{loc}}(\mathbb{R}) \otimes \mathcal{H}$ the

Fréchet space of the $\mathcal{H}$-valued strongly measurable functions $v(\cdot)$ on $\mathbb{R}$ such that $\|v(\cdot)\|$ is locally square integrable there, with the topology induced by the semi-norms $v \mapsto \left(\int_{-T_\ell}^{T_\ell} \|v(t)\|^2 dt\right)^{1/2}$ for a countable set $\{T_\ell\}_{\ell=1}^{\infty}$ of increasing positive numbers accumulating at infinity, $\lim_{\ell\to\infty} T_\ell = \infty$. In a similar way one defines the Fréchet space $L^2_{\mathrm{loc}}([0,\infty);\mathcal{H}) = L^2_{\mathrm{loc}}([0,\infty))\otimes\mathcal{H}$. Under the stated assumptions our main result can be stated as follows

Theorem 1.1. *For every $f \in \mathcal{H}$ and $\varepsilon = \pm 1$ it holds that*

$$(Pe^{-i\varepsilon tH/n}P)^n f \longrightarrow e^{-i\varepsilon tH_P}\, Pf\,, \tag{1.1}$$

$$(P\, e^{-i\varepsilon tH/n})^n f \longrightarrow e^{-i\varepsilon tH_P}\, Pf\,, \tag{1.2}$$

$$(\, e^{-i\varepsilon tH/n}\, P)^n f \longrightarrow e^{-i\varepsilon tH_P}\, Pf\,, \tag{1.3}$$

in the topology of $L^2_{\mathrm{loc}}(\mathbb{R};\mathcal{H})$ as $\to \infty$.

In fact, the proof given in [6] yields a stronger result with P on the left-hand side replaced by the values $P(1/n)$ of a projection-valued family $\{P(t)\}$ defined in a neighborhood of $t = 0$ such that $P(t) \to P(0) = P$ strongly as $t \to 0$, and such that $D[H^{1/2}P(t)] \supset D[H^{1/2}P]$ and $\lim_{t\to 0} \|H^{1/2}P(t)v\| = \|H^{1/2}Pv\|$ holds for $v \in D[H^{1/2}P]$. From the viewpoint of quantum Zeno effect described at the beginning of this section the optimal result would be a strong convergence on $\mathcal{H}$ for a fixed value of the time variable, moreover uniformly on each compact interval of the variable t. Our Theorem 1.1 implies the following weaker result on such a pointwise convergence.

Corollary 1.1. *Under the same hypotheses as in Theorem 1.1, There exist a set $M \subset R$ of Lebesgue measure zero and a strictly increasing sequence n' of positive integers along which we have*

$$(Pe^{-i\varepsilon tH/n'}P)^{n'} f \longrightarrow e^{-i\varepsilon tH_P}\, Pf\,, \tag{1.4}$$

$$(P\, e^{-i\varepsilon tH/n'})^{n'} f \longrightarrow e^{-i\varepsilon tH_P}\, Pf\,, \tag{1.5}$$

$$(\, e^{-i\varepsilon tH/n'}\, P)^{n'} f \longrightarrow e^{-i\varepsilon tH_P}\, Pf\,, \tag{1.6}$$

for every $f \in \mathcal{H}$, strongly in $\mathcal{H}$ for all $t \in \mathbb{R} \setminus M$.

Note that the strong convergence in $\mathcal{H}$ in Theorem 1.1 and Corollary 1.1 holds in fact without restriction to subsequences in the orthogonal complement of the subspace $P\mathcal{H}$, uniformly on each compact t-interval in $\mathbb{R}$.

As we have indicated above, one need not resort to subsequences also in the particular case when the projections involved are finite-dimensional.

Theorem 1.2. *If the orthogonal projection P is of finite dimension, then formulas (1.1)–(1.3) hold in the norm of $\mathcal{H}$, uniformly on each compact interval in the variable $t \in \mathbb{R}$.*

Before proving Theorems 1.1 and 1.2 and Corollary 1.1 let us comment briefly on some other aspects of the result.

Remark 1.1. While the necessity to pick a subsequence makes the pointwise convergence result weaker than desired, let us notice that from the physical point of view the convergence in $L^2_{\rm loc}(\mathbb{R};\mathcal{H})$ can be regarded as satisfactory. The point is that any actual measurement, in particular that of time, is burdened with errors. Suppose thus we perform the Zeno experiment on numerous copies of the system. The time value in the results will be characterized by a probability distribution ϕ
$R_+ \to \mathbb{R}_+$, which is typically a bounded, compactly supported function – in a precisely posed experiment it is sharply peaked, of course. Theorem 1.1 then gives

$$\int \phi(t) \left\| (Pe^{-i\varepsilon tH/n}P)^n f - e^{-i\varepsilon tH_P}\, Pf \right\|^2 dt \to 0 \qquad (1.7)$$

as $n \to \infty$, in other words, the Zeno dynamics limit is valid after averaging over experimental errors, however small they are.

Remark 1.2. The fact that the product formulae require H_P to be densely defined is nontrivial. Recall the example of [5] in which H is the multiplication operator, $(H\psi)(x) = x\psi(x)$ on $L^2(\mathbb{R}_+)$, and P is the one-dimensional projection onto the subspace spanned by the vector ψ_0 $\psi_0(x) = [(\pi/2)(1+x^2)]^{-1/2}$. In this case obviously H_P is the zero operator on the domain $D[H_P] = \{\psi_0\}^{\perp}$. On the other hand, $P e^{-itH} P$ acts on $\operatorname{Ran} P$ as multiplication by the function

$$v(t) = e^{-t} - \frac{i}{\pi}\left[e^{-t}\right.$$

Ei(t) - e^t Ei$(-t)$ $= 1 + \frac{2i}{\pi}\, t \ln t + \mathcal{O}(t)$,where $E_i(-t)$ and $\overline{E}_i(t)$ are exponential integrals [1]; due to the rapid oscillations of the imaginary part as $t \downarrow 0$ a pointwise limit of $v(t/n)^n$ for $n \to \infty$ does not exist. Notice also that different limits may be obtained in this example along suitably chosen subsequences $\{n'\}$.

We will sketch the proof of the theorems in Section 2, and after that we will discuss in Section 3 the example of [7], namely the reduction of a free dynamics to a domain in $\mathbb{R}^d$ by permanent observation.

2. Proof sketch of Theorems 1.1 and 1.2

We deal only with the case for $\varepsilon = 1$; the case $\varepsilon = -1$ can be treated similarly. We restrict ourselves to proving the symmetric product case (1.1), and treat only the convergence in $L^2_{\rm loc}$ in $t \geq 0$ instead of $\mathbb{R}$. The result for negative halfline and the non-symmetric cases (1.2), (1.3) then follow easily – cf. [6]. We shall explain the essential steps of the argument. To begin with, put $F(\zeta,\tau) = P\,e^{-\zeta\tau H}P$ for $\zeta \in \mathbb{C}$ with $\mathrm{Re}\,\zeta \geq 0$ and $\tau > 0$. Furthermore, define

$$S(\zeta,\tau) = \tau^{-1}[I - F(\zeta,\tau)] = \tau^{-1}[I - P\,e^{-\zeta\tau H}P].$$

The key ingredient of the proof is the following lemma.

Lemma 2.1. *$(I + S(it,\tau))^{-1}$ converges to $(I + itH\_P)^{-1}P$ as $\tau \to 0$ strongly in $\$L^2_{\rm loc}([0,\infty);\mathcal{H})$, i.e. for all $f \in \mathcal{H}$ and every finite $T > 0$ we have*

$$\int 0^T|(I + S(it,\tau))^{-1}f - (I + itH\_P)^{-1}Pf|^2 dt \to 0, \quad \tau \to 0.$$

Proof. Our aim is to show that $(I + S(\zeta,\tau))^{-1}$ converges to the indicated limit in $L^2_{\rm loc}([0,\infty);\mathcal{H})$. Note that $S(\zeta,\tau)$ is defined also in the closed halfplane $\mathrm{Re}\,\zeta \geq 0$ except at $\zeta = 0$. The proof is acomplished in three steps. First, we demonstrate the claim for ζ on the positive real axis, next in the open first quarter, and finally we show the desired assertion.

Step 1. For $\zeta = t > 0$ and $S(t,\tau) = \tau^{-1}[I - P\,\mathrm{e}^{-t\tau H}P]$, we have

$$(I + S(t,\tau))^{-1} \longrightarrow (I + tH_P)^{-1}P \quad \text{strongly as} \quad \tau \to 0\,.$$

The proof of this claim is analogous to Kato's argument. Denote $Q = I - P$ and $H(t\tau) = (t\tau)^{-1}[I - \mathrm{e}^{-t\tau H}]$. Given an $f \in \mathcal{H}$, put $\hat{u}(t,\tau) = (I + S(t,\tau)^{-1}f$, so that

$$f = (I + S(t,\tau))\hat{u}(t,\tau) = [I + \tau^{-1}Q + tPH(t\tau)P]\hat{u}(t,\tau)\,.$$

Then we have

$$\langle \hat{u}(t,\tau), f\rangle = \|\hat{u}(t,\tau)\|^2 + \tau^{-1}\|Q\hat{u}(t,\tau)\|^2 + t\|H(t\tau)^{1/2}P\hat{u}(t,\tau)\|^2.$$

Since all the terms on the right-hand side are bounded, to *any fixed* t there exists a subsequence $\{\tau_n(t)\}$, possibly dependent on t, such that $\tau_n(t) \to 0$ as $n \to \infty$, along which the involved vectors converge *weakly in* $\mathcal{H}$,

$$\hat{u}(t,\tau) \longrightarrow \hat{u}(t)\,, \quad \tau^{-1/2}Q\hat{u}(t,\tau) \longrightarrow g_0(t)\,, \quad t^{1/2}H(t\tau)^{1/2}P\hat{u}(t,\tau) \longrightarrow h(t)$$

to some vectors $\hat{u}(t)$, $g_0(t)$ and $h(t)$ in $\mathcal{H}$. One can check that these limits satisfy

$$\hat{u}(t) = P\hat{u}(t)\,, \quad g_0(t) = 0\,, \quad h(t) = t^{1/2}H^{1/2}P\hat{u}(t)\,,$$

and $Pf = \hat{u}(t) + tH_P\hat{u}(t)$, or equivalently $\hat{u}(t) = (I + tH_P)^{-1}Pf$, and show that the convergence is independent of the sequence $\{\tau_n(t)\}$ chosen, and that it is in fact strong. This concludes Step 1. Using Step 1 and mimicking Feldman's argument [8], cf. also [4] and [9], in combination with Vitali theorem on analytic continuation we find that for $\mathrm{Re}\,\zeta > 0$ one has

$$(I + S(\zeta, \tau))^{-1} \longrightarrow (I + \zeta H_P)^{-1}P \quad \text{strongly as} \quad \tau \to 0\,,$$

and therefore on the boundary halfline, $\mathrm{Re}\,\zeta = 0$, or $\zeta = it$ with t real, the convergence holds at least in the following weaker sense

Step 2. For any pair of vectors f, $g \in \mathcal{H}$, the family $\langle g, (I + S(it, \tau))^{-1}f\rangle$ of functions of t in $L^\infty(\mathbb{R})$ converges weakly* to $\langle g, (I + itH_P)^{-1}Pf\rangle$ as $\tau \to 0$.

Step 3. Finally we are going to show the desired assertion. We employ an argument similar to that used in the Step 1 on $\mathcal{H}$, however, this time not on the Hilbert $\mathcal{H}$ but on the Fréchet space $L^2_{\mathrm{loc}}([0, \infty);\, \mathcal{H})$. Given an arbitrary $f \in \mathcal{H}$, put $u(t, \tau) = (I + S(it, \tau))^{-1}f$ and $iH(it\tau) = (t\tau)^{-1}[I - e^{-it\tau H}] = B(t\tau) + iA(t\tau)$ for $t \neq 0$, where $B(t\tau)$ and $A(t\tau)$ are bounded and self-adjoint on $\mathcal{H}$, and $B(t\tau)$ is in addition nonnegative. Then

$$\begin{aligned} f &= (I + S(it, \tau))u(t, \tau) \\ &= [I + \tau^{-1}Q + tP(B(t\tau) + iA(t\tau))P]u(t, \tau)\,, \end{aligned}$$

and therefore

$$\begin{aligned} \langle u(t, \tau), f\rangle &= \|u(t, \tau)\|^2 + \tau^{-1}\|Qu(t, \tau)\|^2 + t\|B(t\tau)^{1/2}Pu(t, \tau)\|^2 \\ &\quad + it\langle Pu(t, \tau), A(t\tau)Pu(t, \tau)\rangle\,. \end{aligned}$$

Inspecting the real part we see that for τ small enough, each of the $\mathcal{H}$-valued families $\{u(t, \tau)\}$, $\{\tau^{-1/2}Qu(t, \tau)\}$ and $\{t^{1/2}B(t\tau)^{1/2}Pu(t, \tau)\}$ is bounded by $\|f\|$ for all $t > 0$. Moreover, they are strongly continuous in t for fixed $\tau > 0$, and locally bounded as $\mathcal{H}$-valued functions of t in $L^2_{\mathrm{loc}}([0, \infty); \mathcal{H})$, uniformly as $\tau \to 0$. Thus there is a sequence $\{\tau_n\}_{n=1}^{\infty}$ with $\tau_n \to 0$ as $n \to \infty$ along which the above families are weakly convergent in $L^2_{\mathrm{loc}}([0, \infty); \mathcal{H})$,

$$\begin{aligned} &u(t, \tau) \xrightarrow{w} u(t)\,, \qquad \tau^{-1/2}Qu(t, \tau) \xrightarrow{w} f_0(t)\,, \\ &t^{1/2}B(t, \tau)^{1/2}Pu(t, \tau) \xrightarrow{w} z(t)\,, \end{aligned}$$

with some limit vectors $u(\cdot)$, $f_0(\cdot)$ and $z(\cdot) \in L^2_{\mathrm{loc}}([0,\infty);\mathcal{H})$. Using the result of Step 2 we can see that $u(t,\tau) = (I+S(it,\tau))^{-1}f$ converges weakly to $u(t) = (I+itH_P)^{-1}Pf$ in $L^2_{\mathrm{loc}}([0,\infty);\,\mathcal{H})$ as $\tau \to 0$. Furthermore, we can also see that the seminorms of $u(t,\tau)$ in the Fréchet space $L^2_{\mathrm{loc}}([0,\infty);\,\mathcal{H})$ converge to those of $u(t)$. Hence it follows that $u(t,\tau)$ converges strongly to $u(t)$ in $L^2_{\mathrm{loc}}([0,\infty);\,\mathcal{H})$, proving thus Lemma 2.1. □

Let us pass to the *proof of Theorem 1.1.* By Lemma 2.1 we can show, by appealing to the separability of our Hilbert space $\mathcal{H}$, the following claim.

Lemma 2.2. *For any sequence $\{m_n\}$ of strictly increasing positive integers, i.e. any subsequence of the sequence of all positive integers, there exist a subsequence $\{n'\}$ of the sequence $\{m_n\}$ and a set $M \subset \mathbb{R}$ of Lebesgue measure zero such that*

$$(I + S(it, 1/n'))^{-1}f \longrightarrow (I + itH_P)^{-1}Pf \tag{2.1}$$

strongly in $\mathcal{H}$ as $n' \to \infty$ for every $f \in \mathcal{H}$ and for each fixed $t \notin M$.

Using then the reasoning from Chernoff [Ch2, Ch1] with Lemma 2.2 we can first demonstrate Corollary 1.1, which implies that $(Pe^{-itH/n'}P)^{n'}f$ converges to $e^{-itH_P}Pf$ in $L^2_{\mathrm{loc}}([0,\infty);\,\mathcal{H})$ as $n' \to \infty$ by the Lebesgue dominated-convergence theorem. The sought result, Theorem 1.1, then follows from a standard argument about subsequences. *Proof of Theorem 1.2.*

If the projection P is of finite dimension, one can show that for any $t, t' \geq 0$ and $0 < \tau \leq 1$ we have

$$\|u(t,\tau) - u(t',\tau)\| \leq C|t-t'|\|f\|$$

with a positive C independent of t, t'. This means that the family $\{u(t,\tau)\}$ is uniformly bounded and equicontinuous in t. Therefore by the Ascoli–Arzelà theorem the subsequence $\{\tau_n\}$ in Step 3 in the proof of of Lemma 2.1 can be so chosen that $\{u(t,\tau_n)\}$ converges strongly to $u(t)$ also pointwise, uniformly on $[0,\infty)$. Then the limit $u(t)$ becomes strongly continuous in $t \geq 0$, and coincides with $(I+itH_P)^{-1}f$ for all $t \geq 0$. Thus we have instead of Lemma 2.1 the following claim,

$$(I + S(it,\tau))^{-1} \longrightarrow (I + itH_P)^{-1}P \tag{2.2}$$

as $\tau \to 0$, strongly on $\mathcal{H}$ and uniformly on each compact interval of the variable t in $[0,\infty)$. Then we can conclude by Chernoff's theorem [Ch2] that the assertion of Theorem 1.2 is valid.

3. An example

Let us return to the situation considered by Facchi et al. [7] which we have mentioned in the introduction. Suppose that we have an open domain $\Omega \subset \mathbb{R}^d$ with a smooth boundary, and denote by P the orthogonal projection on $L^2(\mathbb{R}^d)$ defined as the multiplication operator by the indicator function χ_Ω of the set Ω. Consider further the free Schrödinger particle with the Hamiltonian $H = -\Delta$, i.e. the Laplacian in $\mathbb{R}^d$ which is a nonnegative self-adjoint operator in $L^2(\mathbb{R}^d)$, and on the other hand, the Dirichlet Laplacian $-\Delta_\Omega$ in $L^2(\Omega)$ describing the motion in Ω with a hard-wall constraint; the last named Hamiltonian is defined in the usual way as the Friedrichs extension of the appropriate quadratic form. Let us now consider the Zeno dynamics in the subspace $L^2(\Omega)$ corresponding to a permanent reduction of the wavefunction to the region Ω, which may be identified with the *volume of the detector.* We then claim that the generator of the dynamics in $L^2(\Omega)$ is just the Dirichlet Laplacian $-\Delta_\Omega$. In fact, we can show that the self-adjoint operator

$$(-\Delta)_P = ((-\Delta)^{1/2}P)^*((-\Delta)^{1/2}P) \tag{3.1}$$

is densely defined in $L^2(\mathbb{R}^d)$ and its restriction to the subspace $L^2(\Omega)$ is nothing but the Dirichlet Laplacian $-\Delta_\Omega$ of the region Ω with the domain $D[-\Delta_\Omega] = W_0^1(\Omega) \cap W^2(\Omega)$.

We have, in the sense of the topology of $L^2_{\text{loc}}(\mathbb{R}; L^2(\mathbb{R}^d)) = L^2_{\text{loc}}(\mathbb{R}) \otimes L^2(\mathbb{R}^d)$, which is physically plausible as explained in Remark 1.1,

$$(Pe^{-it(-\Delta/n)}P)^n \longrightarrow e^{-it(-\Delta_\Omega)}P, \quad n \to \infty. \tag{3.2}$$

In this sense therefore our result given in Theorem 1.1 provides one possible abstract version of the result by Facchi et al. [7].

Acknowledgments P.E. and T.I. are respectively grateful for the hospitality extended to them at Kanazawa University and at the Nuclear Physics Institute, AS CR, where parts of this work were done. The research has been partially supported by ASCR and Czech Ministry of Education under the contracts K1010104 and ME482, and by the Grant-in-Aid for Scientific Research (B) No. 13440044 and No. 16340038, Japan Society for the Promotion of Science.

References

1. M.S. Abramowitz and I.A. Stegun, eds. *Handbook of Mathematical Functions*, Dover, New York 1965.
2. J. Beskow and J. Nilsson The concept of wave function and the irreducible representations of the Poincaré group, II. Unstable systems and the exponential decay law, *Arkiv Fys.* **34** (1967), 561-569.
3. P. R. Chernoff Note on product formulas for operator semigroups, *J. Funct. Anal.* **2** (1968), 238–242.
4. P. R. Chernoff *Product Formulas, Nonlinear Semigroups, and Addition of Unbounded Operators*, Mem. Amer. Math. Soc. **140**; Providence, R.I. 1974.
5. P. Exner *Open Quantum Systems and Feynman Integrals*, D. Reidel Publ. Co., Dordrecht 1985.
6. P. Exner and T. Ichinose A product formula related to quantum Zeno dynamics, Preprint 2004.
7. P. Facchi, S. Pascazio, A. Scardicchio, and L.S. Schulman Zeno dynamics yields ordinary constraints, *Phys. Rev.* **A 65** (2002), 012108.
8. J. Feldman On the Schrödinger and heat equations for nonnegative potentials, *Trans. Amer. Math. Soc.* **108** (1963), 251–264.
9. C. Friedman Semigroup product formulas, compressions, and continual observations in quantum mechanics, *Indiana Math. J.* **21** (1971/72), 1001–1011.
10. T. Kato, *Perturbation Theory for Linear Operators*, Springer, Berlin-Heidelberg-New York 1966.
11. T. Kato Trotter's product formula for an arbitrary pair of self-adjoint contraction semigroups, in *Topics in Functional Analysis* (I. Gohberg and M. Kac, eds.), Academic Press, New York 1978; pp.185–195.
12. B. Misra and E.C.G. Sudarshan The Zeno's paradox in quantum theory, *J. Math. Phys.* **18** (1977), 756–763.
13. M. Reed and B. Simon *Methods of Modern Mathematical Physics, IV. Analysis of Operators,* Academic Press, New York 1978.

INVARIANT SUBSPACES AND CONTROL OF DECOHERENCE

P. FACCHI, V.L. LEPORE AND S. PASCAZIO

Dipartimento di Fisica, Università di Bari I-70126 Bari, Italy
E-mail: paolo.facchi@ba.infn.it,
paolo.facchi@ba.infn.it,
saverio.pascazio@ba.infn.it

We discuss three different control strategies, all aimed at countering the effects of decoherence. The first strategy hinges upon the quantum Zeno effect, the second makes use of frequent unitary interruptions ("bang-bang" pulses), and the third of a strong, continuous coupling. Decoherence is suppressed if the frequency N of the measurements/pulses is large enough or if the coupling K is sufficiently strong. However, if N or K are large, but not extremely large, all these control procedures accelerate decoherence.

1. Introduction

The deterioration of the coherence features of quantum systems, due to their interaction with the environment, is known as decoherence [1] and represents the most serious obstacle against the preservation of quantum superpositions and entanglement over long periods of time. The possibility of controlling (and eventually halting) decoherence is a key problem with important applications, e.g. in quantum computation [2]. We focus here on three schemes that have been recently proposed in order to counter the effects of decoherence. The first is based on the quantum Zeno effect (QZE) [3,4,5]), the second on "bang-bang" (BB) pulses and their generalization, quantum dynamical decoupling [6] and the third on a strong, continuous coupling (when this can be viewed as a measurement of some sort [7]). These apparently different methods are in fact related to each other [8] and a sistematic study of their analogies and differencies helps understanding under which circumstances and physical conditions all these controls may *accelerate*, rather than hinder decoherence.

In this paper we will outline the main results of a comparison among these control strategies (the complete proofs can be found in [9]). We stress that the notion of "bang-bang" control is well known in engineering and in connection with spin-echo techniques [10], so that this control can be considered "classical," its revival in quantum-information-related problems being very recent. Moreover, the idea that a strong continuous interaction with an external field or "apparatus" may be viewed as a measurement on the system and can slow its dynamics is not new [7]. (In fact, this turns out to be one of the most efficient control procedures.) For the above-mentioned reasons, the similarities among the three control methods are not surprising and their comparison interesting.

Our main objective is to endeavor to understand in which sense one can *control* decoherence [11] and to outline the key role played by the form factors of the interaction. The method we propose is general and can be applied to diverse situations of practical interest, such as atoms and ions in cavities, organic molecules, quantum dots and Josephson junctions [12].

2. Generalities and notation

Let the total system consist of a target system S and a reservoir B and its Hilbert space be expressed as the tensor product $\mathcal{H}_{\rm tot} = \mathcal{H}_S \otimes \mathcal{H}_B$. The total Hamiltonian

$$H_{\rm tot} = H_0 + H_{SB} = H_S + H_B + H_{SB} \tag{1}$$

is the sum of the system Hamiltonian H_S, the reservoir Hamiltonian H_B and their interaction H_{SB}, which is responsible for decoherence; the operators H_S and H_B act on $\mathcal{H}_S$ and $\mathcal{H}_B$, respectively. H_0 is the free total Hamiltonian.

The dynamics of the total system is conveniently reexpressed in terms of the Liouvillian

$$\mathcal{L}_{\rm tot}\rho \equiv -i[H_{\rm tot}, \rho] = -i\left(H_{\rm tot}\rho - \rho H_{\rm tot}\right) , \tag{2}$$

where ρ is the density matrix. If the Hamiltonian is given by (1), the Liouvillian is accordingly decomposed into

$$\mathcal{L}_{\rm tot} = \mathcal{L}_0 + \mathcal{L}_{SB} = \mathcal{L}_S + \mathcal{L}_B + \mathcal{L}_{SB}, \tag{3}$$

where the meaning of the symbols is obvious.

We assume that the interaction Hamiltonian H_{SB} in (1) can be written as [13]

$$H_{SB} = \sum_m \left(X_m \otimes A_m^\dagger + X_m^\dagger \otimes A_m\right) , \tag{4}$$

where the X_m are the eigenoperators of the system Liouvillian, satisfying

$$\mathcal{L}_S X_m = i\omega_m X_m \qquad (\omega_m \neq \omega_n, \quad \text{for} \quad m \neq n) \tag{5}$$

and A_m are the destruction operators of the bath

$$A_m = A(g_m) = \int d^3k \; g_m^*(\boldsymbol{k}) \; a(\boldsymbol{k}) \; , \tag{6}$$

expressed in terms of bosonic operators $a(\boldsymbol{k})$, with form factors $g_m(\boldsymbol{k})$.

The bare spectral density functions (form factors) read

$$\kappa_m(\omega) = \int d^3k \; |g_m(\boldsymbol{k})|^2 \delta(\omega_k - \omega) \; , \tag{7}$$

with $\kappa_m(\omega) = 0$, for $\omega < 0$, and the thermal spectral density functions $[N(\omega) = 1/(e^{\beta\omega} - 1)$, where β is the inverse temperature]

$$\kappa_m^\beta(\omega) = \kappa_m(\omega)\,(N(\omega)+1) + \kappa_m(-\omega)N(-\omega) = \frac{1}{1 - e^{-\beta\omega}}\,[\kappa_m(\omega) - \kappa_m(-\omega)]$$

extend along the whole real axis, due to the counter-rotating terms, and satisfy the KMS symmetry [14]

$$\kappa_m^\beta(-\omega) = \frac{N(\omega)}{N(\omega)+1}\,\kappa_m^\beta(\omega) = \exp(-\beta\omega)\,\kappa_m^\beta(\omega). \tag{8}$$

We focus on two particular (Ohmic) cases: an exponential form factor

$$\kappa_m^{(E)}(\omega) = g^2\omega\exp(-\omega/\Lambda)\theta(\omega) \tag{9}$$

and a polynomial form factor

$$\kappa_m^{(P)}(\omega) = g^2\frac{\omega}{[1+(\omega/\Lambda)^2]^n}\theta(\omega), \tag{10}$$

where g is a coupling constant, Λ a cutoff and θ the unit step function. In order to properly compare these two cases, we require that the bandwidth be the same

$$W = \frac{\int_{-\infty}^{\infty} d\omega \; |\omega|\kappa_m^{(E)}(\omega)}{\int_{-\infty}^{\infty} d\omega \; \kappa_m^{(E)}(\omega)} = \frac{\int_{-\infty}^{\infty} d\omega \; |\omega|\kappa_m^{(P)}(\omega)}{\int_{-\infty}^{\infty} d\omega \; \kappa_m^{(P)}(\omega)}. \tag{11}$$

We focus on a proper subspace $\mathcal{H}_{\text{comp}} \subset \mathcal{H}_S$, in which quantum computation is to be performed. For this reason we look in detail at the case

$$\mathcal{H}_S = \mathcal{H}_{\text{comp}} \oplus \mathcal{H}_{\text{orth}} \tag{12}$$

and consider for simplicity a single qubit, $\mathcal{H}_{\text{comp}} = \mathbb{C}^2$.

The initial state of the total system $\rho(0)$ is taken to be the tensor product of the system and reservoir initial states

$$\rho(0) = \sigma(0) \otimes \rho_B \ , \tag{13}$$

where the reservoir equilibrium state has an inverse temperature β

$$\rho_B = \frac{1}{Z} \exp(-\beta H_B), \qquad (\mathcal{L}_B \rho_B = 0) \tag{14}$$

and $Z = \mathrm{tr}_B e^{-\beta H_B}$ is the normalization constant.

The system state $\sigma(t)$ at time t is given by the partial trace

$$\sigma(t) \equiv \mathrm{tr}_B \rho(t). \tag{15}$$

There is decoherence when $\sigma(t)$ is not unitarily equivalent to $\sigma(0)$ for a given class of initial states. In the Markov approximation the state of the system (15) satisfies the master equation

$$\dot{\sigma}(t) = (\mathcal{L}_S + \mathcal{L})\, \sigma(t) \ ,$$

where, up to a renormalization of the free Liouvillian $\mathcal{L}_S$ by Lamb and Stark shift terms, $\mathcal{L}$ engenders the dissipation due to the interaction with the bath,

$$\begin{aligned} \mathcal{L}\sigma = \gamma_0 &\left(X_0 \sigma X_0 - \frac{1}{2} \{X_0 X_0, \sigma\} \right) \\ &+ \sum_{m \geq 1} \gamma_m \left(X_m \sigma X_m^\dagger - \frac{1}{2} \{X_m^\dagger X_m, \sigma\} \right) \\ &+ \sum_{m \geq 1} \gamma_{-m} \left(X_m^\dagger \sigma X_m - \frac{1}{2} \{X_m X_m^\dagger, \sigma\} \right) \ , \end{aligned} \tag{16}$$

where

$$\gamma_m = 2\pi \kappa_m^\beta(\omega_m) \tag{17}$$

are the decay rates.

A particular case of the above is the qubit Hamiltonian

$$H_{SB} = \sigma_z \otimes \left[A(g_0) + A^\dagger(g_0)\right] + \sigma_x \otimes \left[A(g_1) + A^\dagger(g_1)\right], \qquad H_0 = \frac{\Omega}{2}\sigma_z \ . \tag{18}$$

This is of the form (4), when one identifies

$$X_0 = \sigma_z, \quad X_{\pm 1} = \sigma_{\mp} = \frac{\sigma_x \mp i\sigma_y}{2}, \qquad \omega_{\pm 1} = \pm\Omega, \quad \omega_0 = 0,$$

hence

$$\mathcal{L}\rho = \gamma_0 (\sigma_z \rho \sigma_z - \rho) + \gamma_{+1} \left(\sigma_- \rho \sigma_+ - \frac{1}{2} \{\sigma_+ \sigma_-, \rho\} \right)$$
$$+ \gamma_{-1} \left(\sigma_+ \rho \sigma_- - \frac{1}{2} \{\sigma_- \sigma_+, \rho\} \right) ,$$

with

$$\gamma_0 = 2\pi \kappa_0^\beta(0), \qquad \gamma_{\pm 1} = 2\pi \kappa_1^\beta(\pm\Omega) . \tag{19}$$

3. Control procedures

3.1. *Quantum Zeno control*

In general, the purpose of the control is to suppress decoherence, as expressed by the "unitarity defect" of the evolution (15). We first look at the Zeno control, by adapting the argument of Ref. [15]. The control is obtained by performing frequent measurements of the system:

$$\rho \to \hat{P}\rho \equiv \sum_n P_n \rho P_n, \tag{20}$$

where $\hat{P}$ is a projection superoperator and $\{P_n\}$ a complete ($\sum_n P_n = \mathbf{1}_S$) set of orthogonal projection operators acting on $\mathcal{H}_S$. We restrict our analysis to a measuring apparatus that does not "select" the different outcomes (nonselective measurement) [16]. The measurement is designed so that

$$\hat{P} H_{SB} = \sum_n P_n H_{SB} P_n = 0 \quad \Leftrightarrow \quad \hat{P} \mathcal{L}_{SB} \hat{P} = 0. \tag{21}$$

We will see that a similar requirement is necessary for the other control procedures, to be analyzed in the next subsections. The Zeno control consists in performing repeated nonselective measurements at times $t = k\tau$ ($k = 0, 1, 2, \ldots$) (we include an initial "state preparation" at $t = 0$). Between successive measurements, the system evolves via H_{tot}. The density matrix after $N+1$ measurements, with an initial state $\rho(0)$, in the limit $\tau \to 0$ while keeping $t = N\tau$ constant, reads

$$\rho(t) = \rho(N\tau) = \left[\hat{P} e^{\mathcal{L}_{\text{tot}} \tau} \hat{P} \right]^N \rho(0)$$
$$= \hat{P} \left[1 + \hat{P} \mathcal{L}_{\text{tot}} \hat{P} \tau + \mathrm{O}\left(\tau^2\right) \right]^{\frac{t}{\tau}} \rho(0) \stackrel{\tau \to 0}{\longrightarrow} \hat{P} e^{\hat{P} \mathcal{L}_{\text{tot}} \hat{P} t} \rho(0) ,$$

where the controlled Liouvillian is

$$\mathcal{L}'_{\text{tot}} = \hat{P} \mathcal{L}_{\text{tot}} \hat{P} = \hat{P} \mathcal{L}_S \hat{P} + \mathcal{L}_B \hat{P} . \tag{22}$$

Hence, as a result of infinitely frequent measurements, the system-reservoir coupling is eliminated and, thus, decoherence is halted. Also, transitions among different sectors of the system Hilbert space, defined by the measurement superoperator $\hat{P}$, become forbidden, yielding a superselection rule and the formation of invariant "Zeno" subspaces [15]. The "decoherence-free" subspace [17] is one of these Zeno subspaces.

We assume for simplicity that $\hat{P}$ commutes with the system Liouvillian

$$\hat{P}\mathcal{L}_S = \mathcal{L}_S\hat{P}, \tag{23}$$

so that

$$\mathcal{L}'_{\text{tot}} = (\mathcal{L}_S + \mathcal{L}_B)\hat{P} \ . \tag{24}$$

The crucial issue is to understand what happens when the interval $\tau = t/N$ between measurements is *finite*. The evolution of the reduced state operator is governed by

$$\dot{\sigma}(t) = [\mathcal{L}_S + \mathcal{L}_Z(\tau)]\,\sigma(t) \ ,$$

where the dissipative part is found to have the explicit form [analogous to Eq. (16)]

$$\begin{aligned}\mathcal{L}_Z(\tau)\sigma = {} & \gamma_0^Z(\tau)\hat{P}\left(X_0\hat{P}\sigma X_0 - \frac{1}{2}\left\{X_0X_0, \hat{P}\sigma\right\}\right) \\ & + \sum_{m\geq 1}\gamma_m^Z(\tau)\hat{P}\left(X_m\hat{P}\sigma X_m^\dagger - \frac{1}{2}\left\{X_m^\dagger X_m, \hat{P}\sigma\right\}\right) \\ & + \sum_{m\geq 1}\gamma_{-m}^Z(\tau)\hat{P}\left(X_m^\dagger\hat{P}\sigma X_m - \frac{1}{2}\left\{X_m X_m^\dagger, \hat{P}\sigma\right\}\right) \ ,\end{aligned} \tag{25}$$

with the controlled decay rates

$$\gamma_m^Z(\tau) = \tau\int_{-\infty}^{\infty} d\omega\ \kappa_m^\beta(\omega)\,\mathrm{sinc}^2\left(\frac{\omega - \omega_m}{2}\tau\right) \ , \tag{26}$$

where $\mathrm{sinc}(x) = \sin(x)/x$. Let us focus on the exponential (9) and polynomial form factors (10). We work in the high-temperature case, which is rather critical from an experimental point of view, because of temperature-induced transitions in two-level systems, and set $\Omega = 0.01W$, $\beta = 50W^{-1}$, so that temperature $= \beta^{-1} = 2\Omega$. Observe that

$$\gamma^Z(\tau) \sim \frac{\tau}{\tau_Z^2}, \quad (\tau \to 0) \tag{27}$$

$$\tau_Z^{-2} = \int_{-\infty}^{\infty} d\omega\ \kappa^\beta(\omega) = \int_0^{\infty} d\omega\ \kappa(\omega)\coth\left(\frac{\beta\omega}{2}\right) \ ,$$

τ_Z being the thermal Zeno time. (We dropped the suffix m for simplicity.) Notice also that

$$\gamma^{Z}(\tau) \to \gamma, \qquad \tau \to \infty , \tag{28}$$

where

$$\gamma = 2\pi\kappa^{\beta}(\Omega) \tag{29}$$

is the natural decay rate (17). The ratio $\gamma^{Z}(\tau)/\gamma$ is the key quantity: decoherence is suppressed (controlled) if $\gamma^{Z}(\tau) < \gamma$, and it is enhanced otherwise. The latter phenomenon is known in the literature as *inverse* Zeno effect [18,19,20]. The key issue is to understand *how small* τ should be in order to get suppression (control) of decoherence (QZE), rather than its enhancement. This ratio $\gamma^{Z}(\tau)/\gamma$ is shown in Figs. 1 and 2 as a function of τ [in units W–the bandwidth defined in Eq. (11)]. The transition between the two regimes takes place at $\tau = \tau^*$, where τ^* is defined by the equation [20]

$$\gamma^{Z}(\tau^*) = \gamma. \tag{30}$$

It is useful to spend a few words on the *physical* meaning of the expressions $\tau \to 0$, $\beta \to \infty$ in the above (and following) formulas. Times and temperatures are to be compared with the bandwidth W (or frequency cutoff Λ). Times (temperatures) are "small" when $\tau \ll W^{-1}$ ($\beta^{-1} \ll W$). For example, when one considers short-time expansions in a Zeno context, the relevant timescale is τ^* [20,21]: the expansion (27) is valid for $\tau \lesssim W^{-1}$ (and not $\tau \lesssim \tau_Z$, as it is sometimes erroneously assumed).

3.2. *Control via quantum dynamical decoupling and "bang-bang" pulses*

We now turn our attention to the so-called quantum dynamical decoupling [6], and in particular to a kicked control ("bang-bang" pulses). In this case, one applies after each time interval τ an *instantaneous* unitary operators $U_{\rm k}$ and gets the following convenient explicit expression of the effective Hamiltonian [8]

$$H'_{\rm tot} = \hat{P} H_{\rm tot} = \sum_n P_n H_{\rm tot} P_n, \tag{31}$$

where the projections P_n arise from the spectral decomposition

$$U_{\rm k} = \sum_n e^{-i\lambda_n} P_n, \qquad (\lambda_n \neq \lambda_m \bmod 2\pi, \quad \text{for } n \neq m) . \tag{32}$$

By assuming again, as in (21) and (23), that $[\hat{P}, \mathcal{L}_S] = 0$ and that $\hat{P}\mathcal{L}_{SB}\hat{P} = 0$ we get the controlled evolution

$$\rho(t) = \left[e^{\mathcal{L}_{\rm k}} e^{\mathcal{L}_{\rm tot}\tau}\right]^{\frac{t}{\tau}} \hat{P}\rho(0) \to e^{\mathcal{L}'_{\rm tot} t} \hat{P}\rho(0), \qquad \tau \to 0 , \tag{33}$$

where $\mathcal{L}_{\rm k}$ is the Liouvillian corresponding to the evolution (32) and

$$\mathcal{L}'_{\rm tot} = \hat{P}\mathcal{L}_{\rm tot}\hat{P} = (\mathcal{L}_S + \mathcal{L}_B)\hat{P}, \tag{34}$$

exactly as in (24). As in the case discussed in the previous subsection, one observes the formation of invariant Zeno subspaces: transitions among different subspaces vanish in the $\tau \to 0$ limit, yielding a superselection rule. In this case, the subspaces are defined by (31)-(32) and are nothing but the ergodic sectors of $U_{\rm k}$. Note also that the controlled Liouvillians for bang bang pulses, (34), and for the Zeno control, (24), coincide when the set of orthogonal projections (20) is chosen equal to the set (32) of eigenprojections of $U_{\rm k}$, namely

$$\mathcal{L}_{\rm k}\hat{P} = 0, \qquad (\hat{P}\mathbf{1}) = \mathbf{1}. \tag{35}$$

Therefore, the two controls are equivalent in the ideal (limiting) case [8]. However, throughout this article, by dynamical decoupling we will refer to a situation where the evolution is coherent (unitary), while by Zeno control to a situation where the evolution involves incoherent (nonunitary) processes, such as quantum measurements.

As in the Zeno control, let us look at the τ-finite situation. Let us consider the two level system (18) with $g_0 = 0$ (spin-flip decoherence). We include an additional third level—that performs the control—and add to (18) the Hamiltonian (acting on $\mathcal{H}_S \oplus \mathrm{span}\{|M\rangle\}$)

$$H_M = -\frac{\Omega}{2}|M\rangle\langle M|, \tag{36}$$

so that $|M\rangle$ is degenerate with $|{\downarrow}\rangle$. The control consists of a sequence of 2π pulses [22] between $|{\downarrow}\rangle$ and $|M\rangle$, given by

$$U_{\rm k} = \exp\left[-i\pi\left(|{\downarrow}\rangle\langle M| + |M\rangle\langle{\downarrow}|\right)\right] = P_{\uparrow} - P_{-1} , \tag{37}$$

where

$$P_{\uparrow} = |\uparrow\rangle\langle\uparrow|, \qquad P_{-1} = P_{\downarrow} + P_M = |\downarrow\rangle\langle\downarrow| + |M\rangle\langle M|, \tag{38}$$

are the eigenprojections of $U_{\rm k}$ (belonging respectively to $e^{-i\lambda_{\uparrow}} = 1$ and $e^{-i\lambda_{-1}} = -1$), that define two Zeno subspaces.

One gets for the decay rate out of state $|\uparrow\rangle$

$$\gamma^{\rm k}(\tau) = \frac{2}{\pi}\sum_{j=0}^{\infty}\frac{1}{\left(j+\frac{1}{2}\right)^2}\left[\kappa^\beta\left(\Omega+\frac{2\pi}{\tau}(j+1/2)\right)+\kappa^\beta\left(\Omega-\frac{2\pi}{\tau}(j+1/2)\right)\right]$$

$$\stackrel{\tau\to 0}{\sim} \frac{2}{\pi}\sum_{j=0}^{\infty}\frac{1}{\left(j+\frac{1}{2}\right)^2}\,\kappa\left(\frac{\pi}{\tau}(2j+1)\right), \tag{39}$$

where in the expansion we assumed that β is not too small (as compared to τ). The key issue, once again, is to understand *how small* τ should be in order to get suppression of decoherence (control), rather than its enhancement.

Notice that

$$\gamma^{\rm k}(\tau) \to \frac{4}{\pi}\,\kappa^\beta\,(\Omega)\sum_{j=0}^{\infty}\frac{1}{\left(j+\frac{1}{2}\right)^2} = \gamma, \qquad \tau\to\infty\,.$$

The ratio $\gamma^{\rm k}(\tau)/\gamma$ is shown in Figs. 1 and 2 as a function of τ. Once again, the transition between the two regimes takes place at $\tau=\tau^*$, where τ^* is defined by the equation

$$\gamma^{\rm k}(\tau^*) = \gamma. \tag{40}$$

Observe that the mechanism of decoherence suppression (39) is not fully determined by $\mathcal{L}_{\rm tot}$ and $\hat{P}$, in contrast to the Zeno case, and depends also on the details of the Liouvillian $\mathcal{L}_{\rm k}$.

3.3. *Control via a strong continuous coupling*

The formulation in the preceding subsections hinges upon instantaneous processes, that can be unitary or nonunitary. However, the basic features of the QZE can be obtained by making use of a continuous coupling, when the external system takes a sort of steady "gaze" at the system of interest. The mathematical formulation of this idea is contained in a theorem [15] on the (large-K) dynamical evolution governed by a *generic* Liouvillian of the type

$$\mathcal{L}_K = \mathcal{L}_{\rm tot} + K\mathcal{L}_{\rm c}. \tag{41}$$

$\mathcal{L}_{\rm c}$ can be viewed as an "additional" interaction Hamiltonian performing the "measurement" and K is a coupling constant. The evolution reads

$$\rho(t) = e^{(K\mathcal{L}_{\rm c}+\mathcal{L}_{\rm tot})t}\hat{P}\rho(0) \to e^{\mathcal{L}'_{\rm tot}t}\hat{P}\rho(0), \qquad K\to\infty\,, \tag{42}$$

[see (33)] where the notation is obvious and

$$\mathcal{L}_c \hat{P} = 0, \qquad (\hat{P}\mathbf{1}) = \mathbf{1} \tag{43}$$

[see (35)].

The above statements can be proved by making use of the adiabatic theorem [23]. Once again, like in the two previous subsections, one observes the formation of invariant Zeno subspaces, that are in this case the eigenspaces of the interaction (43). The links between the quantum Zeno effect and the notion of "continuous coupling" to an external apparatus or environment has often been proposed in the literature of the last 25 years [7]. The novelty here lies in the gradual formation of the Zeno subspaces as K becomes increasingly large. In such a case, they are nothing but the adiabatic subspaces.

In general, as in the BB control but in contrast to the Zeno case, the mechanism of decoherence suppression is not fully determined by H_S and depends on the details of the Hamiltonians H_S *and* H_c. Once again, this can be clarified by looking at a specific example: consider the two level system (18) with $g_0 = 0$ (spin flip decoherence). We add to (18) the Hamiltonian (acting on $\mathcal{H}_S \oplus \mathrm{span}\{|M\rangle\}$)

$$\begin{aligned} H_M &= -\frac{\Omega}{2}|M\rangle\langle M| + KH_c, \\ H_c &= |\downarrow\rangle\langle M| + |M\rangle\langle\downarrow| = P_+ - P_- \,, \end{aligned} \tag{44}$$

where

$$P_\pm = \frac{(|\downarrow\rangle \pm |M\rangle)(\langle\downarrow| \pm \langle M|)}{2} \equiv |\pm\rangle\langle\pm|. \tag{45}$$

The third state $|M\rangle$ is now "continuously" coupled to state $|\downarrow\rangle$, $K \in \mathbb{R}$ being the strength of the coupling. As K is increased, state $|M\rangle$ performs a better "continuous observation" of $|\downarrow\rangle$, yielding the Zeno subspaces [5]. In terms of its eigenprojections, H_c reads

$$H_c = \eta_\uparrow P_\uparrow + \eta_- P_- + \eta_+ P_+, \tag{46}$$

with $P_\uparrow = |\uparrow\rangle\langle\uparrow|$ and $\eta_\uparrow = 0, \eta_\pm = \pm 1$. In the Zeno limit ($K \to \infty$) the subspaces $\mathcal{H}_\uparrow$, $\mathcal{H}_+$ and $\mathcal{H}_-$ decouple due to wildly oscillating phases $\mathrm{O}(K)$. We get

$$\hat{P}H_{SB} = P_\uparrow H_{SB} P_\uparrow + P_- H_{SB} P_- + P_+ H_{SB} P_+ = 0. \tag{47}$$

Therefore in the limit $K \to \infty$, $\gamma_{\pm 1} = 0$ and decoherence is halted.

By diagonalizing the new system Hamiltonian one obtains for the decay rate out of state $|\uparrow\rangle$

$$\gamma^{c}(K) = \frac{\gamma_{+}(K)+\gamma_{-}(K)}{2} = \pi\left(\kappa_1^{\beta}(\Omega-K)+\kappa_1^{\beta}(\Omega+K)\right)$$
$$\sim \pi\kappa(K)\left(1+e^{-\beta K}\right) \sim \pi\kappa(K), \tag{48}$$

for $K \to \infty$. On the other hand,

$$\gamma^{c}(K) \to \gamma, \qquad K \to 0. \tag{49}$$

Notice that the role of K in this subsection and the role of $1/\tau$ in the previous ones are equivalent. This yields a natural comparison [8] between different timescales (τ for measurements and kicks, $1/K$ for continuous coupling).

The ratio $\gamma^{c}(K)/\gamma$ is shown in Figs. 1 and 2 as a function of $2\pi/K$. The transition between these two regimes takes now place at $K = K^{*}$ where K^{*} is defined by the equation

$$\gamma^{c}(K^{*}) = \gamma. \tag{50}$$

3.4. *Comparison among the three control strategies*

There is a clear difference between *bona fide* projective measurements and the other two cases, BB kicks and continuous coupling. In the former case τ^{*} depends on the global features of the form factor (i.e., its integral). By contrast, in the other two cases τ^{*} "pick" some particular ("on-shell") value(s). This important difference is due to the different features of the evolution (non-unitary in the first case, unitary in the latter cases). The different features discussed above yield very different outputs, clearly apparent in Fig. 2, that can be important in practical applications: decoherence can be more easily halted by applying BB and/or continuous coupling strategies. These two methods yield values of τ^{*} (or K^{*}) that are easier to attain. However, this advantage has a price, because BB and continuous coupling yield a larger enhancement of decoherence for $\tau > \tau^{*}$, $K < K^{*}$. The two dynamical methods perform better only when $\tau \lesssim \tau^{*}$, $K \gtrsim K^{*}$. This is apparent in Fig. 1. We notice that a strict comparison between continuous coupling and the other two methods is difficult, as it would involve an analysis of numerical factors of order one in the definition of the relevant conversion factors between the frequency of interruptions τ and the coupling K (this factor has been sensibly–but arbitrarily–set equal to 2π in Figs. 1-2).

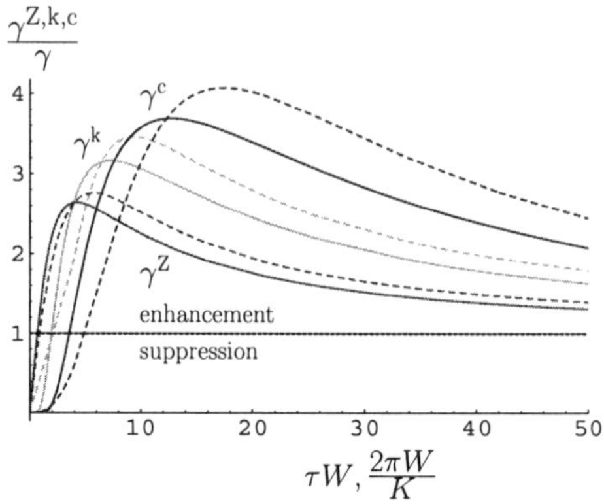

Figure 1. Comparison among the three control methods. The full and dashed lines refer to the exponential and polynomial form factors, respectively. BB kicks and continuous coupling are more effective than *bona fide* measurements for combatting decoherence, as the regime of "suppression" is reached for larger values of τ and K^{-1}.

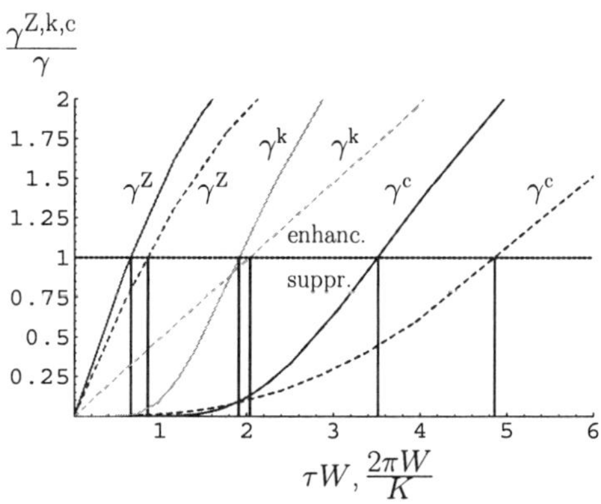

Figure 2. Comparison among the three control methods: small times/strong coupling regions. τ^* and K^* are indicated.

4. The Zeno subspaces

The three different procedures described in the previous section yield, by different physical mechanisms, the formation of invariant Zeno subspaces. This is shown in Fig. 3. If one of these invariant subspaces is the "computational" subspace $\mathcal{H}_{\rm comp}$ introduced in Eq. (12), the possibility arises of inhibiting decoherence in this subspace.

Of course, in principle (in the $\tau, K^{-1} \to 0$ limit), decoherence can be

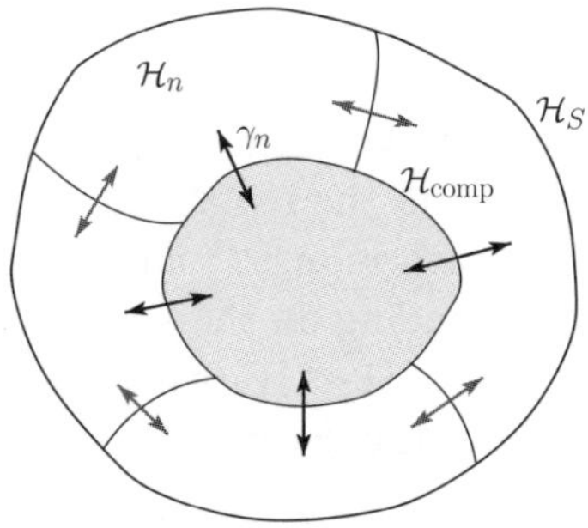

Figure 3. The Zeno subspaces are formed when the frequency τ^{-1} of measurements or BB pulses or the strength K of the continuous coupling tend to ∞. The shaded region represents the "computational" subspace $\mathcal{H}_{\rm comp} \subset \mathcal{H}_S$ defined in Eq. (12). The transition rates γ_n depend on τ or K.

completely *halted*; however, the main objective of our study was to understand *how* the limit is attained and analyze the deviations from the ideal situation. This was done by studying the transition rates γ_n between different subspaces and in particular their τ and K dependence (see Fig. 3). In general, this dependence can be complicated, leading to *enhancement* of decoherence in some cases and *suppression* in other cases. For this reason, the key issue is to understand the *physical* meaning of the expressions $\tau, K^{-1} \to 0$. This point is often sloppily analyzed in the literature.

5. Summary and concluding remarks

We compared three control methods for combatting decoherence. The first is based on repeated quantum measurements (projection operators) and involves a description in terms of nonunitary processes. The second and third methods are both dynamical, as they can be described in terms of unitary evolutions. In all cases, decoherence can be halted by very rapidly/strongly driving or very frequently measuring the system state. However, if the frequency is not high enough or the coupling not strong enough, the controls may accelerate the decoherence process and deteriorate the performance of the quantum state manipulation. The acceleration of decoherence is analogous to the inverse Zeno effect, namely the acceleration of the decay of an unstable state due to frequent measurements [18,19,20].

It is convenient to summarize the main results obtained in this article in the particular case of a two-level system (qubit) with energy difference Ω. If the frequency τ^{-1} of measurements or BB kicks, or the strength K of the coupling tend to ∞, the two-dimensional (Zeno) subspace defining the qubit becomes isolated and decoherence is completely suppressed. However, if τ^{-1} and K are large, but not extremely large, the transition (decay) rates between the qubit subspace and the remaining sector of the Hilbert space display a complicated dependence on τ^{-1} and K, and decoherence can be suppressed or enhanced, depending on the situation.

At low temperatures $\beta^{-1} \ll \Omega \ll W$, where W is the bandwidth of the form factor of the interaction, the decay rates read

$$\begin{cases} \gamma^Z(\tau) \sim \frac{\tau}{\tau_Z^2}, & \tau \to 0 \,, \\ \gamma^{\mathrm{k}}(\tau) \sim \frac{8}{\pi}\kappa\left(\frac{\pi}{\tau}\right), & \tau \to 0 \,, \\ \gamma^{\mathrm{c}}(K) \sim \pi\kappa(K), & K \to \infty \,, \end{cases} \tag{51}$$

where Z, k and c denote (Zeno) measurements, (BB) kicks and continuous coupling, respectively, κ is the form factor and $1/\tau_Z^2 \simeq \int d\omega \kappa(\omega)$ the Zeno time. As we have shown, there is a characteristic transition time τ^* [coupling K^*], such that one obtains:

$$\begin{aligned} &\text{for} \tau < \tau^* \ [K > K^*] \Rightarrow \text{decoherence suppression} : \gamma(\tau) < \gamma \ [\gamma(K) < \gamma], \\ &\text{for} \tau > \tau^* \ [K < K^*] \Rightarrow \text{decoherence enhancement} : \gamma(\tau) > \gamma \ [\gamma(K) > \gamma]. \end{aligned} \tag{52}$$

Therefore, in order to obtain a suppression of decoherence, the interruptions/coupling must be *very* frequent/strong. Notice that both τ^* and $2\pi/K^*$ are not simply related to the inverse bandwidth $2\pi W^{-1}$: they can be in general (much) shorter. For instance, in the Ohmic polynomial case (10), one gets

$$\begin{cases} \tau_Z^* \simeq 2\pi W^{-1}\left(2(n-1)\alpha_n^2 \frac{\Omega}{W}\right) \ll 2\pi W^{-1} \,, \\ \tau_{\mathrm{k}}^* \simeq 2\pi W^{-1} \frac{\alpha_n}{2}\left(\frac{\alpha_n \pi^2}{4}\frac{\Omega}{W}\right)^{\frac{1}{2n-1}} \ll 2\pi W^{-1} \,, \\ K^* \simeq W\alpha_n^{-1}\left(\frac{2}{\alpha_n}\frac{W}{\Omega}\right)^{\frac{1}{2n-1}} \gg W \,, \end{cases} \tag{53}$$

where $\alpha_n = (\sqrt{\pi}/2)\Gamma(n-3/2)/\Gamma(n-1) \leq \pi/2$ is a coefficient of order 1 and n characterizes the polynomial fall off of the form factor (10). The above times/coupling may be (very) difficult to achieve in practice. In fact, we see here that the relevant timescale is not simply the inverse bandwidth $2\pi W^{-1}$, but can be much shorter if $\Omega \ll W$, as is typically the case. These conclusions, summarized here for the simple case of a qubit, are valid in general, when one aims at protecting from decoherence an N-dimensional subspace.

Acknowledgments We thank D. Lidar, H. Nakazato, S. Tasaki and A. Tokuse for many discussions. This work is partly supported by the bilateral Italian-Japanese project 15C1 on "Quantum Information and Computation" of the Italian Ministry for Foreign Affairs.

References

1. D. Giulini, E. Joos, C. Kiefer, J. Kupsch, I.-O. Stamatescu, and H.-D. Zeh, *Decoherence and the Appearance of a Classical World in Quantum Theory* (Springer, Berlin, 1996); M. Namiki, S. Pascazio, and H. Nakazato, *Decoherence and Quantum Measurements* (World Scientific, Singapore, 1997).
2. A. Galindo and M.A. Martin-Delgado, Rev. Mod. Phys. **74**, 347 (2002); D. Bouwmeester, A. Ekert and A. Zeilinger, Eds. *The Physics of Quantum Information* (Springer, Berlin, 2000); M.A. Nielsen and I.L. Chuang, *Quantum Computation and Quantum Information* (Cambridge University Press, Cambridge, 2000).
3. J. von Neumann, *Mathematical Foundation of Quantum Mechanics* (Princeton University Press, Princeton, 1955); A. Beskow and J. Nilsson, Arkiv für Fysik **34**, 561 (1967); L.A. Khalfin, JETP Letters **8**, 65 (1968).
4. B. Misra and E.C.G. Sudarshan, J. Math. Phys. **18**, 756 (1977).
5. P. Facchi and S. Pascazio, Progress in Optics, ed. E. Wolf (Elsevier, Amsterdam, 2001), vol. 42, Chapter 3, p.147.
6. L. Viola and S. Lloyd, Phys. Rev. A **58**, 2733 (1998); L. Viola, E. Knill, and S. Lloyd, Phys. Rev. Lett. **82**, 2417 (1999); *ibid.* **83**, 4888 (1999); *ibid.* **85**, 3520 (2000); P. Zanardi, Phys. Lett. A **258**, 77 (1999); D. Vitali and P. Tombesi, Phys. Rev. A **59**, 4178 (1999); Phys. Rev. A **65**, 012305 (2001); C. Uchiyama and M. Aihara, Phys. Rev. A **66**, 032313 (2002); M.S. Byrd and D.A. Lidar, Quantum Information Processing **1**, 19 (2002); Phys. Rev. A **67**, 012324 (2003).
7. M. Simonius, Phys. Rev. Lett. **40**, 980 (1978); R.A. Harris and L. Stodolsky, Phys. Lett. B **116**, 464 (1982); A. Peres, Am. J. Phys. **48**, 931 (1980); L.S. Schulman, Phys. Rev. A **57**, 1509 (1998); A. Luis and L.L. Sánchez–Soto, Phys. Rev. A **57**, 781 (1998); K. Thun and J. Peřina, Phys. Lett. A **249**, 363 (1998); A.D. Panov, Phys. Lett. A **260**, 441 (1999); J. Řeháček, J. Peřina, P. Facchi, S. Pascazio, and L. Mišta, Phys. Rev. A **62**, 013804 (2000); P. Facchi and S. Pascazio, Phys. Rev. A **62**, 023804 (2000); B. Militello, A. Messina, and A. Napoli, Phys. Lett. A **286**, 369 (2001); A. Luis, Phys. Rev. A **64**, 032104 (2001).
8. P. Facchi, D.A. Lidar, and S. Pascazio, Phys. Rev. A **69**, 032314 (2004).
9. P. Facchi, S. Tasaki, S. Pascazio, H. Nakazato, A. Tokuse, D.A. Lidar, "Control of decoherence: analysis and comparison of three different strategies",

quant-ph/0403205.
10. Y. Takahashi, M.J. Rabins, and D.M. Auslander, *Control and Dynamic Systems* (Addison-Wesley, Reading, MA, 1970); J. Macki and A. Strauss, *Introduction to Optimal Control Theory* (Springer-Verlag, New York, 1982); L. Lapidus and R. Luus, *Optimal Control of Engineering Processes* (Blaisdell Publishing, Waltham, MA, 1967).
11. T. Calarco, A. Datta, P. Fedichev, E. Pazy, and P. Zoller, Phys. Rev. A **68**, 012310 (2003); E. Paladino, L. Faoro, G. Falci, and R. Fazio, Phys. Rev. Lett., **88**, 228304 (2002); G. Falci, A. D'Arrigo, A. Mastellone, and E. Paladino, "Bang-bang suppression of telegraph and 1/f noise due to quantum bistable fluctuators," cond-mat/0312442; A.G. Kofman and G. Kurizki, Phys. Rev. Lett. **87**, 270405 (2001); S. Tasaki, A. Tokuse, P. Facchi, and S. Pascazio, Int. J. Quant. Chem. **98**, 160 (2004).
12. C. Monroe *et al*, Phys. Rev. Lett. **75**, 4714 (1995); R.J. Hughes *et al*, Fortschr. Phys. **46**, 32 (1998); D.G. Cory *et al*, Fortschr. Phys. **48**, 875 (2000); M. Lieven *et al*, Nature **414**, 883 (2001); L. Jacak, P. Hawrylak, and A. Wojs, *Quantum Dots* (Springer, Berlin, 1998); D. Steinbach *et al*, Phys. Rev. B **60**, 12079 (1999); Y. Makhlin, G. Schön and A. Shnirman, Rev. Mod. Phys. **73**, 357 (2001).
13. C.W. Gardiner and P. Zoller, *Quantum Noise* (Springer, Berlin, 2000).
14. H. Spohn and J.L. Lebowitz, Adv. Chem. Physics **38**, 109 (1979).
15. P. Facchi and S. Pascazio, Phys. Rev. Lett. **89** 080401 (2002); "Quantum Zeno subspaces and dynamical superselection rules," in *The Physics of Communication*, Proceedings of the XXII Solvay Conference in Physics, edited by I. Antoniou, V.A. Sadovnichy, and H. Walther (World Scientific, Singapore, 2003) p. 251 (quant-ph/0207030).
16. J. Schwinger, Proc. Natl. Acad. Sci. U.S. **45**, 1552 (1959); *Quantum Kinetics and Dynamics* (Benjamin, New York, 1970).
17. G.M. Palma, K.A. Suominen, and A.K. Ekert, Proc. R. Soc. Lond. A **452**, 567 (1996); L.M. Duan and G.C. Guo, Phys. Rev. Lett. **79**, 1953 (1997); P. Zanardi and M. Rasetti, Phys. Rev. Lett. **79**, 3306 (1997); D.A. Lidar, I.L. Chuang, and K.B. Whaley, Phys. Rev. Lett. **81**, 2594 (1998).
18. A.M. Lane, Phys. Lett. A **99**, 359 (1983); W.C. Schieve, L.P. Horwitz and J. Levitan, Phys. Lett. A **136**, 264 (1989); B. Elattari and S.A. Gurvitz, Phys. Rev. A **62**, 032102 (2000); A.G. Kofman and G. Kurizki, Nature **405**, 546 (2000); K. Koshino and A. Shimizu, Phys. Rev. A **67**, 042101 (2003).
19. P. Facchi and S. Pascazio, Phys. Rev. A **62**, 023804 (2000).
20. P. Facchi, H. Nakazato, and S. Pascazio, Phys. Rev. Lett. **86**, 2699 (2001).
21. I. Antoniou, E. Karpov, G. Pronko, and E. Yarevsky, Phys. Rev. A **63**, 062110 (2001).
22. G.S. Agarwal, M.O. Scully, and H. Walther, Phys. Rev. A **63**, 044101 (2001); M.O. Scully, S.-Y. Zhu, and M.S. Zubairy, Chaos, Solitons and Fractals **16**, 403(2003); K. Shiokawa and D.A. Lidar, Phys. Rev. A **69**, 030302(R) (2004).
23. See, for instance, A. Messiah, *Quantum mechanics* (Interscience, New York, 1961).

CLAUSER-HORNE INEQUALITY FOR ELECTRON COUNTING STATISTICS IN MULTITERMINAL MESOSCOPIC CONDUCTORS

LARA FAORO[(1)], FABIO TADDEI[(1,2)] AND ROSARIO FAZIO[(2)]

[(1)] *ISI Foundation, Viale Settimio Severo, 65, I-10133 Torino, Italy*
E-mail: f.taddei@sns.it

[(2)] *NEST-INFM & Scuola Normale Superiore, I-56126 Pisa, Italy*
E-mail: fazio@sns.it

In this paper we derive the Clauser-Horne (CH) inequality for the full electron counting statistics in a mesoscopic multiterminal conductor and we discuss its properties. We first consider the idealized situation in which a flux of entangled electrons is generated by an *entangler*. Given a certain average number of incoming entangled electrons, the CH inequality can be evaluated for different numbers of transmitted particles. Strong violations occur when the number of transmitted charges on the two terminals is the same ($Q_1 = Q_2$), whereas no violation is found for $Q_1 \neq Q_2$. We then consider two actual setups that can be realized experimentally. The first one consists of a three terminal normal beam splitter and the second one of a hybrid superconducting structure. Interestingly, we find that the CH inequality is violated for the three terminal normal device. The maximum violation scales as $1/M$ and $1/M^2$ for the entangler and normal beam splitter, respectively, $2M$ being the average number of injected electrons. As expected, we find full violation of the CH inequality in the case of the superconducting system.

1. Introduction

Most of the work on entanglement has been performed in optical systems with photons [1], cavity QED systems [2] and ion traps [3]. Only recently attention has been devoted to the manipulation of entangled states in a solid state environment. This interest, originally motivated by the idea to realize a solid state quantum computer [4,5,6], has been rapidly growing and by now several works discuss how to generate, manipulate and detect entangled states in solid state systems. It is probably worth to emphasize already at this point that, differently from the situation encountered in quantum optics, in solid state system entanglement is rather common. What is not trivial is its control and detection (especially if the interaction between the different subsystems forming the entangled state is switched off).

Despite the large body of knowledge developed in the study of optical systems, new strategies have to be designed to reveal the signatures of non-local correlations in the case of electronic states. For mesoscopic conductors, the prototype scheme was discussed in Ref. [7]. In this work it has been shown that the presence of spatially separated pairs of entangled electrons, created by some *entangler*, can be revealed by using a beam splitter and by measuring the correlations of the current fluctuations in the leads. Provided that the electrons injected are in an entangled state bunching and anti-bunching behavior for the cross-correlations of current fluctuations are found depending on whether the state is a spin singlet or a spin triplet. Not only the noise, but the full counting statistics is sensitive to the presence of entanglement in the incoming beam [8]. The distribution of transmitted electrons is binomial and symmetric with respect to the average number of transmitted charges. Moreover, this is important for the problem studied in the present work, the joint probability for counting electrons at different leads unambiguously characterizes the state of the incident electrons if one uses spin-sensitive electron counters. In this case the joint probability cannot be expressed as a product of single-terminal probabilities.

Since Bell's work [9], it is known that a classical theory formulated in terms of a hidden variables satisfying reasonable condition of locality, yields predictions which are different from those of quantum mechanics. These predictions were casted into the form of inequalities which any realistic local theory must obey. Bell inequalities have been formulated for mesoscopic multi-terminal conductors in Refs. [10,11,12,13,14,15] in terms of electrical noise correlations at different terminals [16]. A test of quantum mechanics through Bell inequalities in mesoscopic physics is very challenging and most probably it would be rather difficult, if not impossible, to get around all possible loopholes. Although solid state systems are not the natural arena where to test the foundations of quantum mechanics, it is nevertheless very interesting to have access, manipulate and quantify these non-local correlations.

In this work (see also [17]) we derive a Bell inequality for the full electron counting statistics and discuss its properties. The formulation we follow is based on what is known as the Clauser-Horne (CH) inequality [18,19]. We shall show that the joint probabilities for a given number of electrons to pass through a mesoscopic conductor (in a given time) should satisfy, for a classical local theory, an inequality.

2. CH inequality for the full counting statistics

Recently, in Ref. [12] it has been shown that zero-frequency current cross-correlations, in the tunneling limit, can be used to formulate a Bell inequality. The same Authors, in Ref. [14], have shown that such a result can be extended to arbitrary tunneling rates, since a pair of orbitally entangled electrons is post-selected by the measurement. In this paper we take a different route by resorting to Electron Full Counting Statistics (FCS) for analyzing electronic entanglement. FCS refers to the probability that a given number of electrons has traversed, in a time t, a mesoscopic conductor. In the long time limit the first and the second moment of the probability distribution are related to the average current and noise, respectively.

In its original version [9], the Bell inequality was derived for dichotomic variables. Here we consider the more general formulation due to Clauser and Horne [18]. To this aim, we consider the idealized setup, illustrated in Fig.??, which consists of the following parts. On the left we place an entangler that produces pairs of electrons in a spin entangled state. Each electron propagates respectively into lead 3 and 4 in a superposition of spin states $\uparrow$ and $\downarrow$. (In Section 3 two different situations for the implementation of the entangler are discussed). Two conductors, characterized by some scattering matrix, connect the terminals 3 and 4 of the entangler with the exit leads 1 and 2 so to carry the two particles belonging to each pair into two different spatially separated reservoirs. The electron counting is performed in leads 1 and 2 for electrons with spin aligned along the local spin-quantization axis at angles θ_1 and θ_2. Detection is realized by means of spin-selective counters, *i.e.* by counting electrons with the projection of the spin along a given local quantization-axis. In analogy with the optical case we say that the analyzer is not present when the electron counting is spin-insensitive (electrons are counted irrespective of their spin direction). Since we assume no back-scattering from counters to the entangler, the particles which are not counted are lost and hence there is no communication between the two detectors. In the case where only two entangled electrons are injected, we find a situation similar to that with photons. More generally we discuss the case where a large number of electrons are injected, finding that the CH inequality is violated only when a "coincidence measurement" is performed. In Section 2.1 we present the derivation of the CH inequality for the FCS and in Section 2.2 we resume, for completeness, the relation between FCS and the scattering matrix S.

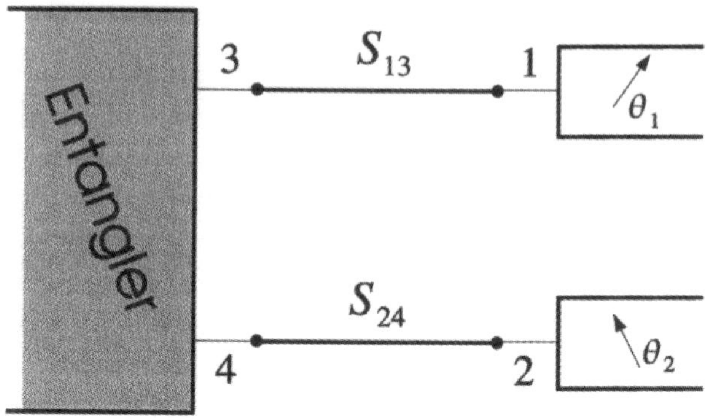

Figure 1. Idealized setup for testing the CH inequality for electrons in a solid state environment. It consists of two parts: an entangler (shaded block) that produces pairs of spin entangled electrons exiting from terminals 3 and 4. These terminals are connected to leads 1 and 2 through two conductors described by scattering matrices S_{13} and S_{24}. Electron counting is performed in leads 1 and 2 along the local spin-quantization axis oriented at angles θ_1 and θ_2.

2.1. *Derivation of the CH inequality*

The basic object for the formulation of the CH inequality is the joint probability $P(Q_1, Q_2)$ for transferring a number of Q_1 and Q_2 electronic charges into leads 1 and 2 over an observation time t. We follow closely the derivation given in Ref. [19].

We now introduce $P^{\theta_1,\theta_2}(Q_1, Q_2)$ as the joint probability for transferring Q_1 and Q_2 electronic charges when both analyzers are present, while $P^{\theta_1,-}(Q_1, Q_2)$ and $P^{-,\theta_2}(Q_1, Q_2)$ are the corresponding joint probabilities when one of the two analyzers is removed. If the condition

$$P^{\theta_\alpha}(Q_\alpha, \tau) \leq P(Q_\alpha, \tau) \tag{1}$$

(known as *no-enhancement assumption*) is verified, it is possible to obtain

$$\begin{aligned} \mathcal{S}_{CH} = {} & P^{\theta_1,\theta_2}(Q_1, Q_2) - P^{\theta_1,\theta_2'}(Q_1, Q_2) + P^{\theta_1',\theta_2}(Q_1, Q_2) + P^{\theta_1',\theta_2'}(Q_1, Q_2) \\ & - P^{\theta_1',-}(Q_1, Q_2) - P^{-,\theta_2}(Q_1, Q_2) \leq 0\,. \end{aligned} \tag{2}$$

Eq.(2) is the CH inequality for the full counting statistics [20], holding for all values of Q_1 and Q_2 which satisfy the no-enhancement assumption. We stress that the no-enhancement assumption, upon which Eq.(2) is based, it is not satisfied in general like its optical version. The quantities that we have to compare are probability distributions, so that Eq.(1) must be checked over the whole range of Q. For a fixed time t and a given mesoscopic system, hence for a given scattering matrix and incident particle state, the no-enhancement assumption is valid only in some range of values of Q. In particular, different sets of system parameters correspond to different such ranges. The quantity $\mathcal{S}_{CH}$ in Eq.(2) depends on Q_1 and Q_2 so that the possible violation, or the extent of it, also depends on Q_1 and Q_2. Given a certain average number M of entangled pairs that have being injected in the time t, one can look for the maximum violation as a function of the transmitted charges Q_1 and Q_2.

2.2. *Scattering approach to the full counting statistics*

The joint probabilities appearing in Eq.(2) can be determined once the scattering matrix S of the mesoscopic conductor is known. The FCS in electronic systems was first introduced by Levitov *et al.* in Ref. [21,22] in the context of the scattering theory and later on the Keldysh Green function method [23] to FCS was developed in Refs.[24] (for a review see Refs. [25]).

The joint probability distribution for transferring $Q_{1\sigma}$ spin-σ electrons in lead 1, $Q_{2\sigma}$ spin-σ electrons in lead 2, etc. is related to the characteristic function χ by the relation [17] (we assume that no polarizers are present):

$$P(Q_{1\uparrow}, Q_{1\downarrow}, Q_{2\uparrow}, ..) = \frac{1}{(2\pi)^{2n}} \int_{-\pi}^{+\pi} d\lambda_{1\uparrow} d\lambda_{1\downarrow} d\lambda_{2\uparrow} .. \chi(\vec{\lambda_\uparrow}, \vec{\lambda_\downarrow}) e^{i\vec{\lambda_\uparrow}\cdot\vec{Q_\uparrow}} e^{i\vec{\lambda_\downarrow}\cdot\vec{Q_\downarrow}}. \tag{3}$$

In the rest of the paper we will consider systems where only two counting terminals are present. In particular, while the counting terminals are kept at the lowest chemical potential, all other terminals are biased at chemical potential eV.

3. Results

The inequality presented in Eq.(2) can be tested in various multi-terminal mesoscopic conductors. In this Section we present several geometries that can be experimentally realized. In order to get acquainted with the informations that can be retrieved from Eq.(2) we start from an ideal case in which the entangled pair is generated by some *entangler* in the same spirit as in the works of Refs. [7,8]. In Section 3.2 we analyze the role of superconductivity in creating spin singlets. In Section 3.3 we shall demonstrate that a normal beam splitter in the absence of interaction is enough to generate entangled pairs of electrons, therefore constituting a simple realization of an entangler.

3.1. *Entangled electrons*

In the setup depicted in Fig.1 we assume the existence of an entangler that produces electron pairs in the Bell state

$$|\psi\rangle = \frac{1}{\sqrt{2}}\left[a^{\dagger}_{3\uparrow}(E)a^{\dagger}_{4\downarrow}(E) \pm a^{\dagger}_{3\downarrow}(E)a^{\dagger}_{4\uparrow}(E)\right] |0\rangle\,, \tag{4}$$

of spin triplet (upper sign) or spin singlet (lower sign) in the energy range $0 < E < eV$. These electrons propagate through the conductors, of transmission probability equal to T, which connect terminals 3 and 4 with leads 1 and 2, as though terminals 3 and 4 were kept at a potential eV with respect to 1 and 2. Our aim is to test the violation of the CH inequality given in Eq.(2) for such maximally entangled states.

For the single terminal probability distributions in leads $i = 1, 2$ we get, in the presence and in the absence of an analyzer, respectively,

$$P^{\theta_i}(Q_i) = \binom{M}{Q_i}\left(\frac{T}{2}\right)^{Q_i}\left(1-\frac{T}{2}\right)^{M-Q_i} \tag{5}$$

$$P(Q_i) = \binom{M}{Q_i}(T)^{Q_i}(1-T)^{M-Q_i}\,, \tag{6}$$

so that the no-enhancement assumption reads:

$$\left(1-\frac{T}{2}\right)^{(M-Q_i)}\left(\frac{1}{2}\right)^{Q_i} \le (1-T)^{(M-Q_i)} \qquad i = 1, 2\,. \tag{7}$$

Note that the probabilities in Eqs. (5) and (6) do not depend on the angles θ_1 and θ_2. As a consequence, the effect of the analyzer is equivalent to a reduction of the transmission probability T by a factor of 2, resulting in a shift of the maximum of the distribution. From Eq.(7) it follows that, for a given number $M = eVt/h$ of entangled pairs generated by the entangler, the no enhancement assumption holds only for certain values of T and of Q_i. Thus the CH inequality of Eq.(2) can be tested for violation only for appropriate values of M, T and Q_1 or Q_2. For example, for a given observation time t (*i.e.* a given M) and a given value of Q, CH inequality can be tested only for transmission T less than a maximum value given by the expression

$$T_{\max} = \frac{2^{\frac{Q_i}{M-Q_i}} - 1}{2^{\frac{Q_i}{M-Q_i}} - \frac{1}{2}}. \tag{8}$$

At the edge of the distribution ($Q_i = M$) the no-enhancement assumption is satisfied for every T. The window of allowed Q_i values where the no-enhancement assumption is satisfied gets wider on approaching the tunneling limit. For large M, $T_{\max} \simeq 2(\log 2)\frac{Q_i}{M}$. The previous inequality can be also interpreted as a limit for the allowed measuring time given a setup at disposal. Alternatively, given a certain transmission, the no-enhancement assumption is verified for points of the distribution such that:

$$\frac{Q_i}{M} \geq \frac{\log \frac{1-T/2}{1-T}}{\log 2 + \log \frac{1-T/2}{1-T}}. \tag{9}$$

It is useful to note here that the joint probabilities with a single analyzer are factorized:

$$\begin{aligned} P^{\theta_1,-}(Q_1,Q_2) &= P^{\theta_1}(Q_1)P(Q_2) \\ P^{-,\theta_2}(Q_1,Q_2) &= P(Q_1)P^{\theta_2}(Q_2), \end{aligned} \tag{10}$$

while joint probabilities with two analyzers are not factorized. Furthermore, all such probabilities have a common factor, $T^{Q_1+Q_2}/2^M$, which leads to an exponential suppression for large M and $Q_1 + Q_2$. We shall address the question of whether this also produces a suppression of $\mathcal{S}_{CH}$ in case of violation.

Let us now analyze the possibility of violation of the CH inequality for different values of Q_1 and Q_2. First consider the situation where the entangler emits a single entangled pair of electrons in which case we find that the CH inequality is maximally violated for the following choice of angles: $\theta_2 - \theta_1 = \theta_2' - \theta_1' = 3\pi/4$. More precisely we obtain:

$$\mathcal{S}_{CH} = T^2 \frac{\sqrt{2}-1}{2} \tag{11}$$

which is equal to the result obtain for an entangled pair of photons [19], where T plays the role of the quantum efficiency of the photon detectors. In the more general case of $Q_1 = Q_2 = M$, for $M \gg 1$, we have

$$\begin{aligned} P^{\theta_1,\theta_2}(M,M) &= \frac{T^{2M}}{2^M}\left[\sin^2\left(\frac{\theta_1 \pm \theta_2}{2}\right)\right]^M \\ P^{\theta_1,-}(M,M) = P^{-,\theta_2}(M,M) &= \frac{T^{2M}}{2^M} \end{aligned} \tag{12}$$

so that the no-enhancement assumption is always satisfied and the quantity $\mathcal{S}_{CH}$ can be easily evaluated:

$$\begin{aligned} \mathcal{S}_{CH} = \frac{T^{2M}}{2^M}\Bigg[&\sin^{2M}\frac{\theta_1 \pm \theta_2}{2} - \sin^{2M}\frac{\theta_1 \pm \theta_2'}{2} + \sin^{2M}\frac{\theta_1' \pm \theta_2}{2} \\ &+ \sin^{2M}\frac{\theta_1' \pm \theta_2'}{2} - 2\Bigg]. \end{aligned} \tag{13}$$

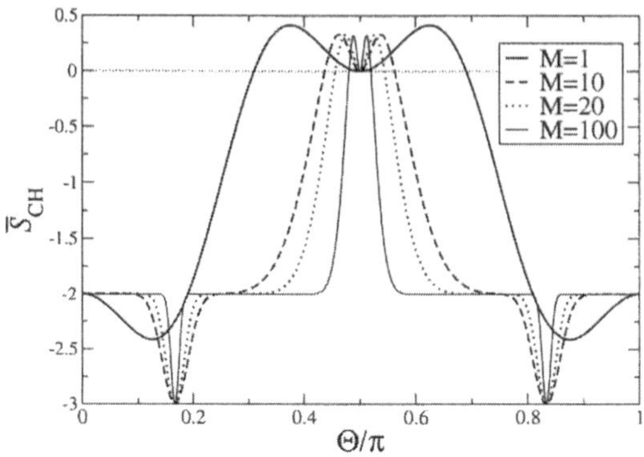

Figure 2. The quantity $\overline{\mathcal{S}}_{CH} = \mathcal{S}_{CH}/(T^{2M}/2^M)$ is plotted as a function of the angle Θ for different numbers M of injected entangled pairs by the entangler. The range of angles relative to positive values shrinks with increasing M, while the value of the maximum slightly decreases.

The rotational invariance makes $P^{\theta_1,-}$ and P^{-,θ_2} independent of angles, and P^{θ_1,θ_2} dependent on the angles through $\frac{\theta_1 \pm \theta_2}{2}$. This allows us, without loss of generality, to define an angle Θ such that $2\Theta = \theta_1 \pm \theta_2 = \theta_1' \pm \theta_2 = \theta_1' \pm \theta_2' = (\theta_1 \pm \theta_2')/3$. As a result Eq.(2) takes the form:

$$\mathcal{S}_{CH} = 3P^{\Theta}_{1,2}(Q_1,Q_2) - P^{3\Theta}_{1,2}(Q_1,Q_2) - P_{1,-}(Q_1,Q_2) - P_{-,2}(Q_1,Q_2) \leq 0 \tag{14}$$

where $P^{\Theta}_{1,2} = P^{\theta_1,\theta_2}$ and $P_{1,-} = P^{\theta_1,-}$. It is useful to define the reduced quantity $\overline{\mathcal{S}}_{CH} = \mathcal{S}_{CH}/(T^{2M}/2^M)$ which is plotted in Fig. 2 as a function of Θ for different values of M (note that since $P^{\theta_1,-}(M,M) = (T^{2M}/2^M)$, $\overline{\mathcal{S}}_{CH}$ is nothing but $\mathcal{S}_{CH}/P^{\theta_1,-}(M,M)$). The violation occurs for every value of M in a range of angles around $\Theta = \pi/2$ (note that $\mathcal{S}_{CH}$ is symmetric with respect to $\pi/2$). The range of angles for which $\overline{\mathcal{S}}_{CH}$ is positive shrinks with increasing M, while the maximum value of $\overline{\mathcal{S}}_{CH}$ decreases very weakly with M (more precisely, $\overline{\mathcal{S}}^{max}_{CH} \propto 1/M$). This means that the effect of the factor $T^{2M}/2^M$ on the value of $\mathcal{S}_{CH}$ is exponentially strong, making the violation of the CH inequality exponentially difficult to detect for large M and $Q_1 = Q_2 = M$. The weakening of the violation is mainly due to the suppression of the joint probabilities. As we shall show later, by optimizing all the parameters it is yet possible to eliminate this exponential suppression.

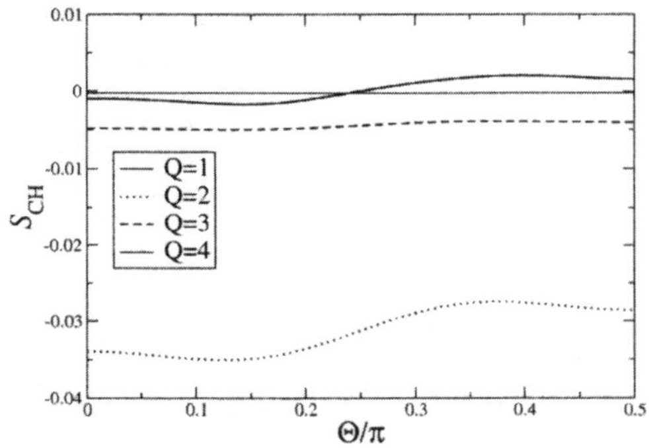

Figure 3. The quantity $\mathcal{S}_{CH}$ is plotted as a function of the angle Θ for $M = 20$ and $T = 0.06917$, which corresponds to the highest value allowed by the no-enhancement assumption for $Q = 1$. The curves are relative to different values of $Q = [1, 4]$. Note that for $Q \geq 4$ the variation of $\mathcal{S}_{CH}$ over the whole range of Θ is small on the scale of the plot. Violations are found only for $Q = 1$ and $Q = 20$.

Let us now consider the violation of the CH inequality as a function of the transmitted charges. We notice that the CH inequality is not violated for the off-diagonal terms of the distributions (when $Q_1 \neq Q_2$), meaning that one really needs to look at "coincidences". Therefore we discuss the case $Q_1 = Q_2 \equiv Q < M$ (remember that the no-enhancement assumption is satisfied only for $T \leq T_{\max}(Q)$). In Fig. 3 we plot the quantity $\mathcal{S}_{CH}$ for $M = 20$ as a function of Θ and different values of Q. The transmission T is fixed at the highest allowed value by the no-enhancement assumption, which corresponds to the smallest Q considered $T_{\max}(Q = 1) = 0.06917$. Fig. 3 shows that the largest positive value of $\mathcal{S}_{CH}$ and the widest range of angles corresponding to positive $\mathcal{S}_{CH}$ occur for $Q = 1$, *i.e.* for a joint probability relative to the detection of a single pair. One should not conclude that, in order to detect the violation of the CH inequality, only very small values of the transmitted charge should be taken. We have in fact considered $T = T_{\max}$ relative to $Q = 1$ and the maximum violation, for given M and Q, always occurs at $T = T_{\max}$. In order to get the largest violation of the CH inequality at a given M and Q one could, in principle, choose the highest allowed value of T for each value of Q ($T = T_{\max}(Q)$).

In Fig. 4 we plot the maximum value of S, with respect to Θ and T, as a function of Q for different values of M. Several observations are in order. For increasing M, the position of the maximum, $Q_{\max}$ is very weakly

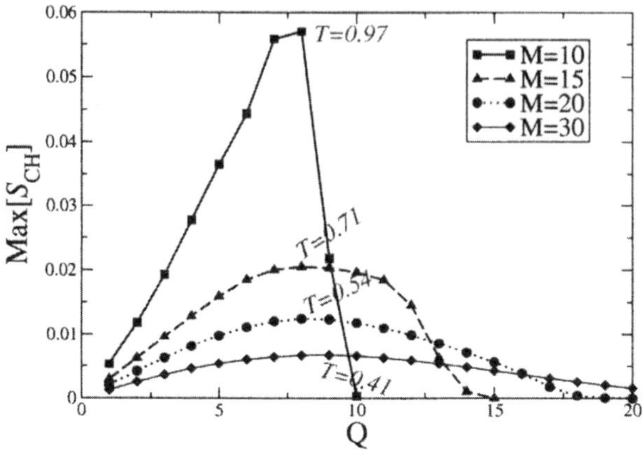

Figure 4. The maximum value of the quantity $\mathcal{S}_{CH}$, evaluated over angles Θ and transmission probabilities T, is plotted as a function of Q. The curves are relative to different values of M ranging from 10 to 30. For points corresponding to the maximum of the curves we indicate the corresponding value of transmission T.

dependent on M. Remarkably, the value of the maximum of the curves does not decreases exponentially, but rather as $1/M^2$. Despite the exponential suppression of the joint probability with M, the extent of the maximal violation scales with M much slowly (polynomially).

If the entangler is substituted with a source that emits factorized states, the CH inequality given in Eq.(2) is never violated. In this case, in contrast to Eq.(4), the state emitted by the source reads: $|\psi\rangle = a^\dagger_{3\uparrow} a^\dagger_{4\uparrow} |0\rangle$. All the previous calculations can be repeated and we find, as expected, that the characteristic functions factorizes, so that the two terminal joint probability distributions are given by the product of the single terminal probability distributions. To conclude, we wish to mention that the CH inequality, Eq.(2) holds for joint probabilities relative to arbitrary observation time, although the FCS requires long observation time, so that $M \gg 1$.

We are now ready to analyze realistic structures by replacing the shaded block in Fig. 1 (which represents the entangler) with a certain system, and discuss the CH inequality along the lines of Section 3.1.

3.2. *Superconducting beam splitter*

In many proposals superconductivity has been identified as a key ingredient for the creation of entangled pairs of electrons. The idea is to extract

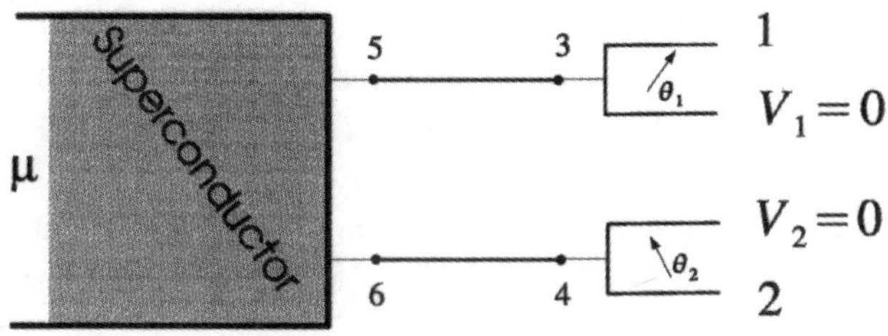

Figure 5. Setup of a realistic system consisting of a superconducting beam splitter (shaded region) for testing the CH inequality. Bold lines represent two conductors of transmission probability T. The superconducting condensate electrochemical potential is set to μ, while terminals 1 and 2 are grounded.

the two electrons which compose a Cooper pair (a pair of spin-entangled electrons) from two spatially separated terminals. We analyze the case of a superconducting beam splitter [26,27] depicted in Fig. 5, which consists of a superconducting lead (with condensate chemical potential equal to μ) in contact with two normal wires of transmission probability equal to T. The wires are then connected to two leads attached to reservoirs kept at zero potential. This is basically what is obtained by replacing the entangler of Fig. 1 by a superconducting lead with two terminals.

The no-enhancement assumption can be calculated along the lines of Eqs. (5-7) and it is easy to check that for $Q_2 = Q_3 = M$ it is always satisfied. Although Andreev processes are fundamental for the injection of Cooper pairs, in the case where Andreev transmissions only are non-zero and $T = 1$ the joint probabilities factorize in a trivial way

$$P^{\theta_1,\theta_2}(Q_1,Q_2) = \delta_{Q_1,2M}\delta_{Q_2,2M} \qquad P^{\theta_1,-}(Q_1,Q_2) = \delta_{Q_1,2M}\delta_{Q_2,4M}\,, \quad (15)$$

in such a way that the CH inequality is never violated. This apparent contradiction is due to the fact that in this situation the scattering processes occur with probability unit, so that the condition of locality is fulfilled. Non-locality can be achieved by imposing $T < 1$. In the limit $T \ll 1$ we obtain the probabilities

$$P^{\theta_1,\theta_2}(M,M) = \left[\frac{2T^2A^6}{[A-T(A-1)]^8}\right]^M \left[\sin^2\left(\frac{\theta_1+\theta_2}{2}\right)\right]^M \quad (16)$$

and

$$P^{-,\theta_2}(M,M) = \left[\frac{2T^2A^6}{[A-T(A-1)]^8}\right]^M \quad (17)$$

for $Q_2 = Q_3 = M$, with $A = 1 + \tau_{he}\tau'^{\star}_{he}$. Eqs. (16) and (17) are equal to Eqs. (12), relative to the case of an entangler, once $2T^2A^6/[A - T(A-1)]^8$ is replaced with $T^2/2$. From this follows that superconductivity leads to violation of the CH inequality For $A = 2$, *i.e.* perfect Andreev transmission, the quantity $2T^2A^6/[A - T(A-1)]^8$ tends to $T^2/2$ in the limit $T \to 0$ so that the analysis of Section 3.3 relative to the case $Q_1 = Q_2 = M$ applies also here.

3.3. *Normal beam splitter*

It is interesting to show that, even in the absence of superconductivity, a normal beam splitter leads to violations of the CH inequality. To this aim, we consider a normal beam splitter (shaded block in Fig. 6) in which lead 3 is kept at a potential eV and leads 1 and 2 are grounded so that the same bias voltage is established between 3 and 1, and 3 and 2. The two conductors, which connect the beam splitter to the leads 1 and 2, are assumed to be normal-metallic and perfectly transmissive, so that the S-matrix of the system for $\theta_1 = \theta_2 = 0$ is equal to the S-matrix of the beam splitter, which reads [28]

$$S = \begin{pmatrix} -(a+b) & \sqrt{\epsilon} & \sqrt{\epsilon} \\ \sqrt{\epsilon} & a & b \\ \sqrt{\epsilon} & b & a \end{pmatrix}. \tag{18}$$

In this parametrization of a symmetric beam splitter $a = \pm(1+\sqrt{1-2\epsilon})/2$, $b = \mp(1-\sqrt{1-2\epsilon})/2$ and $0 < \epsilon < 1/2$. For arbitrary angles θ_1 and θ_2, the S-matrix is obtained rotating the quantization axis in the two conductors independently.

We find [17] that that the characteristic functions for the beam splitter possess the same dependence on the angle difference as the corresponding characteristic functions for the entangler (Section 3.1) but have a different structure as far as scattering probabilities are concerned. In particular, as expected [21], the cross-correlations vanish when the two angles are equal. On the contrary, when the angle difference is π cross-correlations are maximized. Furthermore, when only one analyzer is present the characteristic function shows no dependence on the angle, but it is not factorisable, in contrast to the case of the entangler. As a result, the single terminal probabilities are equal in the two cases provided that ϵ is replaced with $T/2$. The joint probabilities for $Q_1 = Q_2 = M$ are equal in the two cases if ϵ is replaced with $T/\sqrt{2}$ (however, this replacement is not valid in general for

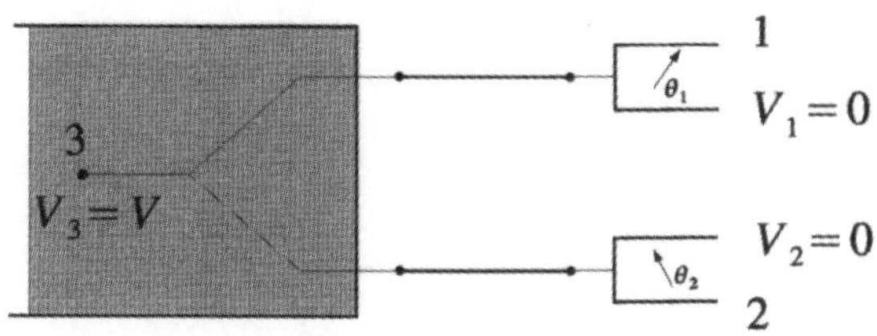

Figure 6. Setup of a realistic system consisting of a normal beam splitter (shaded region) for testing the CH inequality. Bold lines represent two conductors of unit transmission probability. A bias voltage equal to eV is set between terminals 3 and 1 and terminals 3 and 2.

joint probabilities with $Q_1, Q_2 \neq M$):

$$P^{\theta_1,\theta_2}(M,M) = \left[\epsilon^2 \sin^2\left(\frac{\theta_1 - \theta_2}{2}\right)\right]^M \tag{19}$$

$$P^{\theta_1,-}(M,M) = \epsilon^{2M} \tag{20}$$

The no-enhancement assumption is verified when

$$\epsilon \leq \frac{1}{2}\frac{2^{\frac{Q}{M-Q}} - 1}{2^{\frac{Q}{M-Q}} - \frac{1}{2}}, \tag{21}$$

which equals the condition of Eq. (8) once ϵ is replaced with $T/2$. Let us first consider the case for which $Q_1 = Q_2 = M$. We obtain an important result: the CH inequality is violated for the same set of angles found for the case of the entangler, although to a lesser extent, since the prefactors in Eqs. (19) and (20) now varies in the range $0 \leq \epsilon^{2M} \leq \frac{1}{4^M}$. In particular, in the simplest case of $M = 1$, corresponding to injecting a single pair of electrons, the maximum violation corresponds to $\mathcal{S}_{CH} = \frac{\sqrt{2}-1}{4}$, which is a half of the value for the entangler. Furthermore, the plot in Fig. 2 is also valid in the present case with $\overline{\mathcal{S}}_{CH}$ defined as $\overline{\mathcal{S}}_{CH} = \mathcal{S}_{CH}/\epsilon^{2M}$, *i.e.* by replacing $T/\sqrt{2}$ with ϵ. This means that a geometry like that of the beam splitter enables to detect violation of CH inequality without any need to resort to interaction processes to produce entanglement.

Also here we consider the case for which $Q_1 = Q_2 \equiv Q < M$, where interesting differences with respect to the case of the entangler are found. i) We find that the violation of the CH inequality is in general weaker,

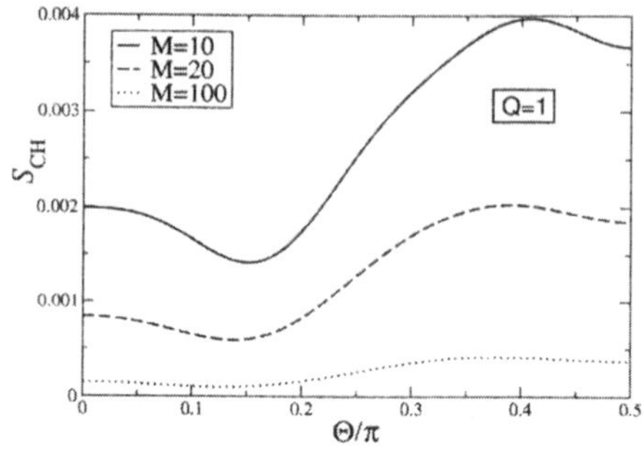

Figure 7. The quantity $\mathcal{S}_{CH}$ for a normal beam splitter is plotted as a function of the angle Θ for three values of $M = eVt/h = 10, 20, 100$ when $Q = 1$. Interestingly, $\mathcal{S}_{CH}$ is positive for every angle and its maximum value decreases like $1/M$.

meaning that the absolute maximum value of $\mathcal{S}_{CH}$ is smaller than in the ideal case of the entangler. ii) The weakening of the violation with increasing M is determined by the suppression of the probability by the prefactor $(\epsilon^2)^{Q1+Q2}$. Remarkably, the maximum value of $\mathcal{S}_{\max}$ decreases like $1/M$, therefore even slower than for the ideal case. iii) Violations occur only for values of Q close to 1, even for large values of M: to search for violations one has to look at single- or few-pair probabilities and therefore, because of the no-enhancement assumption, to small transmissions ϵ. iv) Interestingly, for $Q = 1$ the quantity $\mathcal{S}_{CH}$ is positive for any angles, although the largest values correspond to Θ close to $\pi/2$ (see Fig. 7). We do not find any relevant variation, with respect to the discussion in paragraph 3.1, for probabilities relative to $Q_1 \neq Q_2$.

It is easy to convince oneself that, for an incident state composed of a single pair of particles impinging from the entering arm of the beam splitter, we obtain a final state $|\psi\rangle_{out}$ that contains an entangled part:

$$|\psi\rangle_{out} = \epsilon \left(b^\dagger_{1\uparrow} b^\dagger_{2\downarrow} - b^\dagger_{1\downarrow} b^\dagger_{2\uparrow} \right) |0\rangle + \epsilon\, b^\dagger_{1\uparrow} b^\dagger_{1\downarrow} |0\rangle + \epsilon\, b^\dagger_{2\uparrow} b^\dagger_{2\downarrow} |0\rangle . \qquad (22)$$

In Ref. [29] this fact was already noticed. For mesoscopic conductors, entanglement without interaction for electrons injected from a Fermi sea has been also discussed by Beenakker *et al.* [13]. In the limit of strongly asymmetric beam splitter the state (22) is analogous to the one discussed in Ref.[13].

4. Conclusions

In mesoscopic multiterminal conductors it is possible to observe violations of locality in the whole distribution of the transmitted electrons. In this paper we have derived and discussed the CH inequality for the full counting electron statistics. In an idealized situation in which one supposes the existence of an *entangler*, we have found that the CH inequality is violated for joint probabilities relative to an equal number of electrons that have passed in different terminals. This is related to the intuition that any violation is lost in absence of coincidence measurements. The extent of the violation is suppressed for increasing M (average number of injected pairs), however such a suppression does not scale exponentially with M like the probability, but instead decreases like $1/M^2$. This means that the detection of violation does not become exponentially difficult with increasing M. For fixed transport properties we analyzed the conditions, in terms of M and number of counted electrons, for maximizing the violation of the CH inequality.

The violation of the CH inequality could be achieved in an experiment. Indeed we tested the CH inequality for two different realistic systems, namely a normal beam splitter and a superconducting beam splitter. Interestingly we find a violation even for the normal system, even though weaker with respect to the idealized case of the entangler. In this case the violation is again suppressed for increasing observation time, but scales like $1/M$. We analyzed the superconducting case in the limit of small transmissivity and we also find a violation of the CH inequality to the same extent with respect to the case of the entangler.

It is important to notice that the analyzers should not affect the scattering properties of the system as in the case of ferromagnetic electrodes. In the latter case, in fact, the probability density of the local hidden variables would also depend on the angles θ_1 and θ_2.

Acknowledgments The authors would like to thank M. Büttiker, P. Samuelsson and E. Sukhorukov for helpful discussions and C.W.J. Beenakker for comments on the manuscript. This work has been supported by the EU (IST-FET-SQUBIT, RTN-Spintronics, RTN-Nanoscale Dynamics).

References

1. A. Zeilinger, Rev. Mod. Phys. **71**, S288 (1999).

2. A. Rauschenbeutel, G. Nogues, S. Osnaghi, P. Bertet, M. Brune, J.-M. Raimond, and S. Haroche, Science **288**, 2024 (2000).
3. C.A. Sackett, D. Kielpinski, B. E. King, C. Langer, V. Meyer, C.J. Myatt, M. Rowe, Q. A. Turchette, W.M. Itano, D.J. Wineland, and C. Monroe, Nature **404**, 256 (2000).
4. D. Loss and D.P. DiVincenzo, Phys. Rev. A **57**, 120 (1998).
5. Yu. Makhlin, G. Schön, and A. Shnirman, Rev. Mod. Phys. **73**, 357 (2001).
6. *Semiconductor Spintronics and Quantum Computation*, D.D. Awschalom, D. Loss, and N. Samarth Eds.: Series on Nanoscience and Technology, Springer-Verlag, Berlin, (2002).
7. G. Burkard, D. Loss and E.V. Sukhorukov, Phys. Rev. B **61**, R16303 (2000).
8. F. Taddei and R. Fazio, Phys. Rev. B **65**, 075317 (2002).
9. J.S. Bell, Physics **1**, 195 (1964).
10. S. Kawabata, J. Phys. Soc. Jpn. **70**, 1210 (2001)
11. N.M. Chtchelkatchev, G. Blatter, G.B. Lesovik, T. Martin, Phys. Rev. B **66**, 161320 (2002).
12. P. Samuelsson, E.V. Sukhorukov, M. Büttiker, Phys. Rev. Lett. **91**, 157002 (2003).
13. C.W.J. Beenakker, C. Emary, M. Kindermann, J.L. van Velsen, Phys. Rev. Lett. **91**, 147901 (2003).
14. P. Samuelsson, E.V. Sukhorukov, M. Büttiker, cond-mat/0307473.
15. C.W.J. Beenakker and M. Kindermann, cond-mat/0307103.
16. A central issue of these works is the ability to perform coincidence measurements which can be achieved by performing short-time measurement [10,11], by operating in the tunneling limit [12,13] or even for any transmission rates [14].
17. L. Faoro, F. Taddei, and R. Fazio, cond-mat/0306733 (to appear in Phys. Rev. B).
18. J.F. Clauser and M.A. Horne, Phys. Rev. D **10**, 526 (1974).
19. L. Mandel and E. Wolf, *Optical coherence and quantum optics*, Cambridge University Press (1995).
20. The CH inequality given in Eq.(2) is said to be weak in the sense that it holds only when the no-enhancement assumption, Eq.(1), is satisfied.
21. L.S. Levitov and G.B. Lesovik, JETP Lett. **58**, 225 (1993).
22. L.S. Levitov, H. Lee and G.B. Lesovik, Jour. Math. Phys. **37**, 10 (1996).
23. Yu.V. Nazarov, Ann. Phys. (Leipzig) **8**, Spec. Issue, SI-193 (1999).
24. W. Belzig, Yu.V. Nazarov, Phys. Rev. Lett. **87**, 067006 (2001); W. Belzig, Yu.V. Nazarov, Phys. Rev. Lett. **87**, 197006 (2001).
25. L.S. Levitov, in "Quantum Noise", edited by Yu.V. Nazarov and Ya.M. Blanter (Kluwer); M. Kindermann and Yu.V. Nazarov, *ibidem*; D.A. Bagrets and Yu.V. Nazarov, *ibidem*; W. Belzig, *ibidem.*
26. J. Börlin, W. Belzig, and C. Bruder, Phys. Rev. Lett. **88**, 197001 (2002).
27. P. Samuelsson and M. Büttiker, Phys. Rev. Lett. **89**, 046601 (2002).
28. M. Büttiker, Y. Imry, and M.Ya. Azbel, Phys. Rev. A **30**, 1982 (1984).
29. S. Bose and D. Home, Phys. Rev. Lett. **88**, 050401 (2002).

FIDELITY OF QUANTUM TELEPORTATION MODEL USING BEAM SPLITTINGS

KARL-HEINZ FICHTNER

Friedrich-Schiller-Universität Jena
Fakultät für Mathematik und Informatik
Institut für Angewandte Mathematik
D-07740 Jena Deutschland
E-mail: fichtner@minet.uni-jena.de

TAKAYUKI MIYADERA AND MASANORI OHYA

Department of Information Sciences
Science University of Tokyo
Chiba 278-8510 Japan
E-mail: ohya@is.noda.tus.ac.jp, Tmdella@aol.com

A teleportation model using a beam splitter proposed by Fichtner and Ohya is investigated. Fidelity which represents a perfectness of the model is estimated for general entangled states.

1. Introduction

A model of *Quantum Teleportation* has been first given by Bennett et al.[1,2], in which Alice perfectly sends an unknown state to Bob using the EPR entangled state. In their model, every state is perfectly teleported. The important point to make a perfect teleportation scheme is to use a *maximally* entangled state (EPR state) over Alice and Bob. It is known, however, that preparation of such a maximally entangled states is difficult to realize. Therefore it is important to consider schemes with partially (not maximally) entangled states. As having been pointed out[3], with such an incompletely entangled state one can not obtain a perfect teleportation scheme. In [4,5], protocols employing a partially entangled state constructed by beam splitting technique were introduced to provide examples for both perfect and nonperfect teleportation. In the protocol in nonperfect realistic teleportation, Alice and Bob make *tests* on their own systems and give up the trials when the tests are not passed. If the tests are fortunately passed, the obtained state by Bob perfectly coincides with the original one first possessed by Alice. We calculated the probability to complete successful teleportation, which approaches 1 as the mean energy of the entangled state goes to infinity even in the nonperfect model.

In the present paper, we first give a review on this protocol with tests and next discuss a protocol without tests. In the protocol without tests, let us say it *original* protocol[3,12], we employ the same beam splitting method for the preparation of an entangled state over Alice and Bob. Owing to its non maximal entanglement, the teleportation model becomes nonperfect. We make an estimation on *fidelity* which represents how the model deviates from the ideal perfect one. Furthermore, we give a result for general entangled state prepared by a beam splitter.

In the next section, we review some mathematical notions which are used to construct rigorously a teleportation scheme by beam splittings. In section 3 we give a short review on the teleportation scheme with tests proposed in [4]. Finally we introduce an original teleportation model and discuss how perfect the protocol is by use of a quantity, fidelity.

2. Basic Notions and Notations

First we collect some basic facts concerning the (symmetric) Fock space. We will introduce the Fock space in a way adapted to the language of counting measures. For details we refer to [6,7,8,9,10] and other papers cited in [8].

Let G be an arbitrary complete separable metric space. Further, let μ be a locally finite diffuse measure on G, i.e. $\mu(B) < +\infty$ for bounded measurable subsets of G and $\mu(\{x\}) = 0$ for all singletons $x \in G$. In order to describe the teleportation of states on a finite dimensional Hilbert space through the k–dimensional space $\mathbf{R}^k$, especially we are concerned with the case

$$G = \mathbf{R}^k \times \{1, \ldots, N\}$$
$$\mu = l \times \#$$

where l is the k–dimensional Lebesgue measure and $\#$ denotes the counting measure on $\{1, \ldots, N\}$.

Now by $M = M(G)$ we denote the set of all finite counting measures on G. Since $\varphi \in M$ can be written in the form $\varphi = \sum_{j=1}^{n} \delta_{x_j}$ for some $n =$

$0, 1, 2, \ldots$ and $x_j \in G$ (where δ_x denotes the Dirac measures corresponding to $x \in G$) the elements of M can be interpreted as finite (symmetric) point configurations in G. We equip M with its canonical σ–algebra $\mathfrak{W}$ (cf. [6], [7]) and we consider the measure F by setting

$$F(Y) := \mathcal{X}_Y(O) + \sum_{n \geq 1} \frac{1}{n!} \int_{G^n} \mathcal{X}_Y \left(\sum_{j=1}^{n} \delta_{x_j} \right) \mu^n(d[x_1, \ldots, x_n])(Y \in \mathfrak{W})$$

Hereby, $\mathcal{X}_Y$ denotes the indicator function of a set Y and O represents the empty configuration, i. e., $O(G) = 0$. Observe that F is a σ–finite measure.

Since μ was assumed to be diffuse one can easily check that F is concentrated on the set of a simple configurations (i.e., without multiple points)

$$\hat{M} := \{\varphi \in M | \varphi(\{x\}) \leq 1 \text{ for all } x \in G\}$$

Definition 2.1. $\mathcal{M} = \mathcal{M}(G) := L^2(M, \mathfrak{W}, F)$ is called the (symmetric) Fock space over G.

In [6] it was proved that $\mathcal{M}$ and the Boson Fock space $\Gamma(L^2(G))$ in the usual definition are isomorphic.

For each $\Phi \in \mathcal{M}$ with $\Phi \neq 0$ we denote by $|\Phi\rangle$ the corresponding normalized vector

$$|\Phi\rangle := \frac{\Phi}{\|\Phi\|}$$

Further, $|\Phi\rangle\langle\Phi|$ denotes the corresponding one–dimensional projection, describing the pure state given by the normalized vector $|\Phi\rangle$. Now, for each $n \geq 1$ let $\mathcal{M}^{\otimes n}$ be the n–fold tensor product of the Hilbert space $\mathcal{M}$. Obviously, $\mathcal{M}^{\otimes n}$ can be identified with $L^2(M^n, F^n)$.

Definition 2.2. For a given function $g : G \to \mathbb{C}$ the function $\exp(g) : M \to \mathbb{C}$ defined by

$$\exp(g)(\varphi) := \begin{cases} 1 & \text{if } \varphi = 0 \\ \prod_{x \in G, \varphi(\{x\}) > 0} g(x) & \textit{otherwise} \end{cases}$$

is called exponential vector generated by g.

Observe that $\exp(g) \in \mathcal{M}$ if and only if $g \in L^2(G)$ and one has in this case $\|\exp(g)\|^2 = e^{\|g\|^2}$ and $|\exp(g)\rangle = e^{-\frac{1}{2}\|g\|^2} \exp(g)$. The projection

$|\exp(g)\rangle\langle\exp(g)|$ is called the coherent state corresponding to $g \in L^2(G)$. In the special case $g \equiv 0$ we get the vacuum state

$$|\exp(0)\rangle = \mathcal{X}_{\{0\}} \, .$$

The linear span of the exponential vectors of $\mathcal{M}$ is dense in $\mathcal{M}$, so that bounded operators and certain unbounded operators can be characterized by their actions on exponential vectors.

Definition 2.3. The operator $D : \mathrm{dom}(D) \to \mathcal{M}^{\otimes 2}$ given on a dense domain $\mathrm{dom}(D) \subset \mathcal{M}$ containing the exponential vectors from $\mathcal{M}$ by

$$D\psi(\varphi_1, \varphi_2) := \psi(\varphi_1 + \varphi_2) \quad (\psi \in \mathrm{dom}(D),\ \varphi_1, \varphi_2 \in M)$$

is called compound Hida-Malliavin derivative.

On exponential vectors $\exp(g)$ with $g \in L^2(G)$, one gets immediately

$$D \exp(g) = \exp(g) \otimes \exp(g) \tag{1}$$

Definition 2.4. The operator $S : \mathrm{dom}(S) \to \mathcal{M}$ given on a dense domain $\mathrm{dom}(S) \subset \mathcal{M}^{\otimes 2}$ containing tensor products of exponential vectors by

$$S\Phi(\varphi) := \sum_{\tilde{\varphi} \leq \varphi} \Phi(\tilde{\varphi}, \varphi - \tilde{\varphi}) \quad (\Phi \in \mathrm{dom}(S),\ \varphi \in M)$$

is called compound Skorohod integral.

One gets

$$\langle D\psi, \Phi \rangle_{\mathcal{M}^{\otimes 2}} = \langle \psi, S\Phi \rangle_{\mathcal{M}} \quad (\psi \in \mathrm{dom}(D),\ \Phi \in \mathrm{dom}(S)) \tag{2}$$

$$S(\exp(g) \otimes \exp(h)) = \exp(g + h) \quad (g, h \in L^2(G)) \, . \tag{3}$$

For more details we refer to [11].

Definition 2.5. Let T be a linear operator on $L^2(G)$ with $\|T\| \leq 1$. Then the operator $\Gamma(T)$ called second quantization of T is the (uniquely determined) bounded operator on $\mathcal{M}$ fulfilling

$$\Gamma(T)\exp(g) = \exp(Tg) \quad (g \in L^2(G)) \, .$$

Clearly, it holds

$$\begin{aligned} \Gamma(T_1)\Gamma(T_2) &= \Gamma(T_1 T_2) \\ \Gamma(T^*) &= \Gamma(T^*) \, . \end{aligned} \tag{4}$$

It follows that $\Gamma(T)$ is a unitary operator on $\mathcal{M}$ if T is a unitary operator on $L^2(G)$.

Lemma 2.1. *Let K_1, K_2 be linear operators on $L^2(G)$ with property*

$$K_1^* K_1 + K_2^* K_2 = \mathbf{1} \,. \tag{5}$$

Then there exists exactly one isometry ν_{K_1,K_2} from $\mathcal{M}$ to $\mathcal{M}^{\otimes 2} = \mathcal{M} \otimes \mathcal{M}$ with

$$\nu_{K_1,K_2} \exp\,(g) = \exp(K_1 g) \otimes \exp(K_2 g) \quad (g \in L^2(G)) \tag{6}$$

Further it holds

$$\nu_{K_1,K_2} = (\Gamma(K_1) \otimes \Gamma(K_2))D \tag{7}$$

(at least on $\mathrm{dom}(D)$ *but one has the unique extension).*
The adjoint $\nu^*_{K_1,K_2}$ of ν_{K_1,K_2} is characterized by

$$\nu^*_{K_1,K_2}(\exp\,(h) \otimes \exp\,(g)) = \exp(K_1^* h + K_2^* g) \quad (g, h \in L^2(G)) \tag{8}$$

and it holds

$$\nu^*_{K_1,K_2} = S(\Gamma(K_1^*) \otimes \Gamma(K_2^*)) \tag{9}$$

Remark 2.1. From K_1, K_2 we get a transition expectation $\xi_{K_1 K_2} : \mathcal{M} \otimes \mathcal{M} \to \mathcal{M}$, using ν_{K_1,K_2} and the lifting $\xi^*_{K_1 K_2}$ may be interpreted as a certain splitting (cf. [9]).

Proof. [**Proof of 2.1**] We consider the operator

$$B := S(\Gamma(K_1^*) \otimes \Gamma(K_2^*))(\Gamma(K_1) \otimes \Gamma(K_2))D$$

on the dense domain $\mathrm{dom}(B) \subseteq \mathcal{M}$ spanned by the exponential vectors. Using (1), (3), (4) and (5) we get

$$B \exp\,(g) = \exp\,(g) \quad (g \in L^2(G)) \,.$$

It follows that the bounded linear unique extension of B onto $\mathcal{M}$ coincides with the unity on $\mathcal{M}$

$$B = \mathbf{1} \,. \tag{10}$$

On the other hand, by equation (7) at least on dom (D), an operator ν_{K_1,K_2} is defined. Using (2) and (4) we obtain

$$\begin{aligned} \|\nu_{K_1,K_2}\psi\|^2 &= \langle \nu_{K_1,K_2}\psi, \nu_{K_1,K_2}\psi \rangle \quad (\psi \in \mathrm{dom}\,(D)) \\ &= \langle \psi, B\psi \rangle, \end{aligned}$$

which implies

$$\|\nu_{K_1,K_2}\psi\|^2 = \|\psi\|^2 \quad (\psi \in \mathrm{dom}\,(D)).$$

because of (10). It follows that ν_{K_1,K_2} can be uniquely extended to a bounded operator on $\mathcal{M}$ with

$$\|\nu_{K_1,K_2}\psi\| = \|\psi\| \quad (\psi \in \mathcal{M}).$$

Now from (7) we obtain (6) using (1) and the definition of the operators of second quantization. Further, (7), (3) and (4) imply (9) and from (9) we obtain (8) using the definition of the operators of second quantization and equation (3). □

Here we explain fundamental scheme of beam splitting [8]. We define an isometric operator $V_{\alpha,\beta}$ for coherent vectors such that

$$V_{\alpha,\beta}|\exp(g)\rangle = |\exp(\alpha g)\rangle \otimes |\exp(\beta g)\rangle$$

with $|\ \alpha\ |^2 + |\ \beta\ |^2 = 1$. This beam splitting is a useful mathematical expression for optical communication and quantum measurements [9].

Example 2.1. $\left(\alpha = \beta = 1/\sqrt{2}\text{ above}\right)$ Let $K_1 = K_2$ be the following operator of multiplication on $L^2(G)$

$$K_1 g = \frac{1}{\sqrt{2}}\, g = K_2 g \quad (g \in L^2(G))$$

We put

$$\nu := \nu_{K_1,K_2}$$

and obtain

$$\nu \exp(g) = \exp\left(\frac{1}{\sqrt{2}} g\right) \otimes \exp\,(\frac{1}{\sqrt{2}}\, g) \quad (g \in L^2(G))$$

Example 2.2. Let $L^2(G) = \mathcal{H}_1 \oplus \mathcal{H}_2$ be the orthogonal sum of the subspaces $\mathcal{H}_1, \mathcal{H}_2$. K_1 and K_2 denote the corresponding projections.

We will use Example 2.1 in order to describe a teleportation model where Bob performs his experiments on the same ensemble of the systems like Alice.

Further we will use a special case of Example 2.2 in order to describe a teleportation model where Bob and Alice are spatially separated (cf. section 5).

Remark 2.2. The property (5) implies

$$\|K_1 g\|^2 + \|K_2 g\|^2 = \|g\|^2 \quad (g \in L^2(G)) \tag{11}$$

Remark 2.3. Let U, V be unitary operators on $L^2(G)$. If operators K_1, K_2 satisfy (5), then the pair $\hat{K}_1 = UK_1, \ \hat{K}_2 = VK_2$ fulfill (5).

Our teleportation scheme works for states contained in a subspace of a Fock space. We fix an ONS $\{g_1, \ldots, g_N\} \subseteq L^2(G)$, operators K_1, K_2 on $L^2(G)$ with (5), an unitary operator T on $L^2(G)$, and $d > 0$. We assume

$$TK_1 g_k = K_2 g_k \quad (k = 1, \ldots, N), \tag{12}$$

$$\langle K_1 g_k, K_1 g_j \rangle = 0 \quad (k \neq j;\ k, j = 1 \ldots, N), \tag{13}$$

Using (11) and (12) we get

$$\|K_1 g_k\|^2 = \|K_2 g_k\|^2 = \frac{1}{2}. \tag{14}$$

From (12) and (13) we get

$$\langle K_2 g_k,\, K_2 g_j \rangle = 0 \quad (k \neq j\,;\ k, j = 1, \ldots, N). \tag{15}$$

The state of Alice asked to teleport is of the type

$$\rho = \sum_{s=1}^{N} \lambda_s |\Phi_s\rangle\langle\Phi_s|, \tag{16}$$

where

$$|\Phi_s\rangle = \sum_{j=1}^{N} c_{sj} |\exp{(aK_1 g_j)} - \exp{(0)}\rangle \quad \left(\sum_j |c_{sj}|^2 = 1; s = 1, \ldots, N \right) \tag{17}$$

with $|a|^2 = d$ and

$$\sum_{j=1}^{N} \bar{c}_{sj} c_{kj} = 0 \quad (j \neq k\,;\ j, k = 1, \ldots, N)\,. \tag{18}$$

One easily checks that $(|\exp{(aK_1 g_j)} - \exp{(0)}\rangle)_{j=1}^N$ and $(|\exp{aK_2 g_j)} - \exp{(0)}\rangle)_{j=1}^N$ are ONS in $\mathcal{M}$ and consequently $\{|\Phi_s\rangle\}_{j=1}^N$ forms an ONS. That is, the state of Alice asked to teleport lives in an N-dimensional subspace of the Fock space spanned by the ONS $\{|\exp(aK_1 g_j) - \exp(0)\rangle\}_{j=1}^N$. We write this N-dimensional subspace as $\mathcal{L}$.

Next we introduce an observable measured by Alice. Let $(b_n)_{n=1}^{N}$ be a sequence in $\mathbb{C}^N$,

$$b_n = [b_{n1,\dots,}b_{nN}]$$

with properties

$$|b_{nk}| = 1 \quad (n, k = 1, \dots, N), \tag{19}$$

$$\langle b_n\,,\, b_j\rangle = 0 \quad (n \neq j\,;\; n, j = 1, \dots, N). \tag{20}$$

Projection operators

$$F_{nm} = |\xi_{nm}\rangle\langle\xi_{nm}| \quad (n, m = 1, \dots, N) \tag{21}$$

are given by

$$|\xi_{nm}\rangle = \frac{1}{\sqrt{N}}\sum_{j=1}^{N} b_{nj}|\exp\,(aK_1 g_j) - \exp\,(0)\rangle \otimes |\exp\,(aK_1 g_{j\oplus m}) - \exp\,(0)\rangle, \tag{22}$$

where $j \oplus m := j + m (\text{mod } N)$. One easily checks that $(|\xi_{nm}\rangle)_{n,m=1}^{N}$ is an ONS in $\mathcal{M}^{\otimes 2}$. Because $|\xi_{nm}\rangle$ $(n, m = 1, 2, \cdots N)$ does not form a completly orthonormal system of $\mathcal{M} \otimes \mathcal{M}$, we introduce another projection operator $F_0 := \mathbf{1} - \sum_{nm} F_{nm}$. Thus the measurement of the observable $F = \sum_{nm} z_{nm} F_{nm}$ $(z_{nm} \neq 0)$ distinguishes $\{F_n m\}$'s and F_0, where F_0 corresponds to the case an outcome is zero.
As for unitary operation of Bob, for each $n, m = 1, \cdots, N$ we have U_m, B_n on $\mathcal{M}$ given by

$$\begin{aligned} B_n|\exp\,(aK_1 g_j) - \exp\,(0)\rangle &= b_{nj}|\exp\,(aK_1 g_j) - \exp\,(0)\rangle \quad (j = 1, \dots, N)\\ B_n|\exp(0)\rangle &= |\exp(0)\rangle \end{aligned} \tag{23}$$

$$\begin{aligned} U_m|\exp\,(aK_1 g_j) - \exp\,(0)\rangle &= |\exp\,(aK_1 g_{j\oplus m}) - \exp\,(0)\rangle \quad (j = 1, \dots, N)\\ U_m|\exp(0)\rangle &= |\exp(0)\rangle \end{aligned} \tag{24}$$

where $j \oplus m := j + m(\text{mod } N)$ and define

$$W_{nm} := B_n U_m^* \Gamma(T)^*. \tag{25}$$

In addition we have some arbitrary unitary operator W_0, which we do not specify yet. Further, the state vector $|\xi\rangle$ of the entangled state $\sigma = |\xi\rangle\langle\xi|$ is given by

$$|\xi\rangle = \frac{1}{\sqrt{N}}\sum_k |\exp\,(aK_1 g_k)\rangle \otimes |\exp\,(aK_2 g_k)\rangle \tag{26}$$

which is naturally prepared by use of Beam splitting technique. However, the physical naturalness requires a sacrifice. That is, the state is not maximally entangled state any longer.

3. Teleportation scheme with tests [4]

In this section, we review a teleportation scheme with *tests* which was introduced in [4]. An entangled state between Alice and Bob is prepared by the beam splitter which is described by the isometry ν_{K_1,K_2}. A natural input state for the beam splitter is a superposition of coherent states $\tau_0 := |\phi_0\rangle\langle\phi_0|$;

$$|\phi_0\rangle := \frac{1}{\sqrt{N}}\sum_{j=1}^{N} |\exp g_k\rangle,$$

which is mapped into a partially entangled state $\sigma = |\xi\rangle\langle\xi|$ where

$$|\xi\rangle = \frac{1}{\sqrt{N}}\sum_{k} |\exp(aK_1 g_k)\rangle \otimes |\exp(aK_2 g_k)\rangle \,. \tag{27}$$

On this preparation, the teleportation scheme with tests is performed as follows.

Alice has an unknown quantum state ρ on an N–dimensional subspace $\mathcal{L}$ of a Hilbert space $\mathcal{H}_1$ and she was asked to teleport it to Bob.

For this purpose, we need two other Hilbert spaces $\mathcal{H}_2$ and $\mathcal{H}_3$, $\mathcal{H}_2$ is attached to Alice and $\mathcal{H}_3$ is attached to Bob. By using a beam splitter prearrange the entangled state σ on $\mathcal{H}_2 \otimes \mathcal{H}_3$ having certain correlations and prepare an ensemble of the combined system in the state $\rho \otimes \sigma$ on $\mathcal{H}_1 \otimes \mathcal{H}_2 \otimes \mathcal{H}_3$.

Alice performs a measurement of the observable F, involving only the $\mathcal{H}_1 \otimes \mathcal{H}_2$ part of the system in the state $\rho \otimes \sigma$. Possible outcomes are z_{nm}'s and 0. When Alice obtains z_{nm}, according to the von Neumann rule, after Alice's measurement, the state becomes

$$\rho_{nm}^{(123)} := \frac{(F_{nm} \otimes \mathbf{1})\rho \otimes \sigma(F_{nm} \otimes \mathbf{1})}{\mathrm{tr}_{123}(F_{nm} \otimes \mathbf{1})\rho \otimes \sigma(F_{nm} \otimes \mathbf{1})}$$

where tr_{123} is the full trace on the Hilbert space $\mathcal{H}_1 \otimes \mathcal{H}_2 \otimes \mathcal{H}_3$. The probability to obtain z_{nm} is calculated as

$$p_{nm}(\rho) := \mathrm{tr}_{123}(F_{nm} \otimes \mathbf{1})\rho \otimes \sigma(F_{nm} \otimes \mathbf{1}).$$

On the other hand, when Alice obtains 0, they give up the trial. The probability to quit the experiment in this stage is given by

$$p_0(\rho) := \mathrm{tr}_{123}(F_0 \otimes \mathbf{1})\rho \otimes \sigma(F_0 \otimes \mathbf{1}).$$

Bob is informed which outcome was obtained by Alice. This information is transmitted from Alice to Bob without disturbance and by means of classical tools.

Having been informed an outcome of Alice's measurement, Bob performs a corresponding unitary operation onto his system. That is, if the outcome was z_{nm}, Bob operates a unitary operator W_{nm} and change the state into

$$(\mathbf{1}\otimes\mathbf{1}\otimes W_{nm})\rho_{nm}^{(123)}(\mathbf{1}\otimes\mathbf{1}\otimes W_{nm}^{*}) = \frac{(F_{nm}\otimes W_{nm})\rho\otimes\sigma(F_{nm}\otimes W_{nm}^{*})}{\mathrm{tr}_{123}(F_{nm}\otimes\mathbf{1})\rho\otimes\sigma(F_{nm}\otimes\mathbf{1}}).$$

Bob performs a measurement on his part of the system according to the projection

$$F_{+} := \mathbf{1} - |\exp(0)\rangle\langle\exp(0)|$$

where $|\exp(0)\rangle\langle\exp(0)|$ denotes the vacuum state (the coherent state with density 0). When the result is 1, the trial is regarded as a failure.

In [4], we proved the following theorem:

Theorem 3.1. *When the tests are passed, the resulted state of Bob is exactly same with the original state possessed by Alice. The probability to pass both tests is*

$$P(success) = \frac{\left(1 - e^{-\frac{d}{2}}\right)^2}{1 + (N-1)e^{-d}} \tag{28}$$

which approaches 1 *as* $d \to \infty$.

4. Teleportation and fidelity

In the previous section, we explained that the *tests* can exclude the failed trials and the probability to make successful trials becomes sufficiently high if the parameter d can be large. In this section, we do not employ the scheme with tests but the *original* one and discuss how close the resulted state will be to the state which was possessed by Alice.

Let us describe the scheme employed in this section.

Alice has an unknown quantum state ρ on an N–dimensional subspace $\mathcal{L}$ of a Hilbert space $\mathcal{H}_1$ and she was asked to teleport it to Bob.

For this purpose, we need two other Hilbert spaces $\mathcal{H}_2$ and $\mathcal{H}_3$, $\mathcal{H}_2$ is attached to Alice and $\mathcal{H}_3$ is attached to Bob. By a beam splitter with an input state $\tau_0 := |\phi_0\rangle\langle\phi_0|$ where

$$|\phi_0\rangle := \frac{1}{\sqrt{N}}\sum_{j=1}^{N} |\exp g_k\rangle,$$

prearrange the entangled state σ on $\mathcal{H}_2 \otimes \mathcal{H}_3$ having certain correlations and prepare an ensemble of the combined system in the state $\rho \otimes \sigma$ on $\mathcal{H}_1 \otimes \mathcal{H}_2 \otimes \mathcal{H}_3$.

Alice performs a measurement of the observable F, involving only the $\mathcal{H}_1 \otimes \mathcal{H}_2$ part of the system in the state $\rho \otimes \sigma$. Possible outcomes are z_{nm}'s and 0. When Alice obtains z_{nm}, according to the von Neumann rule, after Alice's measurement, the state becomes

$$\rho_{nm}^{(123)} := \frac{(F_{nm} \otimes \mathbf{1})\rho \otimes \sigma(F_{nm} \otimes \mathbf{1})}{\mathrm{tr}_{123}(F_{nm} \otimes \mathbf{1})\rho \otimes \sigma(F_{nm} \otimes \mathbf{1})}$$

where tr_{123} is the full trace on the Hilbert space $\mathcal{H}_1 \otimes \mathcal{H}_2 \otimes \mathcal{H}_3$. The probability to obtain z_{nm} is calculated as

$$p_{nm}(\rho) := \mathrm{tr}_{123}(F_{nm} \otimes \mathbf{1})\rho \otimes \sigma(F_{nm} \otimes \mathbf{1}).$$

On the other hand, when Alice obtains 0, the state becomes

$$\rho_0^{(123)} := \frac{(F_0 \otimes \mathbf{1})\rho \otimes \sigma(F_0 \otimes \mathbf{1})}{\mathrm{tr}_{123}(F_0 \otimes \mathbf{1})\rho \otimes \sigma(F_0 \otimes \mathbf{1})}.$$

The probability to obtain zero is

$$p_0(\rho) := \mathrm{tr}_{123}(F_0 \otimes \mathbf{1})\rho \otimes \sigma(F_0 \otimes \mathbf{1}).$$

Bob is informed which outcome was obtained by Alice.

Having been informed an outcome of Alice's measurement, Bob performs a corresponding unitary operation onto his system. That is, if the outcome was z_{nm}, Bob operates a unitary operator W_{nm} and change the state into

$$(\mathbf{1} \otimes \mathbf{1} \otimes W_{nm})\rho_{nm}^{(123)}(\mathbf{1} \otimes \mathbf{1} \otimes W_{nm}^*) = \frac{(F_{nm} \otimes W_{nm})\rho \otimes \sigma(F_{nm} \otimes W_{nm}^*)}{\mathrm{tr}_{123}(F_{nm} \otimes \mathbf{1})\rho \otimes \sigma(F_{nm} \otimes \mathbf{1}}).$$

On the other hand, if the outcome was 0, Bob operates another unitary operation W_0 to obtain

$$(\mathbf{1} \otimes \mathbf{1} \otimes W_0)\rho_0^{(123)}(\mathbf{1} \otimes \mathbf{1} \otimes W_0^*) = \frac{(F_0 \otimes W_0)\rho \otimes \sigma(F_0 \otimes W_0^*)}{\mathrm{tr}_{123}(F_0 \otimes \mathbf{1})\rho \otimes \sigma(F_0 \otimes \mathbf{1}}).$$

Thus the state obtained by Bob is

$$\begin{aligned} \Gamma_{nm}^*(\rho) &= \mathrm{tr}_{12}\ (\mathbf{1} \otimes \mathbf{1} \otimes W_{nm})\rho_{nm}^{(123)}(\mathbf{1} \otimes \mathbf{1} \otimes W_{nm}^*) \\ &= \mathrm{tr}_{12}\frac{(F_{nm} \otimes W_{nm})\rho \otimes \sigma(F_{nm} \otimes W_{nm}^*)}{\mathrm{tr}_{123}(F_{nm} \otimes \mathbf{1})\rho \otimes \sigma(F_{nm} \otimes \mathbf{1})} \end{aligned} \tag{29}$$

in case when the outcome is z_{nm} and

$$\begin{aligned}\Gamma_0^*(\rho) &= \mathrm{tr}_{12}\,(\mathbf{1}\otimes\mathbf{1}\otimes W_0)\rho_0^{(123)}(\mathbf{1}\otimes\mathbf{1}\otimes W_0^*)\\ &= \mathrm{tr}_{12}\frac{(F_0\otimes W_0)\rho\otimes\sigma(F_0\otimes W_0^*)}{\mathrm{tr}_{123}(F_0\otimes\mathbf{1})\rho\otimes\sigma(F_0\otimes\mathbf{1})}.\end{aligned} \tag{30}$$

if the outcome was 0.

Once knowing the result of Alice's measurement, the channel becomes nonlinear because of the probabilities $p_{nm}(\rho)$ and $p_0(\rho)$ which appear in the denominator in (29) and (30). We, however, do not know the result of Alice's measurement before the experiment. Therefore it is also important to consider an expected state which is obtained by mixing all possible states with multiplying their probabilities to occur. The teleportation scheme can be written by a linear channel (completely positive map)

$$\Xi^*(\rho) = \sum_{nm}\Xi^*_{nm}(\rho) + \Xi^*_0(\rho),$$

where

$$\begin{aligned}\Xi^*_{nm}(\rho) &:= \ p_{nm}(\rho)\Gamma^*_{nm}(\rho) = \mathrm{tr}_{1,2}(F_{nm}\otimes W_{nm})\rho(F_{nm}\otimes W_{nm})^*\\ \Xi^*_0(\rho) &:= \ p_0(\rho)\Gamma^*_0(\rho) = \mathrm{tr}_{1,2}(F_0\otimes W_0)\rho(F_0\otimes W_0)^*.\end{aligned}$$

We investigate how close the obtained state $\Xi^*(\rho)$ to the original state ρ.

We need some proper quantity (for e.g., [12]) to measure how close these two states are. In this paper we take up *fidelity* [13,14]. The notion of fidelity is frequently used in the context of quantum information, quantum optics and so on. The fidelity of a state ρ with respect to another state σ is defined by

$$F(\rho,\sigma) := \mathrm{tr}[\sqrt{\sigma^{1/2}\rho\sigma^{1/2}}],$$

which possesses some nice properties.

$$0 \le F(\rho,\sigma) \le 1 \tag{31}$$

$$F(\rho,\sigma) = 1 \Leftrightarrow \rho = \sigma \tag{32}$$

$$F(\rho,\sigma) = F(\sigma,\rho) \tag{33}$$

Thus we can say two states ρ and σ are close when the fidelity between them is close to unity. Moreover it satisfies a kind of concavity relation as

$$F(\sum_i p_i\rho_i, \sum_i q_i\sigma_i) \ge \sum_i \sqrt{p_iq_i}F(\rho_i,\sigma_i),$$

where ρ_i's and σ_i's are states and p_i's and q_i's are nonnegative numbers satisfying $\sum_i p_i = \sum_i q_i = 1$. In particular putting $p_j = 1$, one gets

$$F(\rho, \sum_i q_i \sigma) \geq \sqrt{q_j} F(\rho, \sigma_j)$$

for $j = 1, 2, \cdots$.
To estimate $F(\rho, \Xi^*(\rho))$ we begin with a calculation of $\Xi^*(\rho) = \sum_{nm} \Xi^*_{nm}(\rho) + \Xi^*_0(\rho)$.

Lemma 4.1. [4] *For each $n, m, s\,(= 1, \ldots, N)$, it holds*

$$(F_{nm} \otimes \mathbf{1}) \left(|\Phi_s\rangle \otimes |\tilde{\xi}\rangle \right) = \frac{\gamma}{N} \left(1 - e^{-\frac{d}{2}}\right) |\xi_{nm}\rangle \otimes (\Gamma(T) U_m B_n^* |\Phi_s\rangle)$$
$$+ \frac{\gamma}{N} \left(\frac{e^{\frac{d}{2}} - 1}{e^d} \right)^{\frac{1}{2}} \langle b_n, c_s \rangle_{\mathbb{C}^N} \xi_{nm} \otimes |\exp(0)\rangle$$

The following lemma follows immediately.

Lemma 4.2. *For a general state $\rho = \sum_s \lambda_s |\Phi_s\rangle\langle\Phi_s|$, it holds*

$$\begin{aligned} \Xi^*_{nm}(\rho) = {} & \frac{\gamma^2}{N^2}(1 - e^{-d/2})^2 \rho + \frac{\gamma^2}{N^2}(\frac{e^{d/2} - 1}{e^d}) \sum_s \lambda_s |\langle b_n, c_s\rangle|^2 |\exp(0)\rangle\langle\exp(0)| \\ & + \frac{\gamma^2}{N^2}(\frac{e^{d/2} - 1}{e^d})^{1/2}(1 - e^{-d/2})(\sum_s \langle b_n, c_s\rangle^* \lambda_s |\Phi_s\rangle\langle\exp(0)| \\ & + \sum_s \langle b_n, c_s\rangle \lambda_s |\exp(0)\rangle\langle\Phi_s|). \end{aligned} \tag{34}$$

Next we investigate an expression of $\Xi^*_0(\rho)$.

Lemma 4.3. *It holds*

$$(F_0 \otimes \mathbf{1}) \left(|\Phi_s\rangle \otimes |\tilde{\xi}\rangle \right) = |\Phi_s\rangle \otimes |\exp(0)\rangle \otimes \mathrm{e}^{-|a|^2/4} \frac{\gamma}{\sqrt{N}} \sum_{k=1}^{N} |\exp(aK_2 g_k)\rangle$$

Proof. If we let $\mathcal{L}$ a subspace spanned by an ONS $\{|\exp(aK_1 g_k) - \exp(0)\rangle\}$ $(k = 1, \cdots, N)$, $\sum_{nm} F_n m$ is a projection onto $\mathcal{L} \otimes \mathcal{L}$.Therefore we obtain

$$(F_0 \otimes \mathbf{1})(\Phi_s \otimes \tilde{\xi}) = (\mathbf{1} \otimes |\exp(0)\rangle\langle\exp)(0)| \otimes \mathbf{1})(\Phi_s \otimes \tilde{\xi}).$$

Here we used a fact that $|\exp(0)\rangle$ is orthogonal to $\{|\exp(aK_1 g_k) - \exp(0)\rangle\}$'s.
□

Lemma 4.4. *For a general state* $\rho = \sum_s \lambda_s |\Phi_s\rangle\langle\Phi_s|$, *it holds*

$$\Xi_0^*(\rho) = e^{-|a|^2/2}\frac{\gamma^2}{N}\sum_{k=1}^N\sum_{l=1}^N W_0|\exp(aK_2g_k)\rangle\langle\exp(aK_2g_l)|W_0^* \qquad (35)$$

Now let us estimate the fidelity between $\Xi^*(\rho)$ and ρ. We must first compute

$$\begin{aligned}\rho^{1/2}\Xi^*(\rho)\rho^{1/2} &= \sum_{nm}\rho^{1/2}\Xi_{nm}^*(\rho)\rho^{1/2} + \rho^{1/2}\Xi_0\rho^{1/2}\\ &= \gamma^2(1-e^{-d/2})^2\rho^{1/2}\rho\rho^{1/2}\\ &+ \rho^{1/2}e^{-|a|^2/2}\frac{\gamma^2}{N}\sum_{k=1}^N\sum_{l=1}^N W_0|\exp(aK_2g_k)\rangle\langle\exp(aK_2g_l)|W_0^*\rho^{1/2},\end{aligned}$$

where we used the relation $\langle\exp(0)|\Phi_s\rangle = 0$. Because $\Xi_0^*(\rho)$ is positive operator, $\Xi_0^*(\rho)/\mathrm{tr}[\Xi_0^*(\rho)]$ becomes a state and we can rearrange the expression of fidelity as

$$F(\rho,\Xi^*(\rho)) = F(\rho, \gamma^2(1-e^{-d/2})^2\rho + \mathrm{tr}[\Xi_0^*(\rho)]\Xi_0^*(\rho)/\mathrm{tr}[\Xi_0^*(\rho)]). \quad (36)$$

Thanks to the concavity property (**??**) of fidelity, we obtain

$$F(\rho,\Xi^*(\rho)) \geq \gamma(1-e^{-d/2})F(\rho,\rho) = \gamma(1-e^{-d/2}). \qquad (37)$$

Thus we obtain the following theorem.

Theorem 4.1. *For any input state* ρ, *it holds*

$$F(\rho,\Xi^*(\rho)) \geq \sqrt{\frac{(1-e^{-d/2})^2}{1+(N-1)e^{-d}}}. \qquad (38)$$

Therefore the teleportation protocol approaches perfect one as the parameter $|a|$ goes to infinity.

With some additional condition, one can strengthen the above estimate to an equality.

Proposition 4.1. *Let* $L^2(G) = \mathcal{H}_1 \oplus \mathcal{H}_2$ *be the orthogonal sum of the subspaces* $\mathcal{H}_1, \mathcal{H}_2$. K_1 *and* K_2 *denote the corresponding projections and* $W_0 = \mathbf{1}$.

$$F(\rho,\Xi^*(\rho)) = \sqrt{\frac{(1-e^{-d/2})^2}{1+(N-1)e^{-d}}}. \qquad (39)$$

holds for any input state ρ.

Proof. Because $\langle \exp(aK_1 g_k) - \exp(0) | \exp(aK_2 g_l) \rangle = 0$ holds, $\rho^{1/2} \Xi_* (\rho) \rho^{1/2} = \gamma^2 (1 - e^{-d/2})^2 \rho^2$ follows. □

We have considered the fidelity of the teleportation scheme with beam splittings. We showed that as the parameter $|a|$ goes to infinity the fidelity approaches unity and the naive teleportation scheme also approaches a perfect scheme as the teleportation scheme with tests does. In fact the fidelity can be bounded from below by square route of probability to complete successful teleportation with tests. The above result can be extended to the case of general inputs for the beam splitter which produces an entangled state. To get an insight for the generalization, it is instructive to consider the following simple splitting model. Here we put $\mathcal{H}_j \simeq \mathbf{C}^N$ for all $j = 1, 2, 3$ and $\{e_n^{(j)}\}_{n=1}^N$ as its basis. In addition we put input system of the splitter as $\mathcal{H}_0$ which has $\{e_n\}_{n=1}^N$ as its basis. Now we define a simple splitting isometry $J : \mathcal{H}_0 \to \mathcal{H}_2 \otimes \mathcal{H}_3$ by $e_k \mapsto e_k^{(2)} \otimes e_k^{(3)}$ for all k. Then if the input of the simple splitting is $|\phi_0\rangle := \frac{1}{\sqrt{N}} \sum_k e_k$ the resulted state is perfectly entangled, but otherwise not. We put a general input state for the splitter as τ hereafter and the ideal one as $\tau_0 := |\phi_0\rangle\langle\phi_0|$. Let us consider the following teleportation scheme:

Alice has a unknown quantum state $\rho^{(1)}$ (on Hilbert space $\mathcal{H}_1$) and she was asked to teleport it to Bob.

For this purpose, we need three other Hilbert spaces $\mathcal{H}_0, \mathcal{H}_2$ and $\mathcal{H}_3$. $\mathcal{H}_2$ is attached to Alice and $\mathcal{H}_3$ is to Bob. Prearrange a state $J\tau J^* \in \mathcal{S}(\mathcal{H}_2 \otimes \mathcal{H}_3)$ having a certain correlation between Alice and Bob by use of input $\tau \in \mathcal{S}(\mathcal{H}_0)$ for the simple splitter.

Prepare the set of projections $\left\{ F_{nm}^{(12)} \right\}$ and an observable $F^{(12)} := \sum_{nm} z_{nm} F_{nm}^{(12)}$ on a tensor product Hilbert space (system) $\mathcal{H}_1 \otimes \mathcal{H}_2$ defined as: $F_{nm}^{(12)} := |\xi_{nm}\rangle\langle\xi_{nm}|$ with

$$\left| \xi_{nm}{}^{(12)} \right\rangle := \frac{1}{\sqrt{N}} \sum_{j=1}^N b_{nj} e_j^{(1)} \otimes e_{j \oplus m}^{(2)} \qquad \left(\sum_{j=1}^N b_{nj}^* b_{lj} = \delta_{nl}, |b_{nj}| = 1 \right).$$

Alice performs the joint measurement of the observable $F^{(12)}$.

Bob obtained a state $\rho^{(3)}$ due to the reduction of wave packet and he is informed which outcome was obtained by Alice. This information is completely transmitted from Alice to Bob without disturbance (for instance by telephone).

$\rho^{(1)}$ is reconstructed from $\rho^{(3)}$ by using the key which corresponds to the outcome as Bob got from Alice in the above *STEP3*. That is, for z_{nm}, operation W_{nm}. $W_{nm}e^{(3)}_{j\oplus m} := b^*_{nj}e^{(1)}_j$ is employed.

In the above scheme Alice and Bob believe that the teleportation setting (entangled state between them) is perfect and do their own jobs, however the prepared entangled state is not ideal one ($\tau \neq \tau_0$), the experiment yields error. Now the expected state obtained by Bob is written as,

$$\widetilde{\Lambda}^*_\tau(\rho) = \sum_{nm} tr_{1,2}\left(\left(F^{(12)}_{nm} \otimes W_{nm}\right)(\rho \otimes J\tau J^*)\left(F^{(12)}_{nm} \otimes W^*_{nm}\right)\right). \quad (40)$$

Let us estimate the fidelity. We obtain the following theorem.

Theorem 4.2. *The following inequality for the fidelity holds:*

$$(1 \geqslant) F(\rho, \widetilde{\Lambda}^*_\tau(\rho)) \geqslant F(\tau_0, \tau), \quad (41)$$

where $\tau = \tau_0 = |\phi_0\rangle\langle\phi_0|$ *with* $|\phi_0\rangle := \frac{1}{\sqrt{N}}\sum_{k=1}^{N} e_k \in \mathcal{H}_0$ *represents the ideal input state for the splitter to produce a perfect entangled state. In addition, the inequality is strict, that is, there exists an unknown state to be sent which makes the inequality equality.*

Proof. For simplicity, for the unknown state of Alice, we assume pure state $\rho = |\Phi\rangle\langle\Phi|$. The fidelity is written as,

$$F(\rho, \widetilde{\Lambda}^*_\tau(\rho))^2 = \langle\Phi|\,\widetilde{\Lambda}^*_\tau(|\Phi\rangle\langle\Phi|)\,|\Phi\rangle.$$

For $\tau = \sum_k \lambda_k |\varphi_k\rangle\langle\varphi_k|$, it is expressed in a simple form by using $f^k, g \in L^2(\{1, 2, \cdots N\})$ defined by

$$\begin{aligned} f^k(j) &:= \sum_t \langle e_{t\oplus j}|\varphi_k\rangle|\langle e_t|\phi\rangle|^2 \\ g(j) &:= \frac{1}{\sqrt{N}} \quad (j = 1, 2, \cdots N). \end{aligned} \quad (42)$$

We obtain

$$F(\rho, \widetilde{\Lambda}^*_\tau(\rho))^2 = \sum_k \lambda_k \|f_k\|^2 \|g\|^2.$$

Thanks to the Cauchy-Schwarz inequality, it is estimated as,

$$\begin{aligned} F(\rho, \widetilde{\Lambda}^*_\tau(\rho))^2 &\geqslant \sum_k \lambda_k \left|(f^k, g)\right|^2 \quad (43) \\ &= F(\tau_0, \tau)^2 \quad (44) \end{aligned}$$

For the general mixed input case, the proof goes similarly. Since we have just used only Cauchy-Schwarz inequality, the equality is attained with the choice, $|\phi\rangle = \sum_k \frac{e^{i\theta_k}}{\sqrt{N}} |e_k^{(1)}\rangle$. Thus the proof is completed.

□

The theorem shows that closer the input state for the splitter is to the perfect one, the fidelity is closer to one.

Finally we present the generalized result to the generalized beam splitting case:

Theorem 4.3. *For an arbitrary input τ of the generalized beam splitter, the fidelity is bounded by*

$$F(\rho, \Lambda_\tau^*(\rho)) \geqslant \frac{1 - e^{-a^2/2}}{\sqrt{1 + e^{-a^2}}} F(\tau_0, \tau), \tag{45}$$

where $\tau_0 := \left|\tilde{\phi}_0\right\rangle \left\langle \tilde{\phi}_0\right|$ *with*

$$\left|\tilde{\phi}_0\right\rangle := \frac{1}{\sqrt{N}} \sum_{k=1}^{N} |\exp(ag_k) - \exp(0)\rangle. \tag{46}$$

The theorem is proved by combining the above two theorems. (We here omit the proof.) It is seen that when τ is close to τ_0 and $|a|$ is large, the fidelity becomes close to one.

Acknowledgment:
The authors thank SCAT for financial support of our work.

References

1. Bennett, C. H., Brassard, G., Crépeau, C., Jozsa, R., Peres, A. and Wootters, W.: *Teleporting an unknown quantum state viaDual Classical and Einstein-Podolsky- Rosen channels.* Phys. Rev. Lett. **70**, 1895–1899, 1993.
2. Bennett, C.H., G. Brassard, S. Popescu, B. Schumacher, J.A. Smolin, W.K. Wootters, *Purification of noisy entanglement and faithful teleportation via no isy channels*, Phys. Rev. Lett. **76**,722–725 , 1996.
3. Accardi, L. and Ohya, M.: *Teleportation of general quantum states,* Voltera Center preprint,1998.
4. Fichtner,K-H. and Ohya M.: *Quantum Teleportation with Entangled States Given by Beam Splittings*, Commun.Math.Phys.**222**,229–247,2001
5. Fichtner,K-H. and Ohya M.: *Quantum Teleportation and Beam Splittings*, to appear in Commun.Math.Phys.
6. Fichtner, K-H. and Freudenberg, W.: *Pointprocesses and the position distrubution of infinite boson systems.* J. Stat.Phys. **47**, 959–978, 1987.

7. Fichtner, K-H. and Freudenberg, W.: *Characterization of states of infinite Boson systems I.– On the construction of states.* Comm. Math. Phys. **137**, 315–357, 1991.
8. Fichtner, K-H., Freudenberg, W. and Liebscher, V.: *Time evolution and invariance of Boson systems given by beam splittings*, In finite Dim. Anal., Quantum Prob. and related topics I, 511–533, 1998.
9. Accardi L., Ohya M.: *Compound channels, transition expecttations and liftings*, Applied Mathematics & Optimization, **39**, 33–59, 1999.
10. Lindsay, J. M.: *Quantum and Noncausal Stochastic Calculus.* Prob. Th. Rel. Fields **97**, 65–80, 1993.
11. Fichtner, K.-H. and Winkler, G.: *Generalized brownian motion, point processes and stochastic calculus for random fields.* Math. Nachr. **161**, 291–307, 1993.
12. Inoue, K, Ohya, M. and Suyari, H.: *Characterization of quantum teleportation processes by nonlinear quantum mutual entropy*, Physica D, **120**, 117–124, 1998.
13. Uhlmann, A.: *The 'transition probability' in the state spaceof a $*$-algebra.* Rep.Math.Phys,**9**,273–279,1976
14. Jozsa, R.:*Fidelity for mixed quantum states.* J.Mod.Opt.,**41**,2315–2323,1994

QUANTUM LOGICAL GATES REALIZED BY BEAM SPLITTINGS *

WOLFGANG FREUDENBERG

Brandenburgische Technische Universität Cottbus,
Institut für Mathematik, Postfach 101344,
03013 Cottbus, Germany,
E-Mail: freudenberg@math.tu-cottbus.de

MASANORI OHYA

Department of Information Science,
Tokyo University of Science,
Noda City, Chiba 278-8510, Japan,
E-Mail: ohya@is.noda.tus.ac.jp

NATALIYA TURCHINA

Brandenburgische Technische Universität Cottbus,
Institut für Mathematik, Postfach 101344,
03013 Cottbus, Germany

NOBORU WATANABE

Department of Information Science,
Tokyo University of Science,
Noda City, Chiba 278-8510, Japan,
E-Mail: watanabe@is.noda.tus.ac.jp

In [15] Fredkin and Toffoli proposed a logical conservative gate on the base of which Milburn [24] constructed a quantum logical gate. Photon number states were used as the input state for the control gate. The control gate consisted of a Kerr medium. In the present paper we will realize the truth table that has to be satisified by the Fredkin–Toffoli gate solely by using beam splitting procedures. The Kerr medium is replaced by further beam splittings which are applied to both outputs of the first splitting procedure. The information 0 and 1 are encoded by coherent states on a general bosonic Fock space. The aim of the paper is to give examples for quantum channels transmitting some information such that the gate will fulfill the truth table. For the output we get simple explicit expressions.

*Work partially supported by the EU Research Network HPRN-CT-2002-00279

1. Introduction

The aim of the paper is to construct with the aid of quantum channels quantum gates which fulfill the truth table given below. Hereby, the signals 0 and 1 are encoded by two different coherent states. One of these states also could be the vacuum state. The gate is composed of two input gates I_1, I_2, a control gate C and two output gates O_1, O_2. The control gate if switched on will change the information $(0,1)$ into $(1,0)$ but leave the informations $(0,0)$ resp. $(1,1)$ unchanged. The following truth table has to be fulfilled.

Input I_1	Input I_2	Control State	Output O_1	Output O_2
0	0	0	0	0
0	1	0	0	1
1	0	0	1	0
1	1	0	1	1
0	0	1	0	0
0	**1**	**1**	**1**	**0**
1	**0**	**1**	**0**	**1**
1	1	1	1	1

Fredkin and Toffoli proposed in [15] a conservative gate by which any logical gate is realized and it was shown that it is reversible in the sense that there is no loss of information. Such a gate was constructed by Milburn [24] as a quantum gate with quantum input and quantum output. The scheme of such a gate which we call FTM gate is the following:

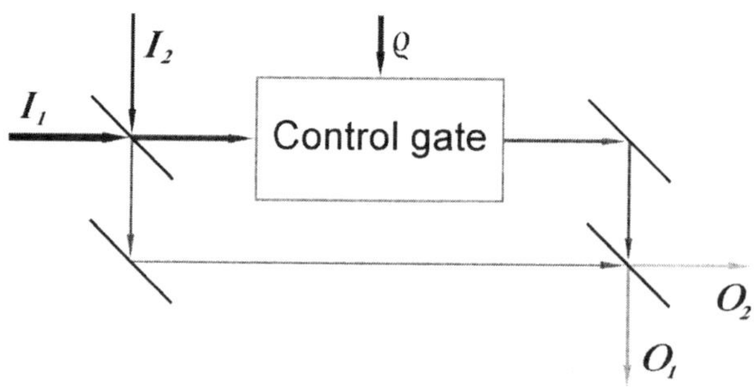

Figure 1. Model of a FTM gate

The two inputs I_1, I_2 come to a first beam splitter and produce two outputs. One of them passes via a control gate to the second beam splitting apparatus, the second output (of the first beam splitting procedure) passes directly to the second beam splitter. After the second beam splitting we will have two final outputs O_1, O_2. This model we discussed in detail in [16, 17].

We will modify the above gate in the following way: As in the above model two inputs I_1, I_2 come to a first beam splitter and produce two outputs. Now, both outputs of the first beam splitting procedure will be splitted separately in the control gate each of them producing two outputs. One pair represents the control gate if switched off, the other pair the control gate if switched on. These pairs produce in the final beam splitting the outputs O_1, O_2. We will consider quantum channels for states on the symmetric Fock space over a general metric space G equiped with a measure ν. In most concrete models G will be the Euclidean space $\mathbb{R}^d$, and ν will be the Lebesgue measure on $\mathbb{R}^d$. The splitting rates α, β are arbitrary measurable functions from G to $\mathbb{C}$ satisfying $|\alpha(x)|^2 + |\beta(x)|^2 = 1$ for all $x \in G$. So we will not assume α, β to be constants. It turns out that for this model the whole process can be described in an explicit way. For the output state we get simple expressions. The information 0 and 1 are encoded by two arbitrary exponential vectors (and multiples of them).

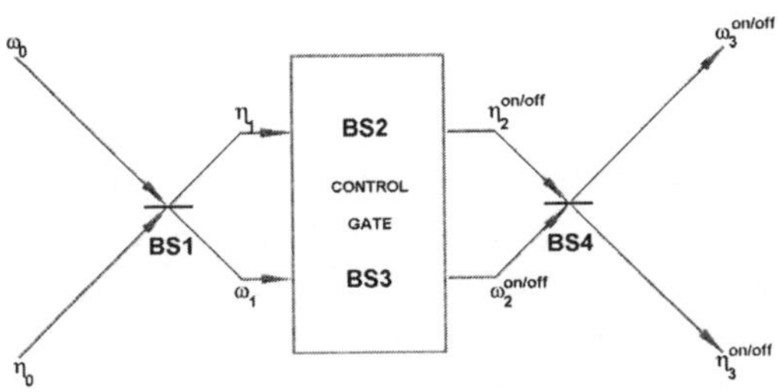

Figure 2. New Model of a FTM gate

In all calculations we have to assume nothing about the two exponential vectors (coherent states) carrying the information 0 or 1. For instance one can take the vacuum state and a coherent (non-vacuum) state. Another possibility is to choose two exponential vectors generated by functions being orthogonal or with disjoint support. So there is a big choice of representatives one can choose for building the logical gate. The concrete choice should depend on the practical possibilities of the construction of such gates.

Our aim is to construct a gate using solely the well-known splitting procedures and coherent states. The proposed model is a first step. In contrary to the model in [16, 17] the operation in the control gate is still a "semi-classical" one and should be replaced by a real quantum switch. For the states appearing in our model we will use the following notations:

ω_0 and η_0	-	the two input states,
ω_1 and η_1	-	the output states after the first beam splitting,
ω_2^{off} and η_2^{off}	-	output states after passing the control gate if control gate is switched *off*,
ω_2^{on} and η_2^{on}	-	output states after passing the control gate if control gate is switched *on*,
ω_3^{off} and η_3^{off}	-	final output states if control gate is switched *off*,
ω_3^{on} and η_3^{on}	-	final output states if control gate is switched *on*.

The present paper continues and generalizes results from [25, 26, 16, 17].

2. The Boson Fock Space

First we will collect some basic facts concerning the Fock space and certain operators acting on it.

Let G be an arbitrary complete separable metric space and $\mathfrak{G}$ its σ-algebra of Borel sets. Further, let ν be a locally finite diffuse measure on $[G, \mathfrak{G}]$, i.e. $\nu(K) < \infty$ for bounded $K \in \mathfrak{G}$ and $\nu(\{x\}) = 0$ for all singletons $x \in G$. Especially, we are concerned with the case where $G = \mathbb{R}^d$ and ν is the $d-$dimensional Lebesgue measure, but with minor changes of notations also the case of atomic measures like the counting measure on $G = \mathbb{Z}^d$ fits into our framework. We avoid these changes for the sake of readability. By M we denote the set of all finite integer-valued measures on $[G, \mathfrak{G}]$, i.e. M is the set of finite counting measures. One can show easily (cf. [22, 3]) that each $\varphi \in M$ can be written in the form $\varphi = \delta_{x_1} + \cdots + \delta_{x_n}$ with $n \in \mathbb{N}$ and $x_j \in G$ (where δ_x denotes the Dirac measure in x). So, the elements of M can be interpreted as finite (symmetric) configurations in G. We equip M with its canonical σ-algebra $\mathfrak{M}$ — the smallest σ-algebra containing all sets of the form $\{\varphi \in M : \varphi(K) = n\}$, $K \in \mathfrak{G}$, $n \in \mathbb{N}$. Observe that $\varphi(K) = n$

means that the configuration φ has exactly n points in the subset K of the phase space G. On $[M, \mathfrak{M}]$ we introduce a measure F by setting for $Y \in \mathfrak{M}$

$$F(Y) := \chi_Y(\mathfrak{o}) + \sum_{n\geq 1} \frac{1}{n!} \int_{G^n} \chi_Y \left(\sum_{j=1}^{n} \delta_{x_j} \right) \nu^n(d[x_1, \ldots, x_n]). \tag{1}$$

Hereby, χ_Y denotes the indicator function of a set Y and $\mathfrak{o}$ is the empty configuration in M, i.e. $\mathfrak{o}(G) = 0$. Observe that F is a σ-finite measure. Since ν was assumed to be diffuse one easily checks that F is concentrated on the set

$$M_s := \{\varphi \in M : \varphi(\{x\}) \leq 1 \text{ for all } x \in G\} \tag{2}$$

of so called simple counting measures (i.e. without multiple points).

Definition 2.1. $\mathcal{M} := L_2(M, \mathfrak{M}, F)$ is called the (symmetric) Fock space over G.

Remark: Usually one defines the symmetric Fock space $\Gamma(\mathcal{H})$ over a Hilbert space $\mathcal{H}$ as the direct sum of the symmetrized tensor products $\mathcal{H}_{\text{symm}}^{\otimes n}$ of the underlying Hilbert space $\mathcal{H}$, i.e. $\Gamma(\mathcal{H}) = \bigoplus_{n=0}^{\infty} \mathcal{H}_{\text{symm}}^{\otimes n}$. We introduced the Fock space in a way adapted to the language of counting measures especially to avoid symmetrization procedures. A further advantage of the definition given above is that the Fock space over a L_2−space again will be a L_2−space allowing very simple calculation of integrals (cf. Lemma 2.1 below). It is very easy to show that $\Gamma(\mathcal{H})$ and $\mathcal{M}$ are isomorphic (cf. [12]). For details we refer to [13, 14, 18] but also e.g. to [20] where a similar definition of the Fock space is given.

Observe that the Hilbert space $\mathcal{M}$ is again separable. Now, for each $n \geq 1$ let $\mathcal{M}^n := \mathcal{M}^{\otimes n}$ be the n-fold tensor product of the Hilbert space $\mathcal{M}$. Obviously, $\mathcal{M}^n$ can be identified with $L_2(M^n, F^n)$ where F^n is the product measure on the cartesian product $M^n = M \times \ldots \times M$ equiped with the usual product σ-algebra.

Basic for the proofs in this paper will be that integration in M^n (with respect to F^n) can be replaced by an integration with respect to F.

Lemma 2.1. *Let $f : M^n \to \mathbb{C}$ be integrable with respect to F^n (or ≥ 0).*

Then

$$
\begin{aligned}
&\int F^n(d[\varphi_1,\ldots,\varphi_n])f(\varphi_1,\ldots,\varphi_n) \\
&= \int F(d\varphi_n) \sum_{\varphi_1\le\varphi_2\le\cdots\le\varphi_n} f(\varphi_1,\varphi_2-\varphi_1,\ldots,\varphi_n-\varphi_{n-1}) \qquad (3) \\
&= \int F(d\varphi) \sum_{\substack{\varphi_1,\ldots,\varphi_n\in M\\ \varphi_1+\cdots+\varphi_n=\varphi}} f(\varphi_1,\ldots,\varphi_n)
\end{aligned}
$$

Hereby, $\hat{\varphi} \le \varphi$ means that $\hat{\varphi}$ is a subconfiguration of φ, i.e. $\varphi - \hat{\varphi} \in M$. The proof of the first part in (3) for the case $n = 2$ one can find e.g. in [11], in [13], or in [21, 23] where the above lemma is called $\sum\!\!\!\!\int$-lemma. A simple induction shows that (3) is valid for all $n \ge 2$. The second part of (3) is just a reformulation of the first one.

Definition 2.2. For a given function $g : G \longrightarrow \mathbb{C}$ the function $\exp_g : M \longrightarrow \mathbb{C}$ defined by

$$
\exp_g(\varphi) = \begin{cases} 1 & \text{if } \varphi = \mathfrak{o}, \\ \prod\limits_{\substack{x\in G\\ \varphi(\{x\})>0}} g(x) & \text{if } \varphi \ne \mathfrak{o} \end{cases} \qquad (4)
$$

is called exponential vector generated by g.

We make use of the following well-known properties of exponential vectors:

Lemma 2.2. *Let f and g be functions from G to $\mathbb{C}$ and $\varphi, \varphi_1, \varphi_2$ be elements from M. Then we have*

$$
\begin{aligned}
\exp_f(\varphi_1+\varphi_2) &= \exp_f(\varphi_1)\cdot\exp_f(\varphi_2), && (5)\\
\exp_{f+g}(\varphi) &= \sum_{\hat{\varphi}\le\varphi} \exp_f(\hat{\varphi})\cdot\exp_g(\varphi-\hat{\varphi}), && (6)\\
\exp_{f\cdot g}(\varphi) &= \exp_f(\varphi)\cdot\exp_g(\varphi). && (7)\\
\|\exp_g\|^2_{\mathcal{M}} &= e^{\|g\|^2_{L_2(G,\nu)}} \qquad (g \in L^2(G,\nu)), && (8)\\
\langle \exp_f, \exp_g\rangle_{\mathcal{M}} &= e^{\langle f,\ g\rangle_{L_2(G,\nu)}} \qquad (f,\ g \in L_2(G,\nu)). && (9)
\end{aligned}
$$

Observe that $\exp_g \in \mathcal{M}$ if and only if $g \in L^2(G,\nu)$. We will make use of the well-known fact that the linear span of exponential vectors from $\mathcal{M}$

is dense in $\mathcal{M}$. Further, observe that the exponential vectors $\exp_f, \exp_g$ generated by orthogonal functions $f,\ g \in L_2(G,\nu)$ are non-orthogonal elements of the Fock space $\mathcal{M}$. Indeed, immediately from (9) it follows that in this case $\langle \exp_f, \exp_g \rangle = 1$. The von-Neumann algebra $\mathcal{L}(\mathcal{M})$ of all bounded linear operators on the Fock space $\mathcal{M}$ we will denote by $\mathcal{A}$. Obviously, we may identify $\mathcal{A}\otimes\mathcal{A}$ and $\mathcal{L}(\mathcal{M}^2)$. The identity in $\mathcal{A}$ we denote by $\mathbb{I}$.

For $g,\ h \in L_2(G,\nu)$ we denote by $B_{g,h}$ the integral operator with kernel $\exp_h \otimes \exp_{\overline{g}}$, i. e.

$$B_{g,h}\Psi(\varphi) = \int_M F(d\hat{\varphi})\exp_h(\varphi)\exp_{\overline{g}}(\hat{\varphi})\Psi(\hat{\varphi}) \qquad (\Psi \in \mathcal{M},\ \varphi \in M). \quad (10)$$

Observe that for $g,\ h \in L_2(G,\nu)$ we have $B_{g,h} \in \mathcal{A}$. Immediately from (9) we get

$$B_{g,h}\exp_f = e^{\langle g,\ f\rangle} \cdot \exp_h \qquad (f,\ g,\ h \in L_2(G,\nu)). \quad (11)$$

Operators of the type $B_{g,h}$ determine a normal state on $\mathcal{A}$ completely, i. e. we have the following result:

Lemma 2.3. *([6, 8]). Let ω_j, $j = 1,\ 2$ be normal states on $\mathcal{A}$. Then $\omega_1 = \omega_2$ if and only if*

$$\omega_1(B_{g,h}) = \omega_2(B_{g,h}) \qquad (g,\ h \in L_2(G,\nu)). \quad (12)$$

3. A General Beam Splitting Model

We start with briefly reviewing the notion of a quantum channel. Let $\mathcal{A}$ denote as above the algebra $\mathcal{L}(\mathcal{M})$ of all bounded linear operators on the symmetric Fock space over the phase space G (cf. Definition 2.1) . By $\mathcal{S}(\mathcal{A})$ we denote the set of normal states on the algebra $\mathcal{A}$.

Definition 3.1. A mapping $\mathcal{E}^* : \mathcal{S}(\mathcal{A}\otimes\mathcal{A}) \longrightarrow \mathcal{S}(\mathcal{A}\otimes\mathcal{A})$ we call a completely positive compound channel if the dual map $\mathcal{E} : \mathcal{A}\otimes\mathcal{A} \longrightarrow \mathcal{A}\otimes\mathcal{A}$ given by the relation

$$\xi(\mathcal{E}(C)) = \mathcal{E}^*(\xi)(C) \qquad (C \in \mathcal{A}\otimes\mathcal{A},\ \xi \in \mathcal{S}(\mathcal{A}\otimes\mathcal{A}))$$

is a linear mapping being completely positive, i. e.

$$\sum_{i,j=1}^{n} D_i^*\mathcal{E}(C_i^*C_j)D_j \geq 0 \qquad (C_i \in \mathcal{A}\otimes\mathcal{A},\ D_i \in \mathcal{A}\otimes\mathcal{A},\ n \in \mathbb{N}) \quad (13)$$

and identity preserving, i. e.

$$\mathcal{E}(\mathbb{I}\otimes\mathbb{I}) = \mathbb{I}\otimes\mathbb{I}. \quad (14)$$

Now, let $\omega_0,\ \eta_0 \in \mathcal{S}(\mathcal{A})$ be the input states. We want to construct logical gates where the inputs carry the information 0 or 1.Therefore, we consider only the case of inputs being "independent", i.e. the compound state (on $\mathcal{A}\otimes\mathcal{A}$) that has to be transformed by $\mathcal{E}^*$ is just the product state $\omega_0\otimes\eta_0$. For a discussion of general compund states in this context we refer to [27]. After the transformation with respect to the channel $\mathcal{E}^*$ we obtain a compound state ζ on $\mathcal{A}\otimes\mathcal{A}$:

$$\zeta = \omega_0\otimes\eta_0 \circ \mathcal{E}.$$

The two "parts" of ζ are given by

$$\omega_1(A) := \zeta(A\otimes\mathbb{I}) = \omega_0\otimes\eta_0 \circ \mathcal{E}(A\otimes\mathbb{I}) \tag{15}$$

$$\eta_1(B) := \zeta(\mathbb{I}\otimes B) = \omega_0\otimes\eta_0 \circ \mathcal{E}(\mathbb{I}\otimes B) \tag{16}$$

with $A,\ B \in \mathcal{A}$.

For the concrete splitting model we will consider in this paper it turns out that starting with the product state of coherent states $\omega_0,\ \eta_0$ the compound state ζ will be again a product state, and one has

$$\zeta(A\otimes B) = \omega_0\otimes\eta_0 \circ \mathcal{E}(A\otimes B) = \omega_1\otimes\eta_1(A\otimes B). \tag{17}$$

We will introduce now a concrete quantum channel constructed with the aid of a beam splitting procedure. The results below in this section are contained at least partially in [8, 7, 9, 10]. However, for completeness and readability of the present paper we will present all proofs.

Let $\alpha,\ \beta : G \longrightarrow \mathbb{C}$ be measurable mappings such that

$$|\alpha(x)|^2 + |\beta(x)|^2 = 1 \qquad (x \in G). \tag{18}$$

We define a linear operator $\mathcal{V}_{\alpha,\beta} : \mathcal{M}^2 \longrightarrow \mathcal{M}^2$ by setting for $\Phi \in \mathcal{M}^2$ and $\varphi_1,\ \varphi_2 \in M$

$$(\mathcal{V}_{\alpha,\beta}\ \Phi)(\varphi_1, \varphi_2) := \sum_{\hat{\varphi}_1\le\varphi_1}\ \sum_{\hat{\varphi}_2\le\varphi_2} \exp_\alpha(\hat{\varphi}_1)\exp_\beta(\varphi_1 - \hat{\varphi}_1)$$
$$\cdot\exp_{-\overline{\beta}}(\hat{\varphi}_2)\exp_{\overline{\alpha}}(\varphi_2 - \hat{\varphi}_2)\Phi(\hat{\varphi}_1 + \hat{\varphi}_2, \varphi_1 + \varphi_2 - \hat{\varphi}_1 - \hat{\varphi}_2). \tag{19}$$

On exponential vectors the operator $\mathcal{V}_{\alpha,\beta}$ has the same form as in the usual beam splitter model of quantum optics (cf. inst. [2]).

Proposition 3.1. *Let $\mathcal{V}_{\alpha,\beta}$ be the operator defined by (19) with α and β satisfying (18). For all $g,\ h \in L_2(G, \nu)$ one has*

$$\mathcal{V}_{\alpha,\beta}\ \exp_g\otimes\exp_h = \exp_{\alpha g+\beta h} \otimes \exp_{-\overline{\beta}g+\overline{\alpha}h}. \tag{20}$$

Proof. For arbitrary $g,\ h \in L_2(G,\nu)$ we have $\exp_g \otimes \exp_h \in \mathcal{M}^2$. Applying several times Lemma 2.2 we get for $\varphi_1, \varphi_2 \in M$

$$
\begin{aligned}
(\mathcal{V}_{\alpha,\beta} \exp_g \otimes \exp_h)(\varphi_1,\varphi_2) &= \sum_{\hat\varphi_1 \le \varphi_1} \sum_{\hat\varphi_2 \le \varphi_2} \exp_\alpha(\hat\varphi_1)\exp_\beta(\varphi_1 - \hat\varphi_1) \\
&\cdot \exp_{-\overline{\beta}}(\hat\varphi_2)\exp_{\overline{\alpha}}(\varphi_2 - \hat\varphi_2)\exp_g(\hat\varphi_1 + \hat\varphi_2)\exp_h(\varphi_1 + \varphi_2 - \hat\varphi_1 - \hat\varphi_2) \\
&= \sum_{\hat\varphi_1 \le \varphi_1} \sum_{\hat\varphi_2 \le \varphi_2} \exp_\alpha(\hat\varphi_1)\exp_g(\hat\varphi_1)\exp_\beta(\varphi_1 - \hat\varphi_1)\exp_h(\varphi_1 - \hat\varphi_1) \\
&\cdot \exp_{-\overline{\beta}}(\hat\varphi_2)\exp_g(\hat\varphi_2)\exp_{\overline{\alpha}}(\varphi_2 - \hat\varphi_2)\exp_h(\varphi_2 - \hat\varphi_2) \\
&= \sum_{\hat\varphi_1 \le \varphi_1} \exp_{\alpha g}(\hat\varphi_1)\exp_{\beta h}(\varphi_1 - \hat\varphi_1) \sum_{\hat\varphi_2 \le \varphi_2} \exp_{-\overline{\beta} g}\exp_{\overline{\alpha} h}(\varphi_2 - \hat\varphi_2) \\
&= \exp_{\alpha g + \beta h}(\varphi_1) \cdot \exp_{-\overline{\beta} g + \overline{\alpha} h}(\varphi_2) = \exp_{\alpha g + \beta h} \otimes \exp_{-\overline{\beta} g + \overline{\alpha} h}(\varphi_1, \varphi_2)
\end{aligned}
$$

what ends the proof. □

Observe that for all $g \in L_2(G,\nu)$

$$
\mathcal{V}_{\alpha,\beta} \exp_g \otimes \exp_0 = \exp_{\alpha g} \otimes \exp_{-\overline{\beta} g} \tag{21}
$$

Splitting procedures of this type (considered as operators from $\mathcal{A}$ to $\mathcal{A} \otimes \mathcal{A}$) were studied in detail for instance in [8]. Similar models were considered in [4, 5]. More general splitting models are discussed in [7, 10].

Proposition 3.2. *The operator $\mathcal{V}_{\alpha,\beta}$ defined by (19) with α and β satisfying (18) is an isometry.*

Proof. For $g,\ h \in L_2(G,\nu)$ we get

$$
\begin{aligned}
||\mathcal{V}_{\alpha,\beta} \exp_g \otimes \exp_h||^2_{\mathcal{M}^2} &= ||\exp_{\alpha g + \beta h}||^2_{\mathcal{M}} \cdot ||\exp_{-\overline{\beta} g + \overline{\alpha} h}||^2_{\mathcal{M}} \\
&= \exp\{||\alpha g + \beta h||^2\} \cdot \exp\{|| - \overline{\beta} g + \overline{\alpha} h||^2\} \\
&= \exp\{|\alpha|^2 \|g\|^2 + |\beta|^2 \|h\|^2 + \overline{\alpha}\beta\langle g, h\rangle + \alpha\overline{\beta}\langle h, g\rangle\} \\
&\quad \cdot \exp\{|\alpha|^2 \|h\|^2 + |\beta|^2 \|g\|^2 - \overline{\alpha}\beta\langle g, h\rangle - \alpha\overline{\beta}\langle h, g\rangle\} \\
&= \exp\{\|g\|^2 + \|h\|^2\} = \|\exp_g \otimes \exp_h\|^2_{\mathcal{M}^2}.
\end{aligned}
$$

Since the linear span of all tensor products $\exp_g \otimes \exp_h$ of exponential vectors is dense in $\mathcal{M}^2$ this proves that $\mathcal{V}_{\alpha,\beta}$ is isometric. □

The operator $\mathcal{V}^*_{\alpha,\beta}$ adjoint to $\mathcal{V}_{\alpha,\beta}$ is of the same type as $\mathcal{V}_{\alpha,\beta}$.

Proposition 3.3. *Let $\mathcal{V}_{\alpha,\beta}$ be given by (19) with α and β satisfying (18). Then*

$$
\mathcal{V}^*_{\alpha,\beta} = \mathcal{V}_{\overline{\alpha},-\beta} \tag{22}
$$

Proof. It is sufficient to show (22) for exponential vectors. For $g,\ h \in L_2(G,\nu)$ we get

$$\begin{aligned}\mathcal{V}_{\overline{\alpha},-\beta}\mathcal{V}_{\alpha,\beta}\exp_g\otimes\exp_h &= \exp_{\overline{\alpha}(\alpha g+\beta h)-\beta(-\overline{\beta}g+\overline{\alpha}h}\otimes\exp_{\overline{\beta}(\alpha g+\beta h)+\alpha(-\overline{\beta}g+\overline{\alpha}h}\\ &= \exp_{|\alpha|^2 g+\overline{\alpha}\beta h+|\beta|^2 g-\overline{\alpha}\beta h}\otimes\exp_{\alpha\overline{\beta}g+|\beta|^2 h-\alpha\overline{\beta}g+|\alpha|^2 h}\\ &= \exp_g\otimes\exp_h\end{aligned}$$

what ends the proof. □

From (18) we get $|\overline{\alpha}(x)|^2+|-\beta(x)|^2=1$ for all $x\in G$. So we conclude from Proposition 3.2 that $\mathcal{V}^*_{\alpha,\beta}=\mathcal{V}_{\overline{\alpha},-\beta}:\mathcal{M}^2\longrightarrow\mathcal{M}^2$ is an isometry too. Consequently, we obtain

Proposition 3.4. *The operator $\mathcal{V}_{\alpha,\beta}:\mathcal{M}^2\longrightarrow\mathcal{M}^2$ with α and β satisfying (18) is unitary.*

From Proposition 3.4 we get immediately that the mapping $\mathcal{E}$: $\mathcal{L}(\mathcal{M}^2)\longrightarrow\mathcal{L}(\mathcal{M}^2)$ defined by

$$\mathcal{E}_{\alpha,\beta}(C):=\mathcal{V}^*_{\alpha,\beta}(C)\mathcal{V}_{\alpha,\beta}\qquad (C\in\mathcal{L}(\mathcal{M}^2)\tag{23}$$

is completely positive and preserves the identity.

Definition 3.2. The completely positive channel $\mathcal{E}^*_{\alpha,\beta}:\mathcal{S}(\mathcal{A}\otimes\mathcal{A})\longrightarrow\mathcal{S}(\mathcal{A}\otimes\mathcal{A})$ the dual map of which is given by (23) (i. e. $\mathcal{E}^*_{\alpha,\beta}\xi(C)=\xi(\mathcal{E}_{\alpha,\beta}(C))$ for $C\in\mathcal{A}\otimes\mathcal{A},\ \xi\in\mathcal{S}(\mathcal{A}\otimes\mathcal{A})$) is called noisy quantum channel with rates $\alpha,\ \beta$.

For more details on beam splittings we refer to [1] and [6, 8, 7, 9, 10], for connections to quantum Markov chains cf. also [19].

Definition 3.3. Let $f\in L_2(G,\nu)$. The coherent state Φ^f corresponding to f is defined by

$$\Phi^f(A):=\langle\exp_f,\ A\exp_f\rangle e^{-\|f\|^2}\qquad (A\in\mathcal{A}).\tag{24}$$

The state Φ^0, i. e. the coherent state corresponding to the function being identical zero is the vacuum state in $\mathcal{S}(\mathcal{A})$.

From (9) in Lemma 2.2 and (11)we get that for an operator $B_{g,h}$ (defined by (10)) and a coherent state Φ^g one has

$$\Phi^f(B_{g,h})=e^{\langle g,\ f\rangle+\langle f,\ h\rangle-\|f\|^2}.\tag{25}$$

Now we want to characterize the transformation of the two coherent normal input states.

Theorem 3.1. *Let* $\omega_0 = \Phi^g$, $\eta_0 = \Phi^h$ *be two coherent states with* g, $h \in L_2(G, \nu)$ *and let* $\mathcal{E}_{\alpha,\beta}$ *be given by (23) wit* α, β *satisfying (18). We set (cf. (15) and(16))*

$$\omega_1 := \omega_0 \otimes \eta_0 \circ \mathcal{E}_{\alpha,\beta}((\cdot)\otimes \mathbb{I}), \quad \eta_1 := \omega_0 \otimes \eta_0 \circ \mathcal{E}_{\alpha,\beta}(\mathbb{I} \otimes (\cdot)).$$

Then

$$\omega_1 = \Phi^{\alpha g + \beta h} \qquad \text{and} \qquad \eta_1 = \Phi^{-\overline{\beta} g + \overline{\alpha} h}. \tag{26}$$

Proof. For arbitrary $A \in \mathcal{A}$ we get

$$\begin{aligned}
&\omega_0 \otimes \eta_0 \circ \mathcal{E}_{\alpha,\beta}(A \otimes \mathbb{I}) \\
&= \langle \mathcal{V}_{\alpha,\beta} \exp_g \otimes \exp_h, (A \otimes \mathbb{I}) \mathcal{V}_{\alpha,\beta} \exp_g \otimes \exp_h \rangle e^{-\|g\|^2 - \|h\|^2} \\
&= \langle \exp_{\alpha g + \beta h}, A \exp_{\alpha g + \beta h} \rangle \cdot \langle \exp_{-\overline{\beta} g + \overline{\alpha} h}, \exp_{-\overline{\beta} g + \overline{\alpha} h} \rangle e^{-\|g\|^2 - \|h\|^2} \\
&= \langle \exp_{\alpha g + \beta h}, A \exp_{\alpha g + \beta h} \rangle \cdot e^{\|-\overline{\beta} g + \overline{\alpha} h\|^2 - \|g\|^2 - \|h\|^2}.
\end{aligned}$$

$\|-\overline{\beta} g + \overline{\alpha} h\|^2 - \|g\|^2 - \|h\|^2 = \|\alpha g + \beta h\|^2$ is just the normalization factor. So we finally obtain $\omega_1 = \Phi^{\alpha g + \beta h}$. Analogously, one can show $\eta_1 = \Phi^{-\overline{\beta} g + \overline{\alpha} h}$. □

We see that coherent signals corresponding to g and h are transformed into coherent ones corresponding to mixtures of these functions with the rates α and β. Since for A, $B \in \mathcal{A}$

$$\begin{aligned}
&\omega_0 \otimes \eta_0 \circ \mathcal{E}_{\alpha,\beta}(A \otimes B) \\
&= \langle \mathcal{V}_{\alpha,\beta} \exp_g \otimes \exp_h, (A \otimes B) \mathcal{V}_{\alpha,\beta} \exp_g \otimes \exp_h \rangle e^{-\|g\|^2 - \|h\|^2} \\
&= \langle \exp_{\alpha g + \beta h}, A \exp_{\alpha g + \beta h} \rangle \cdot \langle \exp_{-\overline{\beta} g + \overline{\alpha} h}, B \exp_{-\overline{\beta} g + \overline{\alpha} h} \rangle e^{-\|g\|^2 - \|h\|^2} \\
&= \omega_1(A) \cdot \eta_1(B)
\end{aligned}$$

we conclude that for coherent states the 'independence' of the two input 'survives', i. e. (cf. (17))

$$\omega_1 \otimes \eta_1 = \omega_0 \otimes \eta_0 \circ \mathcal{E}_{\alpha,\beta}. \tag{27}$$

Immediately from Theorem 3.1 we conclude

Proposition 3.5. *Let* $\omega_0 = \Phi^g$ *be a coherent state with* $g \in L_2(G, \nu)$, *and let the second input* η_0 *be the vacuum state* ($\eta_0 = \Phi^0$). *Further, let* $\mathcal{E}_{\alpha,\beta}$ *be given by (23) wit* α, β *satisfying (18). Setting as above*

$$\omega_1 := \omega_0 \otimes \eta_0 \circ \mathcal{E}_{\alpha,\beta}((\cdot)\otimes \mathbb{I}), \quad \eta_1 := \omega_0 \otimes \eta_0 \circ \mathcal{E}_{\alpha,\beta}(\mathbb{I} \otimes (\cdot)).$$

we get

$$\omega_1 = \Phi^{\alpha g} \qquad \text{and} \qquad \eta_1 = \Phi^{-\overline{\beta} g}. \tag{28}$$

4. The Splitting Procedures

4.1. *The Splitting BS1*

We will denote the splitting functions in this first splitting process **BS1** (cf. Figure 2) by $\alpha_0,\ \beta_0$, i. e. $\alpha_0,\ \beta_0 : G \longrightarrow \mathbb{C}$ and

$$|\alpha_0(x)|^2 + |\beta_0(x)|^2 = 1 \qquad (x \in G). \tag{29}$$

From Theorem 3.1 we obtain for the inputs $\omega_0 = \Phi^g,\ \eta_0 = \Phi^h$ with $g,\ h \in L_2(G,\nu)$ the outputs

$$\omega_1 = \Phi^{\alpha_0 g + \beta_0 h} \qquad \text{and} \qquad \eta_1 = \Phi^{-\overline{\beta_0} g + \overline{\alpha_0} h}. \tag{30}$$

4.2. *The Splittings in the Control Gate*

Now we pass to the interactions in the control gate (cf. Figure 2). There will be two beam splitting procedures **BS2** and **BS3**. The first one (**BS2**) is with splitting functions $\alpha_1,\ \beta_1 : G \longrightarrow \mathbb{C}$ satisfying

$$|\alpha_1(x)|^2 + |\beta_1(x)|^2 = 1 \qquad (x \in G). \tag{31}$$

The two inputs for this splitting are $\omega_1 = \Phi^{\alpha_0 g + \beta_0 h}$ and the vacuum state Φ^0. The second one (**BS3**) is with splitting functions $\alpha_2,\ \beta_2 : G \longrightarrow \mathbb{C}$ satisfying

$$|\alpha_2(x)|^2 + |\beta_2(x)|^2 = 1 \qquad (x \in G). \tag{32}$$

The inputs for this splitting are $\eta_1 = \Phi^{-\overline{\beta_0} g + \overline{\alpha_0} h}$ and the vacuum state Φ^0. Using Proposition 3.5 (see also formula (21)) we obtain as outputs of **BS2** the two states

$$\omega_2^{\text{off}} = \Phi^{\alpha_1(\alpha_0 g + \beta_0 h)} \tag{33}$$

and

$$\omega_2^{\text{on}} = \Phi^{-\overline{\beta_1}(\alpha_0 g + \beta_0 h)}. \tag{34}$$

The splitting procedure **BS3** applied to η_1 and Φ^0 leads to the two outputs

$$\eta_2^{\text{off}} = \Phi^{\alpha_2(-\overline{\beta_0} g + \overline{\alpha_0} h)} \tag{35}$$

and

$$\eta_2^{\text{on}} = \Phi^{-\overline{\beta_2}(-\overline{\beta_0} g + \overline{\alpha_0} h)}. \tag{36}$$

4.3. *The Final Splitting BS4*

For the last beam splitting procedure **BS4** we have to distinguish two cases:

1. The application of **BS4** to the inputs ω_2^{off} and η_2^{off} corresponds to the case the control gate is switched *off*. The outputs we will denote by ω_3^{off} and η_3^{off}.
2. The application of **BS4** to the inputs ω_2^{on} and η_2^{on} corresponds to the case the control gate is switched *on* and the corresponding outputs will be denoted by ω_3^{on} and η_3^{on}.

This last splitting (**BS4**) will be with splitting functions $\alpha_3,\ \beta_3 : G \longrightarrow \mathbb{C}$ satisfying again

$$|\alpha_3(x)|^2 + |\beta_3(x)|^2 = 1 \qquad (x \in G). \tag{37}$$

Using Theorem 3.1 we obtain in the case *1* the outputs

$$\omega_3^{\text{off}} = \Phi^{\alpha_3\alpha_1(\alpha_0 g+\beta_0 h)+\beta_3\alpha_2(-\overline{\beta_0}g+\overline{\alpha_0}h)} \tag{38}$$

and

$$\eta_3^{\text{off}} = \Phi^{-\overline{\beta_3}\alpha_1(\alpha_0 g+\beta_0 h)+\overline{\alpha_3}\alpha_2(-\overline{\beta_0}g+\overline{\alpha_0}h)}. \tag{39}$$

In case *2* (control gate switched *on*) we get the outputs

$$\omega_3^{\text{on}} = \Phi^{-\alpha_3\overline{\beta_1}(\alpha_0 g+\beta_0 h)-\beta_3\overline{\beta_2}(-\overline{\beta_0}g+\overline{\alpha_0}h)} \tag{40}$$

and

$$\eta_3^{\text{on}} = \Phi^{\overline{\beta_3\beta_1}(\alpha_0 g+\beta_0 h)-\overline{\alpha_3\beta_2}(-\overline{\beta_0}g+\overline{\alpha_0}h)}. \tag{41}$$

Now, we still have to find (nontrivial) splitting functions $\alpha_k,\ \beta_k,\ k \in \{0,1,2,3\}$ such that the truth table given at the beginning will be satisfied. This will be done in the next section.

5. Splitting Functions fulfilling the Truth Table

The gate constructed above has to satisfy the truth table mentioned in the introduction. Let g and h be two arbitrary functions from $L_2(G,\nu)$. The information **1** as input is encoded by the coherent state Φ^g the input information **0** will be encoded by the coherent state Φ^h. As an output the information **1** will be represented by each of the coherent states of the form Φ^{cg} where $c \in \mathbb{C}$ is a non-zero constant. The information **0** in the final output is encoded by each of the coherent states of th form Φ^{ch} with $c \in \mathbb{C}$ being different from zero. So we have to require at least that the vectors

g and h are orthogonal. Consequently, the one-dimensional subspaces of $L_2(G,\nu)$ spanned by the vectors g and h will be orthogonal. The concrete choice should depend on the practical possibility to construct such states and to recognize the information **1** or **0** by a measurement. Especially, one could assume g and h to have disjoint support. One also can choose one of these functions as the function being identically zero what corresponds to the choice of the vacuum state as one of the possible inputs.

We have to check case by case all possible inputs ω_0, η_0 with control gate switched off resp. on. This way we will get from case to case more restraints on the splitting functions α_k, β_k. We will see that at the end there remains still a huge number of possibilities for the choice of the splitting functions.

We start with the consideration of the input $(1,0)$, i. e. we consider the case $\omega_0 = \Phi^g$, $\eta_0 = \Phi^h$. In this case there have to exist non-zero constants $c_1, \ldots, c_4$ such that

$$\omega_3^{\text{off}} = \Phi^{c_1 g},\ \eta_3^{\text{off}} = \Phi^{c_2 h},\ \omega_3^{\text{on}} = \Phi^{c_3 h},\ \eta_3^{\text{on}} = \Phi^{c_4 g}. \tag{42}$$

We conclude from (38), (39), (40) and (41) that (42) holds if and only if the following chain of equations is fulfilled:

$$0 = \alpha_3\alpha_1\beta_0 + \beta_3\alpha_2\overline{\alpha_0} \tag{43}$$
$$0 = \overline{\beta_3}\alpha_1\alpha_0 + \overline{\alpha_3}\alpha_2\overline{\beta_0} \tag{44}$$
$$0 = -\alpha_3\overline{\beta_1}\alpha_0 + \beta_3\overline{\beta_2\beta_0} \tag{45}$$
$$0 = \overline{\beta_3\beta_1}\beta_0 - \overline{\alpha_3\beta_2\alpha_0} \tag{46}$$
$$c_1 = \alpha_3\alpha_1\alpha_0 - \beta_3\alpha_2\overline{\beta_0} \tag{47}$$
$$c_2 = -\overline{\beta_3}\alpha_1\beta_0 + \overline{\alpha_3}\alpha_2\overline{\alpha_0} \tag{48}$$
$$c_3 = \alpha_3\overline{\beta_1}\beta_0 + \beta_3\overline{\beta_2\alpha_0} \tag{49}$$
$$c_4 = \overline{\beta_3\beta_1}\alpha_0 + \overline{\alpha_3\beta_2\beta_0} \tag{50}$$

Combining (43) and (44) resp. (45) and (46) we get

$$|\alpha_1| = |\alpha_2| \qquad \text{and} \qquad |\beta_1| = |\beta_2|. \tag{51}$$

Further, (43), (46) and (51) lead to

$$|\alpha_3| = |\beta_3|. \tag{52}$$

Applying the above equalities to to (46) we still get

$$|\alpha_0| = |\beta_0|. \tag{53}$$

Taking into account still the assumption $|\alpha_k|^2 + |\beta_k|^2 = 1$ we can state the following condition:

Proposition 5.1. *If (42) holds the splitting functions α_k, β_k fulfill the following conditions: There exists a function $b : G \longrightarrow [0,1]$ such that for all $x \in G$*

$$b(x) = |\alpha_1(x)| = |\alpha_2(x)| \qquad |\beta_1(x)| = |\beta_2(x)| = \sqrt{1 - b^2(x)} \tag{54}$$

and

$$|\alpha_0(x)| = |\beta_0(x)| = |\alpha_3(x)| = |\beta_3(x)| = \frac{1}{\sqrt{2}}. \tag{55}$$

The first and last splitting have to be done with equal modules of splitting rates. Further restraints arise from (43) to (50). We denote the phases of the splitting functions (being functions from G into $\mathbb{R}$) by a_k, b_k, $k \in \{0, 1, 2, 3\}$:

$$\alpha_k(x) = |\alpha_k(x)| e^{i\, a_k(x)}, \qquad \beta_k(x) = |\beta_k(x)| e^{i\, b_k(x)} \qquad (x \in G) \tag{56}$$

where the moduli have to satisfy (54) and (55). Equations (43) to (46) lead to

$$a_3 - a_2 + a_1 + a_0 - b_3 + b_0 = \pi \tag{57}$$

$$a_3 + a_0 - b_3 + b_2 - b_1 + b_0 = 0. \tag{58}$$

We may add on the right sides even multiples of π. (57) and (58) are equivalent to

$$a_3 + a_0 + b_0 - b_3 = b_1 - b_2 \tag{59}$$

$$\pi + a_2 - a_1 = b_1 - b_2. \tag{60}$$

(47) to (50) still have to be fulfilled (the functions c_k have to be different from zero). Using the above equality (57) we obtain from (47)

$$c_1 = b(x) e^{i(\pi + b_3 - b_0 + a_2)}. \tag{61}$$

Similarly, we get

$$c_2 = b(x) e^{i(\pi - b_3 + b_0 + a_1)}. \tag{62}$$

Using (58), the equalities (49) and (50) are transformed into

$$c_3 = \sqrt{1 - b^2(x)} e^{i(a_3 - b_1 + b_0)}, \tag{63}$$

$$c_4 = \sqrt{1 - b^2(x)} e^{i(-b_3 - b_1 + a_0)}. \tag{64}$$

We see that from (47) to (50) there arise no further constraints. However, if c_k have to be constant necessarily the splitting functions have to be constant. Especially, $b(x) = b$. For the splittings in the control gate there is still a big choice of possibilities.

Obviously, the input $(0,1)$ leads to the same conditions on the splitting rates α_k, β_k. If the input is $(1,1)$ the final output will be

$$\omega_3^{\text{off}} = \Phi^{c_1 g},\ \eta_3^{\text{off}} = \Phi^{c_2 g},\ \omega_3^{\text{on}} = \Phi^{c_3 g},\ \eta_3^{\text{on}} = \Phi^{c_4 g}. \tag{65}$$

If the input is $(0,0)$ the final output will be

$$\omega_3^{\text{off}} = \Phi^{c_1 h},\ \eta_3^{\text{off}} = \Phi^{c_2 h},\ \omega_3^{\text{on}} = \Phi^{c_3 h},\ \eta_3^{\text{on}} = \Phi^{c_4 h}. \tag{66}$$

Example 5.1. Let the information 1 be encoded by the coherent state Φ^g with $\|g\| > 0$. The output 1 has to be detected by any coherent state of the form Φ^{cg} with real $c \neq 0$. The information 0 is encoded in the input and output by the vacuum state Φ^0. We set

$$\begin{aligned}
\alpha_0 &= \beta_0 = 1/\sqrt{2},\\
\alpha_3 &= -\beta_3 = -1/\sqrt{2},\\
\alpha_1 &= 1/2,\ \beta_1 = \sqrt{3}/2\\
\alpha_2 &= 1/2,\ \beta_2 = -\sqrt{3}/2.
\end{aligned}$$

If the input is $(1,0)$ we obtain

$$\omega_3^{\text{off}} = \Phi^{-\frac{1}{2}g},\ \eta_3^{\text{off}} = \Phi^0,\ \omega_3^{\text{on}} = \Phi^0,\ \eta_3^{\text{on}} = \Phi^{\frac{\sqrt{3}}{2}g}.$$

For the input $(0,1)$ one gets

$$\omega_3^{\text{off}} = \Phi^0,\ \eta_3^{\text{off}} = \Phi^{-\frac{1}{2}g},\ \omega_3^{\text{on}} = \Phi^{\frac{\sqrt{3}}{2}g},\ \eta_3^{\text{on}} = \Phi^0.$$

In the case $(0,0)$ all outputs will be equal to the vacuum state Φ^0. Finally, if the input is $(1,1)$ we get as outputs the states

$$\omega_3^{\text{off}} = \Phi^{-\frac{1}{2}g},\ \eta_3^{\text{off}} = \Phi^{-\frac{1}{2}g},\ \omega_3^{\text{on}} = \Phi^{\frac{\sqrt{3}}{2}g},\ \eta_3^{\text{on}} = \Phi^{\frac{\sqrt{3}}{2}g}.$$

We see that the truth table is satisfied. □

Obviously, it is easy to establish a lot of examples with complex-valued splittting rates.

References

1. L. Accardi and M. Ohya. Compound channels, transition expectations and liftings. *Applied Mathematics & Optimization*, 39:33–59, 1999.
2. R.A. Campos, B.E.A. Saleh, and M.C. Teich. Quantum-mechanical lossless beam splitter: $su(2)$ symmetry and photon statistics. *Phys.Rev. A*, 40(3):1371–1384, 1989.
3. D.J. Daley and D. Vere-Jones. *An Introduction to the Theory of Point Processes*. Springer Series in Statistics. Springer-Verlag, New York, Berlin, Heidelberg, 1988.

4. Karl-Heinz Fichtner and Masanori Ohya. Quantum teleportation with entangled states given by beam splittings. *Comm. Math. Phys.*, 222:229–247, 2001.
5. Karl-Heinz Fichtner and Masanori Ohya. Quantum teleportation and beam splitting. *Comm. Math. Phys.*, 225:67–89, 2002.
6. K.-H. Fichtner, W. Freudenberg, and V. Liebscher. Beam Splittings and Time Evolutions of Boson Systems. *Forschungsergebnisse der Fakultät für Mathematik und Informatik*, Math/Inf/96/39:105 pages, 1996.
7. K.-H. Fichtner, W. Freudenberg, and V. Liebscher. Non-independent Splittings and Gibbs States. *Mathematical Notes*, 64(3-4):518 – 523, 1998.
8. K.-H. Fichtner, W. Freudenberg, and V. Liebscher. Time Evolution and Invariance of Boson Systems Given by Beam Splittings. *Infinite Dimensional Analysis, Quantum Probability and Related Topics*, 1(4):511 – 531, 1998.
9. K.-H. Fichtner, W. Freudenberg, and V. Liebscher. Characterization of Classical and Quantum Poisson Systems by Thinnings and Splittings. *Math. Nachr.*, 218:25–47, 2000.
10. K.-H. Fichtner, W. Freudenberg, and V. Liebscher. On exchange mechanisms for bosons. *Random Operators and Stochastic Equations.* volume 12, No.4:331 - 348, VSP Leiden, The Netherlands, 2004.
11. K.-H. Fichtner and W. Freudenberg. Point processes and normal states of boson systems. *Naturwiss.-Technisches Zentrum NTZ*, Leipzig, 1986. 56 p.
12. K.-H. Fichtner and W. Freudenberg. Point processes and the position distribution of infinite boson systems. *J. Stat. Phys.*, 47:959–978, 1987.
13. K.-H. Fichtner and W. Freudenberg. Characterization of states of infinite boson systems I. On the construction of states of boson systems. *Commun. Math. Phys.*, 137:315 – 357, 1991.
14. K.-H. Fichtner and W. Freudenberg. Remarks on stochastic calculus on the Fock space. In L. Accardi, editor, *Quantum Probability and Related Topics VII*, pages 305 – 323, Singapore, New Jersey, London, Hong Kong, 1991. World Scientific Publishing Co.
15. E. Fredkin and T. Toffoli. Conservative logic. *International Journal of Theoretical Physics*, 21:219–253, 1982.
16. W. Freudenberg, M. Ohya, and N. Watanabe. On beam splittings and mathematical construction of quantum logical gate. *Surikaisekikenkyusho Kokyuroku*, 1142:23–35, 2000. Mathematical aspects of quantum information and quantum chaos.
17. W. Freudenberg, M. Ohya, and N. Watanabe. On quantum logical gates on a general Fock space. *QP-PQ: Quantum Probability and White Noise Analysis*, volume 18:252 - 268, World Scientific, N.J., 2005
18. W. Freudenberg. Characterization of states of infinite boson systems II. On the existence of the conditional reduced density matrix. *Commun. Math. Phys.*, 137:461 – 472, 1991.
19. W. Freudenberg. On a class of quantum Markov chains on the Fock space. In L. Accardi, editor, *Quantum Probability and Related Topics*, volume IX, pages 215–237, Singapore, New Jersey, London, Hong Kong, 1994. World Scientific Publishing Co.

20. A. Guichardet. *Symmetric Hilbert Spaces and Related Topics*, volume 231 of *Lecture Notes in Mathematics.* Springer-Verlag, Berlin, Heidelberg, New York, 1972.
21. J.M. Lindsay and H. Maassen. An integral kernel approach to noise. In L. Accardi and W. von Waldenfels, editors, *Quantum Probability and Applications III*, volume 1303 of *Lecture Notes in Mathematics*, pages 192 –208, Heidelberg, 1988. Springer-Verlag.
22. K. Matthes, J. Kerstan, and J. Mecke. *Infinitely Divisible Point Processes.* Wiley, Chichester, 1978.
23. P.A. Meyer. *Quantum Probability for Probabilists*, volume 1538 of *Lecture Notes in Mathematics.* Springer-Verlag, Berlin, Heidelberg, 1993.
24. G.J. Milburn. Quantum optical Fredkin gate. *Physical Review Letters*, 62:2124–2127, 1989.
25. M. Ohya and H. Suyari. An application of lifting theory to optical communication process. *Reports on Mathematical Physics*, 36:403 – 420, 1995.
26. M. Ohya and N. Watanabe. On mathematical treatment of Fredkin-Toffoli-Milburn gate. *Physica D*, 120:206–213, 1998.
27. M. Ohya. Some aspects of quantum information theory and their applications to irreversible processes. *Reports on Mathematical Physics*, 27:19 – 47, 1989.

INFORMATION DIVERGENCE FOR QUANTUM CHANNELS

S.J. HAMMERSLEY AND V.P. BELAVKIN

Mathematics Department, University of Nottingham University Park, Nottingham NG7 2RD, United Kingdom
E-mail: Viacheslav.Belavkin@nottingham.ac.uk

Many definitions of relative information (often called relative entropy) have been proposed in order to characterise the 'distance' as information divergence between states. One such class of relative informations are the quasi-entropies proposed by Petz in [66] (of which the standard Araki-Umegaki form, [53], is a particular example). This paper axiomatises the idea of relative information and then proceeds to build a large class of examples that satisfy these axioms. By using the Radon-Nikodym derivative for completely positive maps as an 'operational density' for channels, along with a theorem by Belavkin and Ohya, [57], for taking the supremum over local couplings these ideas are then lifted beyond the realm of states into that of quantum channels. This method for extending information quantities to channels was pioneered by V.P. Belavkin, with respect to fidelity, [61] (see [60] for an in-depth presentation of the subject).

1. Introduction

The relative information $\mathsf{I}(\varrho;\varsigma)$, in a generalized sense, is a type of contrast (information distinguishability) measure on the set of quantum states described by positive trace class operators $\varrho, \varsigma \in \mathcal{B}_{\mathrm{T}}(\mathfrak{h}_A)$ on an input ("Alice") Hilbert space $\mathfrak{h}_A$. In quantum thermodynamics, in which the states are given by positive operators normalized to the total mass $\operatorname{Tr}\varrho = M$, finite for finite volume, such a function plays the role of negaentropy $\mathsf{I}(\varrho;\varsigma) = -\mathsf{S}(\varrho;\varsigma)$, where $\mathsf{S}(\varrho;\varsigma)$ is one of the variants of relative entropy, for example Araki-Umegaki entropy

$$\mathsf{S}_{\mathrm{a}}(\varrho;\varsigma) := \operatorname{Tr}\varrho(\ln\varsigma - \ln\varrho),$$

which is positive whenever $\varrho \leq \varsigma$. (Usually relative entropy is taken with the opposite sign and identified directly with entropic relative information $\mathsf{I}(\varrho;\varsigma)$, although relative information with respect to the usual trace $\mathsf{I}(\varrho;\tau) = \operatorname{Tr}\varrho\ln\varrho$, corresponding to $\varsigma = I$, is obviously opposite to von

Neumann entropy $\mathsf{S}(\varrho) := \mathsf{S}(\varrho; I)$ of the state ϱ). Having in mind the thermodynamical origin of the entropy and entropic information, we shall not restrict the function $\mathsf{I}(\varrho; \varsigma)$ only to trace one operators ϱ and ς (although the examples are given in this setting it is only for convenience rather than any technical difficulty).

Although Araki-Umegaki entropy is a very important type of relative information it is by no means the only one. Nor is it alone in its importance for example, Sanov's theorem in classical statistics proves that the probability of error in distinguishing n copies of two different states falls exponentially with a rate function given by the relative entropy. However, attempts to generalise this theorem to distinguishing two *quantum* states have only been partially successful. Unlike the classical case the Araki-Umegaki relative entropy only provides an upper-bound for the rate function, it is still an open problem to find a quantum relative information that will fit into a quantum Sanov theorem. Therefore it remains a worthwhile venture to explore other possible variants. To this end the first section of this paper discusses a list of axioms that a relative information functional should satisfy. Following on from this a substantial, but non-exhaustive, class of examples of relative information are characterised. Finally the whole theory is lifted from the domain of states into the more general setting of quantum channels.

2. Axiomatic Properties of Relative Information

The information content of ϱ relative to ς should be positive if $\operatorname{Tr}\varrho \geq \operatorname{Tr}\varsigma$, homogeneous under simultaneous rescaling $\varrho, \varsigma \mapsto \lambda\varrho, \lambda\varsigma$, additive under simultaneous direct decompositions $\varrho = \oplus\varrho^i, \varsigma = \oplus\varsigma^i$, and not increase under channel transformations. Channel transformations are described by linear trace-preserving maps T with completely positive (CP) duals $\Phi = \mathrm{T}^*$, where the dual is with respect to the Hilbert-Schmidt product,

$$\langle\varrho|\Phi(A)\rangle = \operatorname{Tr}\varrho\Phi(A) = \langle\mathrm{T}(\varrho)|A\rangle.$$

Thus the relative information can be defined as any functional $\mathsf{I}(\varrho; \varsigma)$ on the set of positive trace class densities, $\{\varrho, \varsigma\}$, that satisfies the properties:
$\mathsf{I}(\varrho; \sigma) \geq 0$ if $\operatorname{Tr}\varrho \geq \operatorname{Tr}\varsigma$ with $\mathsf{I} = 0 \Leftrightarrow \varrho = \varsigma$.
$\mathsf{I}(\lambda\varrho; \lambda\varsigma) = \lambda\mathsf{I}(\varrho; \varsigma)$ for any $\lambda > 0$.
$\mathsf{I}\left(\oplus\varrho^i, \oplus\varsigma^i\right) = \sum\mathsf{I}\left(\varrho^i, \varsigma^i\right)$ (*direct additivity*).
$\mathsf{I}(\mathrm{T}(\varrho); \mathrm{T}(\varsigma)) \leq \mathsf{I}(\varrho; \varsigma)$ for every channel T (*monotonicity*).

In addition, we say that $\mathsf{I}(\varrho; \varsigma)$ is *semi-finite* if,
$\mathsf{I}(\rho; \sigma) < \infty$ if $\varrho \leq \lambda\varsigma$ for a positive λ.

That it is *differentiable* if,

The function $\mathsf{i}(s,t) = \mathsf{I}(\varrho + sA; \varsigma + tB)$ is differentiable.

That it is *monotone w.r.t.* ς if,
$\mathsf{I}(\varrho;\varsigma_1) \geq \mathsf{I}(\varrho;\varsigma_2)$ for every ϱ and $\varsigma_2 \geq \varsigma_1$.

That it is *relative negaentropy* if,
$\mathsf{I}(\varrho;\varsigma) \leq 0$ for every $\varrho \leq \varsigma$.

Direct additivity for any finite number of orthogonal components $\varrho_i, \varsigma_i \in \mathcal{B}_{\mathsf{T}}(\mathfrak{h}_i)$ of $\varrho = \oplus\varrho_i$ and $\varsigma = \oplus\varsigma_i$ follows from direct 2-additivity

$$\mathsf{I}\left(\varrho^1 \oplus \varrho^2, \varsigma^1 \oplus \varsigma^2\right) = \mathsf{I}\left(\varrho^1, \varsigma^1\right) + \mathsf{I}\left(\varrho^2, \varsigma^2\right).$$

If the monotonicity property is satisfied, then direct additivity implies subadditivity

$$\mathsf{I}\left(\sum \varrho^i, \sum \varsigma^i\right) \leq \sum \mathsf{I}\left(\varrho^i, \varsigma^i\right).$$

If in addition property two is satisfied then this implies joint convexity with respect to both input states:

$$\mathsf{I}\left(\sum \lambda^i \varrho_i, \sum \lambda^i \varsigma_i\right) \leq \sum \lambda^i \mathsf{I}\left(\varrho_i, \varsigma_i\right),$$

for any positive $\varrho_i, \varsigma_i \in \mathcal{B}_{\mathsf{T}}(\mathfrak{h}_A)$ and $\lambda^i \geq 0$, $\sum \lambda^i = 1$.

The natural direct additivity condition is so strong that it determines the structure of $\mathsf{I}(\varrho;\varsigma)$ for commuting ϱ and ς, up to a scalar function $f(\mu,\nu)$ on $\mathbb{R}_+$:

$$[\varrho,\varsigma] = 0 \;\Rightarrow\; \mathsf{I}(\varrho;\varsigma) = \sum_i f\left(\mu^i, \nu^i\right).$$

Here μ_i and ν_i are respectively the eigen-values of ϱ and ς defining the simultaneous Schatten orthogonal decompositions $\varrho = \oplus\mu^i$, $\varsigma = \oplus\nu^i$, and $f(\mu,\nu) = \mathsf{I}(\mu;\nu)$ is the contrast function for the thermodynamical states (masses) $\mu,\nu \geq 0$ in the one-dimensional Hilbert spaces $\mathfrak{h}_i = \mathbb{C}$. Moreover, condition **2.** implies that $f(\mu,\nu) = g(\mu/\nu)$, where $g(r) = f(r\nu,\nu)$ for any $\nu > 0$. As follows from **1.**, the function g should satisfy the strict positivity condition

$$r \geq 1 \Rightarrow g(r) \geq 0 \text{ with } g = 0 \Leftrightarrow r = 1,$$

and condition **4.** is equivalent to its convexity

$$\lambda^i > 0, \sum \lambda^i = 1 \Rightarrow g\left(\sum \lambda^i r_i\right) \leq \sum \lambda^i g(r_i).$$

The other conditions, **5.**–**8.**, mean that g satisfies: $g(r) < \infty$ for all $r \in \mathbb{R}_+$; it is differentiable; $\tilde{g}(s_1) \geq \tilde{g}(s_2)$ if $s_2 \geq s_1$ where $\tilde{g}(s) = sg(1/s)$ and finally that $g(r) \leq 0$ if $r \leq 1$. It is also easy to see that this function $\tilde{g}$ is convex, finite and differentiable if g is; however, it is not strictly positive but strictly negative,

$$s \geq 1 \Rightarrow \tilde{g}\,(s) \leq 0 \text{ with } \tilde{g} = 0 \Leftrightarrow s = 1,$$

$\tilde{g}\,(s) \geq 0$ if $1 \geq s > 0$. Unlike convexity it is not true that monotonicity of g or $\tilde{g}$ implies monotonicity of the other one, as demonstrated by the logarithmic information function. Indeed $g\,(r) = r \ln r$ is convex but not monotone, whereas the corresponding function $\tilde{g}\,(s) = -\ln s$, is both convex and monotone. The logarithmic function $\tilde{g} = -\ln$ is uniquely defined as a convex function with monotonicity condition given by the equation,

$$\tilde{g}\,(s_2) - \tilde{g}\,(s_1) = \ln \frac{s_1}{s_2}.$$

3. Examples of Relative Information for States

To the best of our knowledge it is an open problem as to how to characterise all possible relative informations in the non-commutative case. The best that we can do at present is to characterise a subset of possible forms.

3.1. *Mathematical Preliminaries*

Consider a C*-algebras, and suppose it has a representation such that its weak closure is the von Neumann algebras $\mathcal{A}$. Furthermore assume that this algebra is semi-finite in that there exists a faithful, semi-finite trace λ defined on it. The property of semi-finiteness is defined such that the two sided ideal[a]

$$\mathcal{A}_\lambda := \{A \in \mathcal{A} : \lambda(A) < \infty\},$$

must be weakly dense in the algebra.

Ordinarily normal (ultra-weakly continuous) states on the algebras would be represented using positive trace one operators as densities. However, in this paper the density operators will be defined in the transposed algebra, one of the advantages of this is that the channels can then identified with *positive* density operators instead of the rather cumbersome 'partially-transposed positive' operators. In order to make this clear densities that live in the transposed algebra compared to the algebra on which the state is defined will be called *contravariant* densities. Whereas the more standard

[a]It is a two-sided ideal because if $A \in \mathcal{A}_\lambda$ then for all $X \in \mathcal{A}$ then $\lambda(AX) \leq \lambda(A)||X|| < \infty$, because $\mathcal{A}$ is the algebra of *bounded* operators.

densities that lives in the same algebra as the state will be called *covariant* densities.

Define $\mathcal{A}_*$ as the completion of $\mathcal{A}_\lambda$ with respect to the predual norm $||A||_* = \lambda\left(\sqrt{A^\dagger A}\right)$, conventionally the density operators corresponding to normal states would live in $\mathcal{A}_*$. As mentioned, in order to extend this idea to channels it is more convenient to house the densities in the opposite algebra. To define the transpose algebra the Hilbert space of the representation of $\mathcal{A}$ is assumed to be equipped with an isometric involution operator J, such that $\tilde{A} := JA^\dagger J \equiv \overline{A^\dagger}$. The operator J simply performs complex conjugation (in a particular basis) on the underlying Hilbert space. Now define $\mathcal{A}_\intercal := J\mathcal{A}_*^\dagger J \equiv J\mathcal{A}_* J$, it is this algebra that we take as the predual of $\mathcal{A}$.

In order to link normal states with *contravariant* densities it is necessary to define the functional with respect to the pairing between $\mathcal{A}$ and $\mathcal{A}_\intercal$:

$$\langle \rho, A \rangle_\lambda \equiv \lambda(\bar{\rho}A) = \lambda(\tilde{\rho}A).$$

The subscript λ will usually be omitted unless it becomes ambiguous. For the purposes of clarity we will adopt the notation that the normal states ρ and σ on $\mathcal{A}$ have corresponding contravariant densities given by the same notations, i.e.

$$\rho(A) \equiv \langle \rho, A \rangle,$$
$$\sigma(A) \equiv \langle \sigma, A \rangle.$$

Whereas the covariant densities for these states will be represented by their variants ϱ and ς,

$$\rho(A) \equiv \langle \varrho | A \rangle,$$
$$\sigma(A) \equiv \langle \varsigma | A \rangle,$$

where $\langle \cdot | \cdot \rangle$ is the Hilbert-Schmidt product using the trace λ. Obviously from the definition $\tilde{\rho} = \varrho = \bar{\rho}$ and similarly for σ. Note that these covariant densities are actually contravariant densities for states defined on $\bar{\mathcal{A}}$, if necessary the functionals will then be denoted by ϱ and ς. Using $J^2 = I$ and the cyclicity of the trace, then the semi-finite trace λ defined above can easily be adapted to work on the opposite algebra, i.e. define $\tilde{\lambda}(\tilde{A}) := \lambda(A)$ on $\overline{\mathcal{A}}$. So, if required, the states on the opposite algebra can be written with respect to the pairing,

$$\varrho(\tilde{A}) = \left\langle \varrho, \tilde{A} \right\rangle_{\tilde{\lambda}}.$$

All of these different types of density operator may seem superfluous at the minute, indeed if attention is restricted to states then it is. However, the necessity of this extra complication will hopefully become apparent when we come to deal with channels.

3.2. *A-Type Relative Information*

The most naive extension of of the relative information formula

$$\mathsf{I}(\varrho;\varsigma)=\sum\nu_i g\left(\frac{\mu_i}{\nu_i}\right)$$

from the commuting $\oplus\mu^i$, $\oplus\nu^i$ to non-commuting ϱ, ς is to define the operator-valued function $f(\varrho,\varsigma)=\varsigma g(\varrho/\varsigma)$ on the tensor product $\mathfrak{h}_A\otimes\mathfrak{h}_{\bar{A}}$ with the opposite (complex-conjugate copy) Hilbert space $\mathfrak{h}_{\bar{A}}$ as the function of two commuting operators $\varsigma\otimes I_{\bar{A}}$ and $I_A\otimes\tilde{\varrho}$, where $\tilde{\varrho}=\bar{\varrho}$ is the transposed density operator representing ϱ on $\mathfrak{h}_{\bar{A}}$. Then one can easily evaluate an Araki variant of the generalized relative information as the quantum expectation

$$\mathsf{I}^g_{\mathrm{a}}(\varrho;\varsigma)=\left\langle\sqrt{\varsigma}|g\left(\varsigma^{-1}\otimes\tilde{\varrho}\right)|\sqrt{\varsigma}\right\rangle,$$

where $|\sqrt{\varsigma}\rangle\in\mathfrak{h}_A\otimes\mathfrak{h}_{\bar{A}}$ is the standard entangling state vector for two anti-isomorphic systems A and $\bar{A}$, corresponding to the Hilbert-Schmidt operator $\sqrt{\varsigma}$. Using Hilbert-Schmidt space instead of $\mathfrak{h}_A\otimes\mathfrak{h}_{\bar{A}}$ one can write this in the form

$$\mathsf{I}^g_{\mathrm{a}}(\varrho;\varsigma):=\lambda\left(\sqrt{\varsigma}g(\mathbf{L}_\varsigma^{-1}\mathbf{L}'_\varrho)\sqrt{\varsigma}\right),$$

where $\mathbf{L}_\varsigma^{-1}\chi=\varsigma^{-1}\chi$ is the operator of left multiplication by ς^{-1} isomorphic to the operator $\varsigma^{-1}\otimes I_{\bar{A}}$, and $\mathbf{L}'_\varrho\chi=\chi\varrho$ is the right multiplication by ϱ isomorphic to the operator $I_A\otimes\tilde{\varrho}$. It is easy to prove that this g-information satisfies all of the necessary properties **1.** – **4.** for any strictly positive function g at $r\geq 1$ with $g(1)=0$ which is operator-convex[b]; i.e. satisfying

$$g(\lambda X+(1-\lambda)Y)\geq\lambda g(X)+(1-\lambda)g(Y),$$

for any positive operators X, Y and any $\lambda\in\mathbb{R}$. It also satisfies the other properties **5.**– **8.** if the real function g satisfies the appropriate conditions on $\mathbb{R}_+$. This class of relative informations includes the Araki-Umegaki relative negaentropy

$$\mathsf{I}_{\mathrm{a}}(\varrho;\varsigma):=\mathrm{Tr}\,\varrho(\ln\varrho-\ln\varsigma)$$

[b] An operator convex function is necessarily a convex function on $\mathbb{R}_+$ but the converse is not always true.

as a special case given by $g(r) = r \ln r$, where $0 \ln 0 := 0$. Indeed, $\mathsf{I}^g_{\mathrm{a}}(\varrho;\varsigma) = \mathsf{I}_{\mathrm{a}}(\varrho;\varsigma)$ since $g\left(\varsigma^{-1} \otimes \tilde{\varrho}\right) = \varsigma^{-1} \otimes \tilde{\varrho} \ln \tilde{\varrho} - \varsigma^{-1} \ln \varsigma \otimes \tilde{\varrho}$,

$$g\left(\varsigma^{-1} \otimes \tilde{\varrho}\right) |\sqrt{\varsigma}\rangle = |\sqrt{\varsigma^{-1}}\,(\ln \varrho)\, \varrho\rangle - |\sqrt{\varsigma^{-1}}\,(\ln \varsigma)\, \varrho\rangle,$$

and $\langle \sqrt{\varsigma} | \chi \rangle = \mathrm{Tr} \sqrt{\varsigma} \chi$.

This form of generalized relative information was suggested by Petz [66]. In Petz's paper the generalized informations considered as quasi-distances were called quasi-entropies, and our function g corresponds to Petz's function $f = \tilde{g}$ via $g(s) = sf(s^{-1})$. It can be generalized to arbitrary von Neumann algebras in the spirit of Araki, see [62].

3.3. *B and γ Type Relative Information*

There are many other candidates for logarithmic and generalized relative information. Another, more natural class of such informations is based on the Radon-Nikodym derivative ρ_ς of the state ϱ with respect to the state ς on the opposite algebra $\overline{\mathcal{A}}$. It is given as

$$\rho_\varsigma = \sqrt{\sigma^{-1}} \rho \sqrt{\sigma^{-1}}$$

by the transposed operators $\rho = \tilde{\varrho}$ and $\sigma = \tilde{\varsigma}$ representing ϱ and ς on the opposite Hilbert space $\mathfrak{h}_{\bar{A}}$. It defines the expectations $\lambda(\varrho A)$ in the form of the standard ς-pairing,

$$\left\langle \rho_\varsigma, A \right\rangle_\varsigma := \lambda\left(\tilde{\rho}_\varsigma \sqrt{\varsigma} A \sqrt{\varsigma}\right) = \lambda\left(\varrho_\sigma \sqrt{\varsigma} A \sqrt{\varsigma}\right) \equiv \left\langle \varrho_\sigma | A \right\rangle_\sigma,$$

given in terms of the ς-Hilbert-Schmidt product $\langle C | A \rangle_\varsigma = \mathrm{Tr}\left[C^* \sqrt{\varsigma} A \sqrt{\varsigma}\right]$ with $C^* = \tilde{\rho}$. Defining $g\left(\rho_\varsigma\right)$ as a function of the self-adjoint operator ρ_ς one can obtain a simpler B-type relative information written as the trace

$$\mathsf{I}^g_{\mathrm{b}}\left(\varrho;\varsigma\right) = \left\langle \hat{1}_A, g\left(\varrho_\sigma\right)\right\rangle_\varsigma = \lambda\left[\sqrt{\varsigma} g(\varrho_\sigma) \sqrt{\varsigma}\right].$$

Here $\varrho_\sigma = \sqrt{\varsigma^{-1}} \varrho \sqrt{\varsigma^{-1}} = \widetilde{\rho_\varsigma}$ is the Radon-Nikodym derivative of ϱ with respect to the functional σ. In the case of $g\left(r\right) = r \ln r$ this gives the B-type logarithmic formula

$$\mathsf{I}_{\mathrm{b}}\left(\varrho;\varsigma\right) = \lambda\left[\sqrt{\varsigma} \varrho_\sigma \ln(\varrho_\sigma) \sqrt{\varsigma}\right] = \lambda\left[\sqrt{\varrho}\left(\ln \varsigma_\rho^{-1}\right)\sqrt{\varrho}\right],$$

which was defined on the general von Neumann algebras in [73] (see also Petz's book [54c]). Here $\varsigma_\rho^{-1} = \sqrt{\varrho} \varsigma^{-1} \sqrt{\varrho}$ and we have used the identity

$$\sqrt{\varsigma} g\left(\varrho_\sigma\right) \sqrt{\varsigma} = \sqrt{\varrho} \tilde{g}\left(\varsigma_\rho\right) \sqrt{\varrho},$$

[c] Ohya and Petz were able to show that, at least in finite dimensions and with faithful states, B-type logarithmic relative information is greater than A-type logarithmic information, see p126 of their book.

in terms of $\tilde{g}(s) = sg(1/s)$ for the invertible ς_ρ. It is easily seen that $\mathsf{I}_{\mathrm{b}}^g$ satisfies all four necessary properties if g is an operator-convex function, strictly positive on $r \geq 1$ with $g(1) = 0$, and I_{b} satisfies the further four properties if g obeys the corresponding properties on $\mathbb{R}_+$; and is definable as long as care is taken when using singular operators.

Taking into account that B-type g-information can be written in the Hilbert-Schmidt form as

$$\mathsf{I}_{\mathrm{b}}^g(\varrho;\varsigma) = \lambda\left[\sqrt{\varsigma}g(\mathbf{L}'_{\varrho_\sigma})\sqrt{\varsigma}\right],$$

one can consider also an interpolation between these A and B type informations. Rather than using the Radon-Nikodym derivative between ρ and σ we define Radon-Nikodym derivatives with respect to an arbitrary, not necessarily normalised, density $\gamma \in \mathcal{A}_\mathsf{T}$ (i.e. a positive linear functional on $\mathcal{A}$),

$$\varrho_\gamma = \tilde{\gamma}^{-1/2}\varrho\tilde{\gamma}^{-1/2},$$
$$\varsigma_\gamma = \tilde{\gamma}^{-1/2}\varsigma\tilde{\gamma}^{-1/2}.$$

Then define the γ-type relative information as,

$$\mathsf{I}_\gamma^g(\varrho;\varsigma) := \lambda\left(\sqrt{\varsigma}g(\mathbf{L}_{\varsigma_\gamma}^{-1}\mathbf{L}'_{\varrho_\gamma})\sqrt{\varsigma}\right).$$

(Is there a nicer expression for this in terms of some sort of γ product?) Note that this type of relative information includes the other two as special cases: if $\gamma = I$ it gives A-type and if $\gamma = \sigma$ it gives B-type. γ-type relative g-information clearly satisfies all the necessary properties **1.** – **4.** for an operator-convex function $g(r)$ that is strictly positive at $r \geq 1$ with $g(1) = 0$. It also satisfies the other properties **5.** – **8.** for any appropriate function g.

Despite the apparent generality of the γ-type relative g-information this is in no way exhaustive of all possibilities (unless the algebra is commutative). In fact there are some very simple relative informations that aren't expressible in this form. For example the trace distance $\mathsf{I}_{\mathrm{tr}}(\varrho,\varsigma) = \lambda(|\varrho - \varsigma|)$ and the fidelity distance $\mathsf{I}_{\mathrm{fid}}(\varrho,\varsigma) = 1 - \lambda(|\sqrt{\varrho}\sqrt{\varsigma}|)$ aren't of γ-type. In fact it is hard to see how, even these simple expressions, can be incorporated into a general characterisation.

4. Relative Information for Channels

The notion of relative information can be extended to channels, $\Phi_\mathsf{T}, \Psi_\mathsf{T} : \mathcal{A}_\mathsf{T} \to \mathcal{B}_\mathsf{T}$, that take input states from the Banach space of trace-class operators $\mathcal{A}_{\mathsf{T}}$ into an output algebra $\mathcal{B}_{\mathsf{T}}$ (the algebras $\mathcal{B}$ and $\mathcal{B}_{\mathsf{T}}$ are defined in an analogous way to $\mathcal{A}$ and $\mathcal{A}_{\mathsf{T}}$). First it can be done in the following obvious but naive way: for the channels Φ_{T} and Ψ_{T} and a given input state ρ compute the relative information between the output states, $\Phi_{\mathsf{T}}(\rho)$ and $\Psi_{\mathsf{T}}(\rho)$; then take the supremum over all input states,

$$\mathsf{I}(\Phi_*;\Psi_*) := \sup_{\rho} \{\mathsf{I}(\Phi_*(\varrho);\Psi_*(\varrho)) : \varrho \geq 0, \mathrm{Tr}\varrho = 1\}.$$

However, as unintentionally pointed out by Einstein, Podolsky and Rosen in their famous 'paradox' quantum channels can transmit information in an unorthodox way, i.e. via entanglement. Consider an input algebra, $\mathcal{A}$, for a channel, then the standard way of passing information along the channel is to let each of the input states correspond to a piece of information, i, mathematically corresponding to parameterising the density matrices, ρ_i. This is equivalent to *coupling* the input algebra to a classical commutative algebra $\mathcal{C}(I)$ - continuous functions on the indexing set I, i.e. instead of transmitting states on $\mathcal{A}$ the channel is viewed as transmitting states on $\mathcal{A} \otimes \mathcal{C}(I)$. So the naive version of the relative information can be written,

$$\begin{aligned}&\sup_{\varrho}\sup_{\{\varrho_i\}}\left\{\mathsf{I}(\oplus_i\Phi_*(\varrho^i);\oplus_i\Psi_*(\varrho^i)) : \sum\varrho^i = \varrho\right\}\\ &= \sup_{\rho}\sup_{\{\rho_i\}}\left\{\sum_i \mathsf{I}(\Phi_*(\varrho_i);\Psi_*(\varrho_i)) : \sum\varrho^i = \varrho\right\},\end{aligned}$$

where ρ_i are the states normalized to the probabilities $\lambda^i = \mathrm{Tr}\rho^i$ on I. From the convexity property this appears to be greater than the supremum of $\mathsf{I}^g(\Phi_*(\varrho);\Psi_*(\varrho))$ over all states ϱ, but in fact it coincides with $\mathsf{I}(\Phi_*;\Psi_*)$ since

$$\sup_{\{\lambda_i \geq 0\}}\left\{\sum_i \lambda^i \mathsf{I}(\Phi_*(\varrho_i);\Psi_*(\varrho_i)) : \sum\lambda^i = 1\right\} = \sup_i \mathsf{I}(\Phi_*(\rho_i);\Psi_*(\rho_i)).$$

Where this follows because the supremum is achieved on an extremal probability distribution λ^i, i.e. on a delta distribution. However, in quantum mechanics many other *couplings* are possible, in general the input algebra can be coupled to any other system, $\mathcal{A} \otimes \mathcal{C}$, where $\mathcal{C}$ can be any von Neumann algebra. The difference between classical and quantum couplings is that *q-coupling* allows for exotic encodings of information such as sending one half of an entangled pair of particles through the channel. Therefore a more operational definition of the relative information is to take the supremum over *all* possible local couplings as well as over the input states. In this way the relative information will reflect the true information carrying capability of the channels, as it was shown in [57] for the quantum capacity of a single channel. Also this definition of the relative information can be viewed as more comparable with the standard interpretation of the state relative information. Consider two different pure states, then the relative information between them is always infinite no matter how similar they are. This is because the relative information is a measure of how distinguishable the states are *in principle.* That is, not how easy the states are to distinguish in a normal experiment but how distinguishable they are if the experiment ran perfectly. Therefore if the two channels can be distinguished using a q-coupling with an external algebra then the relative information should reflect this.

4.1. *Mathematical Preliminaries*

There are two areas that needed to be covered before the channel relative g-informations can be formulated. The first is to extend the notion of density operators to channels, it is here that we will derive advantage from our seemingly unnecessarily complicated exposition for states. The second issue is to formulate what is meant by local couplings and then to prove the existence of an optimal coupling in the sense of maximising monotone functionals.

4.1.1. *Channel Densities*

A quantum channel is understood as a linear map, mapping the predual space $\mathcal{A}_{\mathsf{T}}$ of an input von Neumann algebra $\mathcal{A}$ into $\mathcal{B}_{\mathsf{T}}$, the predual space of a von Neumann algebra $\mathcal{B}$, such that its dual (w.r.t. the pairing defined previously), $\Phi : \mathcal{B} \to \mathcal{A}$ is completely positive (CP) and identity preserving (unital). Where complete positivity is defined as:

$$\sum_{ik} (h_i|\Phi(B_i B_k^\dagger)|h_k) \geq 0,$$

B_i is a sequence of operators in $\mathcal{B}$ and $|h_i)$ a sequence of vectors in the representing Hilbert space for the algebra $\mathcal{A}$. It was shown in [6] that subject to a technical constraint 'any' CP map can be represented by,

$$\Phi(B) = (\mathrm{id} \otimes \tilde{\mu}) \left[\Phi_\mu \left(I_{\mathfrak{f}} \otimes \tilde{B}\right)\right] \equiv \langle \Phi_\mu, B \rangle_{\tilde{\mu}} \,,$$

where $\Phi_\mu \in (\mathcal{A} \otimes \mathcal{B}_\mathsf{T})_+$ is called the *density* of the channel[d]. The technical constraint refers to a limitation as to which CP maps can be represented by the particular reference trace μ. For the purposes of this paper a strong version of this constraint will be used which also forces the density to be a *bounded* operator. This constraint requires that the CP map is *completely majorised* by the map $T_\mu : B \mapsto \mu(B) I_\mathfrak{f}$, this means that there must exist a $\chi < \infty$ such that $\chi T_\mu - \Phi$ is completely positive[e]. If this is satisfied then not only does the density Φ_μ exist but it is bounded, $||\Phi_\mu|| \leq \chi$.

The dual form, or Schrödinger representation is given by, $\Phi_\mathsf{T} : \mathcal{A}_\mathsf{T} \to \mathcal{B}_\mathsf{T}$, defined as the solution to,

$$\begin{aligned}
\langle \Phi_\mathsf{T}(\rho), B \rangle_\mu &:= \langle \rho, \Phi(B) \rangle_\lambda , \\
&= \lambda \left[\left((\mathrm{id} \otimes \tilde{\mu})[(I_\mathfrak{f} \otimes \tilde{B})\Phi_\mu] \right) \varrho \right] , \\
&= (\lambda \otimes \tilde{\mu})(I_\mathfrak{f} \otimes \tilde{B})\Phi_\mu(\varrho \otimes I_\mathfrak{g}), \\
&= \langle (\lambda \otimes \mathrm{id})[\Phi_\mu(\varrho \otimes I_\mathfrak{g})], B \rangle_{\tilde{\mu}} , \\
&= \langle (\lambda \otimes \mathrm{id})[\Phi_\mu(\varrho \otimes I_\mathfrak{g})], B \rangle_\mu .
\end{aligned}$$

Interchanging between the traces and their transposed forms was used to avoid having to take the transpose of the density, as this would require transposition on one algebra only. Obviously this isn't a necessity but it makes it easier to keep track of the argument. The final line follows trivially in this case, but be careful, it is not true in general, for example $\langle \Phi_\mu, B \rangle_{\tilde{\mu}} \neq \langle \Phi_\mu, B \rangle_\mu$ because the result of the pairing is not transpose invariant. It is now easy to read off the form for the channel in the Schrödinger picture,

$$\Phi_\mathsf{T}(\rho) = (\lambda \otimes \mathrm{id})[(\varrho \otimes I_\mathfrak{g})\Phi_\mu] \equiv \langle \rho, \Phi_\mu \rangle_\lambda .$$

---

[d]If, for illustrative purposes, the density is defined in a sort of covariant sense, i.e. $\phi \in \mathcal{A} \otimes \mathcal{B}$ rather than $\mathcal{A} \otimes \tilde{\mathcal{B}}$, then $\Phi(B) \equiv \mu((I \otimes B)\phi)$. But this implies that

$$\sum (h_i | \Phi(B_i B_k^\dagger) | h_k) = \sum \mu \left(\left((h_i| \otimes B_k^\dagger \right) \phi (|h_k) \otimes B_i) \right)$$

and therefore Φ is CP iff the density is 'partially-transposed positive'. This is why it is natural, although not initially obvious, to consider the density as an element of the algebra $\mathcal{A} \otimes \tilde{\mathcal{B}}$.

[e]The weaker version of the constraint requires that Φ is completely absolutely continuous (CAC) with respect to the map T_μ, but the densities may then be unbounded operators, see [6] for details.

4.1.2. *Local Couplings*

Now consider a third algebra, represented by the von Neumann algebra $\mathcal{C}$, that can be *coupled* to the input algebra of the channel. The term 'coupling' means that the two systems are allowed to interact with each other, this interaction could leave the two systems uncorrelated or fully entangled[f].

Definition 4.1. A *q-coupling* is defined as a CP map $\pi : \mathcal{C} \to \mathcal{A}_{\mathsf{T}}$, such that $\pi(I_{\mathcal{C}}) = \rho$ where ρ is a density operator in $\mathcal{A}_{\mathsf{T}}$.

The coupling ϖ can be used to define the linear functional $\varpi : \mathcal{A}\otimes\mathcal{C} \to \mathbb{C}$ via

$$\varpi(A \otimes C) := \langle \pi(C), A \rangle_{\lambda}.$$

It was proven by Belavkin and Ohya ([57] theorem 1) that ϖ is a positive functional iff π is CP. So if π is CP then ϖ is a (compound) state on the tensor product algebra, moreover all states, subject to $\varpi(A \otimes I) = \langle A \rangle_{\rho}$, can be written in this form. However, it is also possible to use π to define a state on $\mathcal{B} \otimes \mathcal{C}$ by utilising the channel that connects $\mathcal{A}$ to $\mathcal{B}$. Define the functional,

$$\varpi_{\Phi}(B \otimes C) := \langle \Phi_{\mathsf{T}}(\pi(C)), B \rangle_{\mu}.$$

As the composition of two CP maps is CP then this functional is also positive, therefore ϖ_{Φ} is a state on the tensor product algebra of the output and the external system.

It can easily be seen that for any CP map, π, that satisfies $\pi(I_{\mathcal{C}}) = \rho$ then there exists a UCP map $\Pi : \mathcal{C} \to \tilde{\mathcal{A}}$, such that

$$\pi(C) = \sqrt{\rho}\Pi(C)\sqrt{\rho}.$$

With this form in mind we can define the *standard coupling* $\varpi^0 : \tilde{\mathcal{A}} \to \mathcal{A}_{\mathsf{T}}$ as the map

$$\pi^0(\tilde{A}) := \sqrt{\rho}\tilde{A}\sqrt{\rho}.$$

This allows π to be written as $\pi = \pi^0 \circ \Pi$, where Π is the channel defined above (and $\circ$ denotes composition). Now take any monotone functional, $\mathcal{F}$, of π. Where monotonicity is defined as

$$\mathcal{F}(\pi) \geq \mathcal{F}(\pi \circ \Lambda),$$

---

[f] In the literature the term 'entanglement' is often used instead of 'couplings', however in my opinion entanglements suggests that the joint state is entangled which may not be the case.

for any channel Λ from a fourth algebra $\mathcal{D}$, $\Lambda : \mathcal{D} \to \mathcal{C}$. However by choosing $\Lambda = \Pi$ (and choosing the algebras $\mathcal{D}$ and $\mathcal{C}$ accordingly) this shows that,

$$\mathcal{F}(\pi^0) \geq \mathcal{F}(\pi).$$

Which implies that

$$\sup_{\pi} \mathcal{F}(\pi) = \mathcal{F}(\pi^0).$$

Therefore, the supremum over all encodings is actually achieved on the standard encoding π^0 which couples $\mathcal{A}$ to its transpose, rather than on some fancy encoding coupling $\mathcal{A}$ with some exotic algebra $\mathcal{C}$. All of this work on encodings is contained in the Belavkin and Ohya paper, it is presented here as a summary so that notation can be established and continuity maintained. Note that the standard coupling is non-classical in the sense that the supremum over classical couplings (encodings coupling $\mathcal{A}$ with a classical commutative algebra) is strictly less than that achieved by the standard coupling, except when $\mathcal{A}$ is itself classical, in which case the suprema obviously coincide.

Using the standard coupling representation the output state, ϖ_{Φ}, for a general coupling is defined as,

$$\begin{aligned}
\left\langle \Phi_{\mathsf{T}}(\pi^0 \circ \Pi(C)), B \right\rangle_{\mu} &= (\mathrm{id} \otimes \tilde{\mu})(I \otimes \tilde{B})(\lambda \otimes \mathrm{id}) \\
&\quad [\Phi_{\mu}\{(\sqrt{\varrho} \otimes I)(\widetilde{\Pi(C)} \otimes I)(\sqrt{\varrho} \otimes I)\}], \\
&= (\lambda \otimes \tilde{\mu})(\widetilde{\Pi(C)} \otimes \tilde{B})(\sqrt{\varrho} \otimes I)\Phi_{\mu}(\sqrt{\varrho} \otimes I), \\
&= (\tilde{\lambda} \otimes \mu)(\Pi(C) \otimes B)[(\sqrt{\rho} \otimes I)\tilde{\Phi}_{\mu}(\sqrt{\rho} \otimes I)], \\
&\equiv \langle (\sqrt{\varrho} \otimes I)\Phi_{\mu}(\sqrt{\varrho} \otimes I), \Pi(C) \otimes B \rangle_{\tilde{\lambda} \otimes \mu}, \\
&\equiv \langle \Pi(\pi)_{\mathsf{T}}(\sqrt{\varrho} \otimes I)\Phi_{\mu}(\sqrt{\varrho} \otimes I), C \otimes B \rangle_{\tilde{\lambda} \otimes \mu}.
\end{aligned}$$

The last line is used here only in a symbolic way so that the density can be written explicitly as

$$\Phi_{\mu}^{\rho}(\pi) := \Pi(\pi)_{\mathsf{T}}[(\sqrt{\varrho} \otimes I)\Phi_{\mu}(\sqrt{\varrho} \otimes I)].$$

In this paper only the standard coupling is needed, in which case $C = \tilde{A}$ and $\Pi = \mathrm{id}$, therefore only the penultimate line is required to generate an expression for the density. However, the final line can be justified, if necessary, assuming $\mathcal{C}$ is semi-finite with a corresponding reference trace which completely majorizes Π_{T}, then Π_{T} is calculated in a similar way to Φ_{T}. In the simple case using the standard coupling the output state is a compound state on $\mathcal{C} \otimes \mathcal{B} = \tilde{\mathcal{A}} \otimes \mathcal{B}$, with density operator $(\sqrt{\varrho} \otimes I)\Phi_{\mu}(\sqrt{\varrho} \otimes I) =: \varpi_{\Phi}^{\rho} \in \mathcal{A} \otimes \tilde{\mathcal{B}}$.

4.2. γ-*Type Channel Relative Information*

As mentioned at the beginning the definition of channel relative g-information proposed in this paper is the supremum of the relative information of the outputs over all possible input states *and* any local couplings. However, we have just proven that the supremum over local couplings is achieved on the standard coupling providing the functional is monotone (which it is by definition). Therefore, expressions for the channel relative g-information can automatically be written down by substituting the compound densities ϖ_{Φ}^{ρ} and ϖ_{Ψ}^{ρ} for the state densities ρ and σ. For γ-type relative informations the only difference is that the Radon-Nikodym derivative is now between the compound densities and a, not necessarily normalised, (???do we need to stick to a product density???) product density on the tensor product, $\gamma \in \tilde{\mathcal{A}} \otimes \mathcal{B}$. Adopting a slight change of notation define,

$$\varpi_{\gamma}^{\rho}(\Phi) := \tilde{\gamma}^{-1/2}(\sqrt{\varrho} \otimes I)\Phi_{\mu}(\sqrt{\varrho} \otimes I)\tilde{\gamma}^{-1/2}.$$

The γ-type relative g-information can now be written,

$$\mathsf{I}_{\gamma}^{g}(\Phi;\Psi) := \sup_{\rho} \lambda \otimes \tilde{\mu}\left(\sqrt{\varpi_{\gamma}^{\rho}(\Psi)}\, g\left(\mathbf{L}_{\varpi_{\gamma}^{\rho}(\Psi)}^{-1} \mathbf{R}_{\varpi_{\gamma}^{\rho}(\Phi)}\right) \sqrt{\varpi_{\gamma}^{\rho}(\Psi)}\right).$$

As before $\gamma = I \otimes I$ gives A-type and $\gamma = \sigma \otimes I$ gives B-type relative information. However, it is also interesting to note that $\gamma = \rho \otimes I$ gives the naive channel relative information.

At present it remains unknown how to evaluate the remaining supremum over input states in all but fairly trivial cases, for example unitary channels. However, if the input algebra is assumed to be commutative then the problem greatly simplifies because by commuting the input density through the expression it can easily be seen to depend convexly on it. Therefore, the supremum is satisfied on an extremal probability distributions and the problem reduces to a simple function maximisation. For general channels between non-commutative algebras convexity with respect to the input state is not guaranteed, consequently formulating non-Bayesian relative informations would severely restrict the number of possible forms. However, for some functions g convexity can be maintained, for example so called A-type relative quadratic-informations, i.e. with $g = (x-1)^2$. From the theory of operator convex functions it can be shown that any operator convex g s.t. $g(1) = 0$ can be expressed in integral form as,

$$g(x) = a(x-1) + b(x-1)^2 + c\frac{(x-1)^2}{x} + \int_0^{\infty} \frac{(x-1)^2}{(x+s)} d\nu(s),$$

where $b, c \geq 0$ and ν is a positive measure with finite mass, see [63]. Lesniewski and Ruskai also show that for symmetric A-type informations there exists a maximal and a minimal relative information (called relative g-entropy in their paper),

$$\mathsf{I}^{\text{sym}}_{(w-1)^2}(\rho;\sigma) \geq \mathsf{I}^{\text{sym}}_{g}(\rho;\sigma) \geq \mathsf{I}^{\text{sym}}_{\frac{(w-1)^2}{w+1}}(\rho;\sigma).$$

The more ubiquitous Umegaki relative information also has a fairly simple A-type form,

$$\mathsf{I}_{\text{ume}}(\Phi;\Psi) = \sup_{\rho} \tau\left(\rho\Phi_\mu\left(\ln\rho\Phi_\mu - \ln\rho\Psi_\mu\right)\right).$$

Unfortunately the supremum over input states appears to be difficult to calculate in this case.

References

1. Belavkin, V.P., Generalised uncertainty relations and efficient measurement in quantum systems, Theoretical and Mathematical Physics (USA), 26, 213–222, 1976.
2. Petz, D., Covariance and Fisher information in quantum mechanics, Journal of Physics, A. Mathematical and General, 35, 929–939, 2002.
3. Petz, D., Monotone metrics on matrix spaces, Linear Algebra and its Applications, 244, 81–96, 1996.
4. Helstrom, C.W., Minimum mean square estimation in quantum statistics, Physics letters A, 25, 101-102, 1967.
5. Kadison, R.V., A generalised Schwarz inequality and algebraic invariants for operator algebras, Annals of Mathematics, 56, No.3, 494–503, 1952.
6. V.P. Belavkin and P. Staszewski, A Radon-Nikodym theorem for completely positive maps, Reports on mathematical physics, 24, 1986.
7. Lance, E.L., Hilbert C* modules: A toolkit for operator algebraists, Cambridge University Press, 1995.
8. Stinespring, W.F., Positive functions on C*-algebras, Proceedings of the American Mathematical Society, 6, No.2, 211–216, 1955.
9. von-Neuman, J., Mathematical Foundations of Quantum Mechanics, Princeton University Press, 1955.
10. Matsumoto, S., A New Approach to the Cramer-Rao-type Bound of the Pure-State Model, J. Phys A: Math. Gen., 35, 3111-3123, 2002.
11. R.D. Gill and S. Massar, State Estimation for Large Ensembles, Physical Review A, 61, 2000.
12. Completely Bounded Maps and Dilations, Longman Wiley, New York, 1986.
13. S.L. Braunstein and C.M. Caves, Statistical Distance and the Geometry of Quantum States, Physical Review Letters, 72, No.22, 3439-3443, 1994.
14. O.E. Barndorff-Nielsen and R.D. Gill, Fisher Information in Quantum Statistic, quant-ph/9808009, 2000.

15. Chirbella, D'Ariano, Perinotti and Sacchi, Covariant Quantum Measurment which Maximise the Liklihood, quant-ph/0403083, 2004.
16. G. Vidal, J.I. Latorre, P. Pascual and R. Tarrach, Optimal Minimal Measurement of MIxed States, Physical Review A, 60, No.1, 126-135, 1998.
17. J. Rehacek and Z. Hradil, Invariant Information and Quantum State Estimation, Physical Review Letters, 88, No.13, 2002.
18. P.B. Slater, Incresed Efficiency of Quantum State Estimation Using Non-Seperable Measurments, quant-ph/0006009, 2004.
19. K. Matsumoto, A New Approach to the Cramer-Rao Type Bound of the Pure State Model, quant-ph/9711008, 1997.
20. K. Banaszek, G.M. D'Ariano, M.G.A. Paris and M.F. Sacchi, Maximum-Likelihood Estimation of the Density Matrix, quant-ph/9909052, 2003.
21. M. Keyl and R.F. Werner, Estimating the Spectrum of a Density Operator, Physical Review A, 64, 2001.
22. M. Hayashi, Asymptotic Quantum Estimation Theory for the Displaced Thermal States Family, quant-ph/9809002, 1998.
23. D.G. Fischer and M. Freyberger, Estimating Mixed Quantum States, quant-ph/0005090, 2000.
24. K. Usami, Y. Nambu, Y. Tsuda, K. Matsumoto and K. Nakamura, A New Strategy of Quantum State Estimation for Achieving the Cramer-Rao Bound, quant-ph/0209074, 2002.
25. O.E. Barndorff-Nielsen and R.D. Gill, Fisher Information in Quantum Statistics, Journal of Physics A, 33, 4481-4490, 2000.
26. Z. Hradil and J. Summhammer, Quantum Theory of Incompatible Observations, quant-ph/9911067, 1999.
27. O.E. Barndorff-Nielsen and R.D. Gill, An Example of Non-Attainability of Expected Fisher Information, http://www.math.uu.nl (follow links to R. Gill preprints), 1999.
28. M. Freyberger, P. Bardroff, C. Leichtle, G. Schrade and W. Schleich, The Art of Measuring Quantum States, Physics World, 10, No.11, 41-45, 1997.
29. R.D. Gill, Asymptotics in Quantum Statistics, http://www.math.uu.nl (follow links to R. Gill preprints), ?1999?.
30. O.E. Barndorff-Nielsen, R.D. Gill and P.E. Jupp, On Quantum Statisitical Inference, http://www.math.uu.nl (follow links to R. Gill preprints), 2001.
31. Z. Hradil and J. Rehacek, Uncertainty Relations on the Slit and Fisher Information, quant-ph/0309184, 2004.
32. M. Reginatto, Derivation of the Equations of Non-relativistic Qauntum Mechanics Using the Principle of Minimum Fisher Information, Physical Review A, 58, No.3, 1775-1778, 1998.
33. J. Summhammer, Structure of Probabilistic Information and Quantum Laws, quant-ph/0102099, 2001.
34. B.R. Frieden adn B.H. Soffer, Lagrangians of Physics and the Game of Fisher-Information Transfer, Physical Review E, 52, No.3, 2274-2286, 1995.
35. H. Cramer, Mathematical Methods of Statistics, Princeton University, Princeton, NJ, 1946.
36. E.A. Morozova and N.N. Chentsov, Markov Invariant Geometry on State

Manifolds (in Russian), Itogi Nauki i Tekhniki, 36, 69-102, 1990.
37. B.R. Frieden, Physics from Fisher Information, Cambridge University Press, Cambridge, 1999.
38. H.P. Yuen and M. Lax, Trans. IEEE Info. Th., 19, No.740, 1973.
39. V.P. Belavkin and B.A. Grishnin, Probl. Pereachi Inf., 4, No.44, 1973.
40. D.T. Smithey, M. Beck, M.G. Raymer and A. Faridani, Physical Review Letters, 70, No.1244, 1993.
41. T.J. Dunn, I.A. Walmsey and S. Mukamel, Physical Review Letters, 74, No.884, 1995.
42. D. Leibfried et al, Physical Review Letters, 77, No.4281, 1996.
43. C. Kurtsiefer, T. Pfau and J. Mlynek, Nature, 386, No.150, 1997.
44. S. Masser and S. Popescu, Optimal Extraction of Information from Finite Quantum Ensembles, Physical Review Letters, 74, No.8, 1259-1263, 1993.
45. D.F.V. James, P.G. Kwait, W.J. Munro and A.G. White, Measurement of Qubits, Physical Review A, 64, No.052312, 2001.
46. R.Tarrach and G. Vidal, Universality of Optimal Measurement, Physical Review A, 60, No.5, 1999.
47. D.G. Fischer, H. Kienle and Matthias Freyberger, Quantum-state Estimation by Self-learning Measurements, Physcial Review A, 61, No.032306, 2000.
48. J.I. Latorre, P. Pascual and R. Tarrach, Minimal Optimal Generalised Quantum Measurements, Physical Review Letters, 81, No.7, 1351-1354, 1998.
49. R. Derka, V.Buzek and A.K. Ekert, Universal Algorithm for Optimal Estimation of Quantum States from Finite Ensembles via Realizable Generalised Measurement, Physical Review Letters, 80, No.8, 1571-1575, 1998.
50. A.S. Holevo, Probabilistic and Statistical Aspects of Quantum Theory, North-Holland, Amsterdam, 1982.
51. Z. Hradril, Quantum-state Estimation, Physical Review A, 55, No.3, R1561, 1996.
52. J.D. Malley and J. Hornstein, Quantum Statistical Inference, Statistical Science, 8, No.4, 433-457, 1993.
53. H. Umegaki, Conditional Expectations in an Operator Algebra IV, Kodai Mathematical Seminar Reports, 14, 59-85, 1962 (This is where Umegaki proposed the difference in logarithms form of the relative entropy).
54. M. Ohya and D. Petz, Quantum Entropy and Its Use, Springer-Verlag, Berlin, 1993.
55. O. Klein, Z. Phys, 72, 767-775, 1931 (First proof of positivity for relative entropy on q. states).
56. M.A. Nielsen and I.L. Chuang, Quantum Computation and Qauntum Information, Cambridge university press, Cambridge, 2000.
57. V.P. Belavkin and M. Ohya, Quantum Entropy and Information in Discrete Entangled States, Infinite Dimensional Analysis, Quantum Probability and Related Topics, 4, 2, 137-160, 2000.
58. G. Linblad, Completely Positive Maps and Entropy Inequalities, Communications in Mathematical Physics, 40, 147-151, 1975 (First noting of monotonicity for (at Umegaki) relative entropy).
59. J.R. Klauder and E.C.G. Sudarshan, Fundamentals of Quantum Optics,

W.A. Benjamin, inc. New York, 1968 (Contains stuff on Coherent Vectors).
60. V.P. Belavkin, G.M. D'Ariano and M. Raginsky, Operational Distance and Fidelity for Quantum Channels, quant-ph/0408159, 2004.
61. V.P. Belavkin, Contravariant Densities, Operatoional Distances and Quantum Channel Fidelities, Presented at the sixth annual Quantum Communication, Measurement and Computing (QCMandC) conference, Boston, July 2002, 2002 (First presnetation of generalised fidelity idea).
62. H. Araki, Relative Entropy for States of von Neumann algebras, Publ. RIMS Kyoto University, 11, 809-833, 1976 (generalised Relative entropy for states on von Neumann algebras).
63. A. Lesniewski and M.B. Ruskai, Monotone Riemannian metrics and Relative Entropy on Noncommutative probability spaces, Journal of Mathematical Physics, 40, No.11, 5702-5724, 1999.
64. A. Uhlmann, Optimizing Entropy Relative to a Channel or a Subalgebra, quant-ph/9701014, 1997.
65. K. Krauss, States, Effects and Operations: Fundamental Notions of Quantum Theory, Springer, New York, 1983 (Reference for Krauss decompositions of CP maps).
66. D. Petz, Quasi-Entropies for Finite Quantum Systems, Reports on Mathematical Physics, 23, 57-65, 1984.
67. T. Ando, Concavity of Certain Maps on Positive Definite Matricies and Applications to Hadamard Products, Linear Algebra and its Applications, 26, 203-241, 1979 (Discussion of operator monotone functions).
68. S.J. Hamemrsley and V.P. Belavkin, Operational Relative g-Entropies for Quantum Channels, unpublished, 2004.
69. D. Petz, Monotonicity of Quantum Relative Entropy Revisited, quant-ph/0209053, 2002.
70. D. Petz, Sufficiency of Channels over von Neumann Algebras, Quartely Journal of Mathematics, 39, No.2, 97-108, 1986.
71. S. Stratila, Modular Theory in Operator Algebras, Abacus Press, Tunbridge Wells, 1981 (haven't seen book, quoted in Petz paper [72]).
72. D. Petz, Structure of Sufficient Coarse-Grainings, quant-ph/0312221, 2003.
73. V.P. Belavkin and P. Staszewski, C*-algebraic generalisation of relative entropy and entropy, Ann. Inst. Henri Poincare, 37, Sec. A, 51-58, 1982.

ON THE UNIQUENESS THEOREM IN QUANTUM INFORMATION GEOMETRY

HIROSHI HASEGAWA

Institute of Quantum Science, CST, Nihon University,
Chiyoda-ku, Tokyo 101-8308, Japan
E-mail: h-hase@mxj.mesh.ne.jp

0. Introduction

The Chentsov Theorem(1972) [1] provided *uniqueness of the Fisher information metric* among all possible metrics on the set of probability simplexes $\mathcal{P}(\chi); \chi = \{0, 1, ..n\} n \in I$ i.e. $\sum_i p_i = 1$ under *markov invariance.* It was reformulated by Amari [3] by using *invariant divergences* for two probabilities p and q defined by

$$D_f(p,q) = \int f\left(\frac{q}{p}\right) pd\mu, \quad f(1) = 0,\ f''(1) > 0:$$

"The Fisher information is the only metric introduced by invariant divergences, and $\pm\alpha$-connections are the only connections which are dual to each other with respect to this metric, where α is given by $\alpha = 2\frac{f'''(1)}{f''(1)} + 3$."
Explicitly,

$$\textbf{I} \quad f''(1) g_{ij}^{Fisher}(\theta) = \partial_i \partial_j D_f(p(\theta), p(\theta'))\Big|_{\theta'=\theta} = -\partial_i \partial'_j D_f(p(\theta), p(\theta'))\Big|_{\theta'=\theta}$$

$$\textbf{II} \quad f''(1)\Gamma_{ij,k}^{(\alpha)}(\theta) = -\partial_i \partial_j \partial'_k D_f(p(\theta), p(\theta'))\Big|_{\theta'=\theta};$$

$$f''(1)\Gamma_{ij,k}^{(-\alpha)}(\theta) = -\partial_{i'} \partial'_j \partial_k D_f(p(\theta), p(\theta))\Big|_{\theta'=\theta},$$

where $\partial_i \equiv \partial/\partial\theta^i, \partial_{i'} \equiv \partial/\partial\theta^{i'}$ etc. for the parameter sets $\{\theta^i\}$ and $\{\theta'^i\}$ ($p(\theta)$ is an abbreviation of $p(x, \{\theta^i\})$) to define a given statistical model. The purpose of this paper is to extend the theorem to a quantum framework. Our strategy is as follows.

(a) f-divergence $D_f(p,q)$ is replaced by the quasi-entropy$S_f(\omega_\sigma, \omega_\rho)$ introduced by Petz[4] with p vs q being replaced by two invertible density matrices ρ vs σ i.e. $S_F(\rho,\sigma) = \mathrm{Tr}\rho^{1/2}F(\Delta_{\sigma,\rho})\rho^{1/2}$, and the convexity of f-function is replaced by the operator convexity of the (operator) function F restricted to a certain class.

(b) ∂_i is interpreted to be the respective Fréchet partial derivative

(c) $g^{Fisher}(\theta)$ in **I** is replaced by the Wigner-Yanase-Dyson metrics $g^{WYD}(\theta)$ indexed by $\pm\alpha$ with restriction $|\alpha| \leq 3$ due to the operator-convexity of the quasi-entropy.

With these modifications, we prove that the satisfaction of **I** and **II** is true if and only if the metric is identical to one of the $g^{WYD}(\theta)$'s.

Historical remark 1: the Wigner-Yanase-Dyson metrics
The original definition of the skew information:

$$-\frac{1}{2}\mathrm{Tr}[\rho^{1/2}, k]^2 \quad \text{Wigner-Yanase(1963)[11], where}$$

Dyson suggested that it could be generalized to the one exponent ϑ and the other $1-\vartheta$ for the power of ρ. Lieb[12] expressed this as

$$I_p(\rho,\Delta) \equiv c_p \times \mathrm{Tr}[\rho^p,\Delta][\rho^{1-p},\Delta] \quad \Delta \in \{\text{skew hermitians}\},\ \text{i.e.}\, i\Delta \in \mathcal{M}^s.$$

This adds to the classical Fisher term to yield a full metric form in terms of a Fréchet differential denoted by D_ρ(a Fréchet derivation on a matrix function $\varphi(\rho)$ is written as $D_\rho\varphi(\rho)(A)$, where A is an arbitrarily chosen traceless hermitian–a tangent vector.)

$$\mathrm{Tr}A^{c\dagger}g^F A^c + I_p(\rho,\Delta_A)\ (g^F = \rho^{-1}) = \mathrm{Tr}D_\rho(p^{-1}\rho^p)(A)D_\rho((1-p)^{-1}\rho^{1-p})(A):$$

The quantum correspondent to the Fisher metric has an additional term which could be derived from the corresponding quantum divergence[8].

Remark 2 on a later development: Morozova-Chenzov and Petz axiomatic approach to the uniqueness theorem by means of monotone metrics: i.e. by denoting a Riemannian metric form on matrix spaces as $K(B,A) \equiv \mathrm{Tr}B^\dagger\mathbf{K}(A)$, where

(a) $(A,B) \mapsto K_\rho(A,B)$ is sesquilinear

(b) $K_\rho(A,A) \geq 0$ and the equality holds if and only if $A = 0$

(c) $\rho \mapsto K_\rho(A,A)$ is continuous on $\mathcal{M}_n^+$ for every fixed A

(d) monotonicity condition: $K_{T(\rho)}(T(A),T(A)) \leq K_\rho(A,A)$ for every stochastic(linear, completely positive and trace-preserving)map $T: \mathcal{M}_n(\mathbf{C}) \mapsto \mathcal{M}_m(\mathbf{C}), T\mathcal{M}_n \subset \mathcal{M}_m$ and for every $\rho \in \mathcal{M}_n^{++}$ and

$A \in \mathcal{M}_n$. Instead of condition (d) we could require the weaker condition(Chentsov's markovian invariance!)

(d') $K_{U^*\rho U}(U^*AU, U^*AU) = K_\rho(A, A)$

$$K'_\rho(B, A) \equiv \frac{1}{2}(K_\rho(B, A) + K_\rho(A, B)) = K'_\rho(A, B) \quad A, B \in \mathcal{M}_n^s.$$

By taking the ρ-diagonalized basis, it can be expressed in terms of a two-variable function $c(\lambda, \mu)$ on $\mathbf{R}^+ \times \mathbf{R}^+$ and a single variable one $c(\lambda) \equiv c(\lambda, \lambda)$ in the above bilinear form as

$$K'_\rho(B, A) = \sum_i c(\lambda_i) B_{ii} A_{ii} + 2 \sum_{i<j} c(\lambda_i, \lambda_j) B^*_{ij} A_{ij} \qquad c(\mu, \lambda) = c(\lambda, \mu).$$

1. Extended Chentsov Theorem for noncommutative (finite-dimensional matrix) manifolds

Theorem 1.1 (Morozova-Chentsov (1990)[2]). *Suppose that (a), (b), (c) and (d') hold for a real, bilinear form $K'_\rho(A, B)$ on self-adjoint elements in $\mathcal{M}_n$. Then,Morozova-Chentsov(MC)-function to represent the symmetric metric $K'_\rho(A, A)$ satisfies:*

(i) *$c(\lambda)$, $c(\lambda, \mu)(= c(\mu, \lambda))$ are continuous, positive functions*
(ii) $\lim_{\mu \to \lambda} c(\lambda, \mu) = c(\lambda) = \lambda^{-1}$, $c = 1$*(the Fisher term of the metric)*
(iii) *$c(\lambda, \mu)$ is homogeneous of order -1 in λ and μ, implying $c(x\lambda, x\mu) = x^{-1} c(\lambda, \mu)$ for any $x > 0$.*

Petz's representation of a real symmetric monotone metric $f(x); \mathbf{R}^+ \mapsto \mathbf{R}^+$

$$\mathrm{K}_\rho = R_\rho^{-1/2} f(L_\rho R_\rho^{-1})^{-1} R_\rho^{-1/2} \text{ where } L_\rho, R_\rho : L_\rho R_\rho^{-1} A \equiv \rho A \rho^{-1}.$$

Theorem 1.2 (Petz (1996)[5]). *There exists a one-to-one correspondence between the MC function $c(\lambda, \mu)$ subject to (i),(ii),(iii) for a symmetric monotone metric $K'_\rho(A, A)$ and a metric-characterizing function $f(x); f(1) = 1$ with the monotonicity as follows.*

$$\textbf{i)} \quad f(x) = \frac{1}{c(x, 1)} \ (xf(1/x) = f(x)), \quad \textbf{ii)} \quad c(\lambda, \mu) = \frac{1}{\mu f(\lambda/\mu)},$$

and every normalized monotone function $f(x)$ lies in a narrow range

$$\frac{1+x}{2}(min.\ metric) \geq f(x) \geq \frac{2x}{1+x}(max.\ metric) \text{ (Fig.1)}.$$

MC-function and Petz's f-function for the WYD metrics

$$c(\lambda,\mu) = \frac{1}{p(1-p)}\frac{(\lambda^p-\mu^p)(\lambda^{1-p}-\mu^{1-p})}{(\lambda-\mu)^2} \qquad -1 \le p \le 2.$$

The corresponding metric-characterizing function is given from **i)** by

$$f_p(x) = p(1-p)\frac{(1-x)^2}{(1-x^p)(1-x^{1-p})}.$$

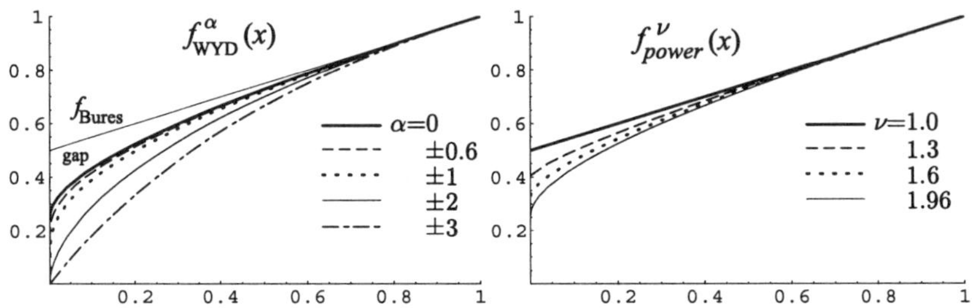

Fig.1 Order structure of monotone metrics on matrix spaces in terms of Petz's f-functions lying in a narrow range

$$f^{\alpha}_{WYD}(x) = \frac{1-\alpha^2}{4}\frac{(1-x)^2}{(1-x^{\frac{1-\alpha}{2}})(1-x^{\frac{1+\alpha}{2}})} \qquad |\alpha| \le 3$$

Theorem 1.3 (Hasegawa-Petz(1997), Hasegawa(2003)[9]). *The WYD skew information is a monotone metric defined on matrix spaces. Conversely, if a metric defined on matrix spaces of the form* $\mathrm{Tr}D_\rho\varphi(\rho)D_\rho\chi(\rho)$ *with* D_ρ *being a Fréchet derivative is monotone, then it is identical to one of the WYD metrics.*

In classical(commutative) information geometry, the F-divergence function(redenoted by F to avoide confusion with Petz's f) conditioned by $F(1) = 0$ and $F''(x) > 0$, was shown to be entitled all the way to yield α-divergence with any real number α [3]. One of the remarkable outcomes of the extended Chentsov Theorem for matrix manifolds is the limited order structure of the monotone metrics. Namely,

the quantum α-divergence may exist only in the indicated narrow range

(i) the real number α is restricted as $-3 \le \alpha \le 3$, or $|\alpha| \le 3$, and for $|\alpha| > 3$ the F-divergence must be forbidden for the violation of *operator convexity* of the F-function, but to be curious enough

(ii) there exists another region of forbiddenness which we may call "gap region"that is, there exists a blank between the maximum of the f_{WYD}-functions that corresponds to $\alpha = 0$ and the $f_{Bures} = \frac{1+x}{2}$(the so-called Bures metric). It has been shown in [9] that this gap can be filled by another class of Petz's functions called "power-mean". Namely, $f^{\nu}_{power}; 1 \leq \nu < 2(f^{\nu=1}_{power} = f_{Bures}$ and $f^{\nu=2}_{power} = f^{\alpha=0}_{WYD})$: roughly speaking, all Petz's functions $f(x)$ with $f(1) = 1$ are covered by the combination $\{f_{power}\} \cup \{f_{WYD}\}$ which forms a linearly ordered set, as in Fig.1.

Accordingly, a quantum version of the uniqueness theorem requires the quantum α-divergence to be distinguished from those in the gap region. This can be done by exploiting the fact that no dual structure of divergence exists in this region.

2. Derivation and Fréchet differentiation

$$d\varphi(\rho) = \varphi'(\rho)d^c\rho + [\varphi(\rho), \Delta] \text{ under the hypothesis}$$
$$d\rho = d^c\rho + [\rho, \Delta], \; d^c\rho \in \mathcal{C}_\rho \; \textit{commutant of } \rho. \tag{1}$$

Let $\varphi(z)$ denote an analytic function on the complex z-plane. By writing $\varphi(\rho)$(which we call "analytic function of ρ"), it can be regarded as a C^∞ function on $\mathcal{M}$. Its 1st order derivation, which has been shown to arise by a contour integration of the resolvent for $\varphi(z)$[9], $\varphi(\rho + d\rho) - \varphi(\rho)$, gives rise to the above decomposition. It can be regarded as the Hilbert-Schmidt orthogonal decomposition: $T_\rho\mathcal{M} = T_\rho\mathcal{C}_\rho \oplus T_\rho\mathcal{C}^\perp_\rho$, since $\mathrm{Tr} d^c\varphi(\rho) d^\perp\varphi(\rho) = 0$.

(1) **Tangent space at a fixed density matrix** $T_\rho\mathcal{M} = \{A \in \mathcal{B}_h; \mathrm{Tr}A = 0\}, \mathrm{dim} T_\rho\mathcal{M}^n = n^2 - 1$. It constitutes all analytic functions of ρ to form a commutative subalgebra, and is denoted by $\mathcal{A}_\rho$. An element of $T_\rho\mathcal{M}$ denoted by $A, B, ..$ is called a tangent vector: this is shown to be related to $\Delta_{A,B..}$ in *derivation* (eq.(4) below).

(2) **Commutant** $\mathcal{C}_\rho$ $\mathcal{C}_\rho = \{X \in \mathcal{B}(H^n); [\rho, X] = 0\}$, $d^c\varphi(\rho) \in \mathcal{C}_\rho$, and $\mathcal{A}_\rho \subset \mathcal{C}_\rho \subset T_\rho\mathcal{M}$.

(3) **Fréchet derivatives**

$$D_\rho(\varphi(\rho))(A) \equiv \lim_{t\to 0} \frac{\varphi(\rho + tA) - \varphi(\rho)}{t}; \quad \text{which satisfies}$$

$$\lim_{h\to 0} \frac{\| \varphi(\rho + h) - \varphi(\rho) - D_\rho(\varphi(\rho))(h) \|}{\| h \|} = 0 \quad \text{for any } A \in T_\rho\mathcal{M}. \tag{2}$$

(4) **Two kinds of a tangent vector** A **and** Δ_A: $[\rho, \Delta_A] = A, A \in T_\rho\mathcal{M}$ and $\Delta_A \in T_\rho\phi\circ\varphi$, we can express the map(the nonparametric tangent space $\mapsto$ the parametrized tangent space denoted by $\phi\circ\varphi$) as

$$A \mapsto \Delta_A = \sum_{i\neq j} \frac{A_{ij}}{\lambda_i - \lambda_j} e_{ij} \ \in T_\rho\phi\circ\varphi \subset T_\rho\mathcal{M}; (\rho = \sum_i \lambda_i e_{ii}). \tag{3}$$

Then, $D_\rho(\varphi(\rho))(A) = D^c_\rho\varphi(\rho)(A^c) + [\varphi(\rho), \Delta_A]$ $(A^c = \text{diag}A \in \mathcal{C}_\rho)$

$$= \sum_{i,j} \frac{\varphi(\lambda_i) - \varphi(\lambda_j)}{\lambda_i - \lambda_j} A_{ij} e_{ij} \in T_\rho\mathcal{M} \quad (\varphi'(\lambda_i)e_{ii} \quad \text{for } i = j)$$

$$= \sum_{i,j} \varphi^{[1]}(\lambda_i, \lambda_j) A_{ij} e_{ij}, \ \varphi^{[1]}(\lambda_i, \lambda_j) \tag{4}$$

called *1st divided difference.*

3. Classical and quantum α-divergence

$$D_\alpha(p, q) \equiv \frac{4}{1-\alpha^2} \int (1 - q(x)^{\frac{1+\alpha}{2}} p(x)^{-\frac{1+\alpha}{2}}) p(x) d\mu \quad \alpha \in \mathbf{R}, \neq \pm 1$$

$$S_\alpha(\rho, \sigma) \equiv \frac{4}{1-\alpha^2} \text{Tr}(1 - \sigma^{\frac{1+\alpha}{2}} \rho^{-\frac{1+\alpha}{2}}) \rho \, (\text{quasi entropy}) \tag{5}$$

$$\alpha \in [-3, 3], \neq \pm 1;$$

for $\alpha = \pm 1$ the classical divergence is identified with the Kullback-Leibler divergence: including this, $D_\alpha(p, q)$ is of the form $\int f(q/p) p d\mu$, that is **f-divergence**, and

$$\text{for } \alpha = 1 \qquad D_1(p, q) \equiv \int (\log q - \log p) q d\mu$$

$$\text{for } \alpha = -1 \quad D_{-1}(p, q) \equiv \int (\log p - \log q) p d\mu. \tag{6}$$

The corresponding quantum $\alpha = \pm 1$ divergence is given by the *Umegaki entropy*:

$$\text{for } \alpha = 1 \qquad S_1(\rho, \sigma) \equiv \text{Tr}(\log \sigma - \log \rho) \sigma$$

$$\text{for } \alpha = -1 \quad S_{-1}(\rho, \sigma) \equiv \text{Tr}(\log \rho - \log \sigma) \rho. \tag{7}$$

3.1. *Lesniewski-Ruskai representation of the quasi-entropy*

Theorem 3.1(Lesniewski-Ruskai [10]). *The quasi-entropy on finite quantum states $S_F(\rho,\sigma) = \langle \rho^{1/2} F(\Delta_{\sigma,\rho})\rho^{1/2}\rangle$ (ρ,σ: both invertible density matrices and $\mathrm{Tr}\rho,\sigma = 1$), where $F(u)$ is operator-convex and conditioned by $F(u) \geq 0$($u \in \mathcal{M}^+$ and the equality holds if and only if $u = 1$) admits a general representation*

$$S_F(\rho,\sigma) = \mathrm{Tr}(\sigma-\rho)F(\Delta_{\sigma,\rho})R_\rho^{-1}(\sigma-\rho) \tag{8}$$

with constants b, c and a measure $\nu(\geq 0)$,

$$F(u) = b(u-1)^2 + c\frac{(u-1)^2}{u} + \int_0^\infty \frac{(u-1)^2}{u+s}d\nu(s), \tag{9}$$

$$\int_0^\infty d\nu(s) < \infty.$$

General properties of $S_F(\rho,\sigma)$ are as follows.

(a) $S_F(\rho,\sigma) \geq 0, = 0$ if and only if $\sigma = \rho$
(b) $S_F(\lambda\rho,\lambda\sigma) = \lambda S_F(\rho,\sigma)$ (homogenuity of order 1 with respect jointly to ρ and σ)
(c) $S_F(T\rho,T\sigma) \leq S_F(\rho,\sigma)$ with every completely, trace preserving(i.e. stochastic) map T. Equivalently, $S_F(\rho,\sigma)$ is jointly convex in ρ and σ.
(d) $S_F(\rho,\sigma)$ is Fréchet differentiable with respect independently to ρ and σ.

3.2. *Classification of quasi-entropies into selfdual and non-selfdual classes*

Let us recall that $F^{dual}(\Delta_{\sigma,\rho})$ denotes the dual of the function $F(\Delta_{\sigma,\rho})$: $\Delta_{\sigma,\rho}F(\Delta_{\sigma,\rho}^{-1})$, and then $S_{F^{dual}}(\rho,\sigma) = S_F(\sigma,\rho)$ holds. We have

$$\begin{aligned} F^{dual}(u) &= uF(u^{-1}) \\ &= c(u-1)^2 + b\frac{(u-1)^2}{u} + \int_0^\infty \frac{(u-1)^2}{u+s}d\bar{\nu}(s), \end{aligned} \tag{10}$$

$$\int_0^\infty d\bar{\nu}(s) < \infty,$$

where

$$\bar{\nu}(s) = s\nu(1/s). \tag{11}$$

Theorem 3.2(Lesniewski-Ruskai [10]). *There exists one-to-one correspondence between a monotone metric K_ρ with operator-monotone decrasing function(denoted by $k(u) = 1/f(u), f$: Petz[5]), and a symmetrized quasi-entropy $S_F(\rho,\sigma) + S_{F^{dual}}(\rho,\sigma)$, which is written as*

$$k(u) = \frac{F(u) + F^{dual}(u)}{(u-1)^2} \quad \textit{for the monotone metric, i.e.} \qquad (12)$$

$$K_\rho(A,B) = -D_\sigma D_\rho S_F(\rho,\sigma)(A,B)|_{\sigma=\rho} = \langle A, R_\rho^{-1} k(\Delta_{\rho,\rho})(B)\rangle, \quad (13)$$

where $S_{F^{dual}}(\rho,\sigma) = S_F(\sigma,\rho)$ holds.

Definition 3.1

(1) Selfdual class: $F^{dual}(u) = F(u)$ and the quasi-entropy $S_{s.dual}(\rho,\sigma) = S_{s.dual}(\sigma,\rho)$. In terms of the measure representation (8),(9), both $b = c$ and $\bar{\nu}(s) = \nu(s)$ hold.

(2) Non-selfdual class: $F^{dual}(u) \neq F(u)$ and $S_{ns.dual}(\rho,\sigma) \neq S_{ns.dual}(\sigma,\rho)$.

A pair of smooth(Fréchet differentiable) operator functions $\varphi(u)$ and$\chi(u)$ is called *dual-pair*, if it is of the form $(\varphi(u) = u^p, \chi(u) = u^{1-p})^7$. This is defined by the pair of functions $(\varphi(u) = A\varphi_0(u) + B, \chi(u) = C\chi_0(u) + D)$ conditioned by $AC = 1$. It yields an equivalent class for deriving the metric by means of Fréchet differentiation on each function of the pair $(\varphi(u), \chi(u))$ i.e. $\langle D_\rho\varphi(\rho)(X), D_\rho\chi(\rho)(Y)\rangle = K_\rho(X,Y)$ irrespective of (A,B,C,D), provided $AC = 1$.

A typical example of *selfdual* and *non-selfdual* class for quasi-entropy is that of *Bures metric* and that of *Wigner-Yanase-Dyson(WYD) metric* $\alpha \neq 0$, respectively, given by

$$F_{Bures}(u) = \frac{2(1-u)^2}{1+u} \quad \text{(expandable in a power series for } |u| < 1\text{)}, \quad (14)$$

and

$$F^{\alpha}_{WYD}(u) = \varphi_\alpha(u)\chi_\alpha(u); \varphi_\alpha(u) = \frac{2}{1-\alpha}(1 - u^{\frac{1-\alpha}{2}}),$$

$$\chi_\alpha(u) = \frac{2}{1+\alpha}(1 - u^{\frac{1+\alpha}{2}}) \qquad (15)$$

(an equivalent class of the single dual pair$(\varphi_\alpha, \chi_\alpha)$).

Theorem 3.3(Hasegawa[9]).*Let $\mathcal{F}, \mathcal{F}_{power}, \mathcal{F}_{WYD}$ denote the set of all Petz functions f for monotone metrics, the set of those for power metrics indexed by $\nu\,(1 \le \nu < 2)$, and the set of those for Wigner-Yanase-Dyson metrics indexed by $\alpha\,(0 \le |\alpha| \le 3)$, respectively, normalized such that $f(1) = 1$. Then, $\mathcal{F}_{power} \cup \mathcal{F}_{WYD}$ forms a linearly ordered set which is everywhere* *dense in $\mathcal{F}$(by a topology for $\mathcal{F}$ unspecified at present).*

Quasi-entropy for the power metrics

$$F^{\nu}_{power}(u) = \frac{2^{\nu}(1-u)^2}{(1+u^{1/\nu})^{\nu}} \tag{16}$$

selfdual, $1 \leq \nu < 2$ "gap region"(Fig.1),

$$S_{F^{\nu}}(\rho,\sigma) = S_{F^{\nu}}(\sigma,\rho) = \sum_n^{\infty} c_n \mathrm{Tr}\sigma^{n/\nu}\rho^{1-n/\nu}, \quad ||\sigma\rho^{-1}|| < 1, \tag{17}$$

specifically, for the Bures case $\nu = 1$

$$S_{F_{Bures}}(\rho,\sigma) = \sum_n^{\infty} 2\mathrm{Tr}(-1)^n\sigma^n\rho^{1-n}(1-\sigma\rho^{-1})^2 \qquad ||\sigma\rho^{-1}|| < 1.$$

In the gap region no members of the WYD metrics exists (except the special member $\alpha = 0$ which is to be included in the selfdual class). However, this does not mean that the selfdual and non-selfdual classes are topologically separate: the two classes may coexist in theWYD region.

Quasi-entropy for the WYD metrics

$$S_{F_\alpha}(\rho,\sigma) \quad \text{with } F_\alpha(u) = \frac{4}{1-\alpha^2}(1-u^{\frac{1+\alpha}{2}}) \ \alpha \neq \pm 1, \text{and} \begin{cases} u\log u & \alpha = 1 \\ -\log u & \alpha = -1. \end{cases}$$

Theorem 3.4 (Characterization of $S_{F_\alpha}(\rho,\sigma)$ Hasegawa[9]). *Given a Riemannian metric on matrix spaces derived from a quasi-entropy by* $K_\rho(A,B) = -D_\sigma D_\rho S_F(\sigma,\rho) \times$
$(A,B)|_{\sigma=\rho}$ *in eq.(11), which is of the form*

$K_\rho(A,B) = \mathrm{Tr}D_\sigma\varphi(\sigma)D_\rho\chi(\rho)(A,B)|_{\sigma=\rho}$ *(metric for a pair of functions)*

and is monotone with respect to every stochastic mappings. Then, the original quasi-entropy must be identical to one of the quantum α-divergence i.e. S_{F_α} including $\alpha = \pm 1$ with dual pair $(\log u, u)$ *in order for it to satisfy Lesniewski-Ruskai relation (10).*

4. Proof of the uniqueness for matrix manifolds–Main theorem in the nonparametric version

$$\mathbf{I'} \quad D^2_\rho S_F(\rho,\sigma)(A,B)\Big|_{\rho=\sigma} = -D_\rho D_\sigma S_F(\rho,\sigma)(A,B)\Big|_{\rho=\sigma} = F''(1)K_\rho$$

$$\mathbf{II'} \quad -D^2_\rho D_\sigma S_F(\rho,\sigma)(C,A;B)\Big|_{\rho=\sigma} = F''(1)\Gamma^{(\alpha)}(C,A;B)$$

$$-D_\rho D_\sigma^2 S_F(\rho,\sigma)(C,B;A)\Big|_{\rho=\sigma} = F''(1)\Gamma^{(-\alpha)}(C,B;A).$$

Let the convex operator function F in the above equalities be restricted to the non-selfdual class which only applies to the WYD region (with $\alpha \neq 0$) in Fig.1. Then, **I**$'$ *and* **II**$'$ *hold if and only if the metric K_ρ in* **I**$'$ *is identical to one of the WYD metrics K_ρ^{WYD} classified by $|\alpha|$(including $|\alpha| = 1$), $S_F(\rho,\sigma) = F''(1)S_{F_\alpha}(\rho,\sigma)$, where*

$$F_\alpha(u) = \frac{4}{1-\alpha^2}(1-u^{\frac{1+\alpha}{2}}) \ (\alpha \neq \pm 1)\,, \text{ and } \begin{cases} u\log u \ (\alpha = +1) \\ -\log u \ (\alpha = -1) \end{cases}$$

$$F_\alpha''(1) = 1, \tag{18}$$

and **II**$'$ *the associated dual $\pm\alpha$-connections with identity $\alpha = 3 + \frac{F'''(1)}{F''(1)}$.*

Here, we restrict ourselves to those quasi-entropies which correspond to the metrics of paired functions $\mathrm{Tr}(D_\rho\varphi(\rho)D_\rho\chi(\rho))$. Then, Theorem 3.4 ensures that the only possible quasi-entropy of non-selfduality class which derives a monotone metric is the quantum α-divergence, more precisely the single dual pair(13) $(\varphi_\alpha, \chi_\alpha)$, and the corresponding metric is the WYD indexed by $|\alpha|$. The remaining question, the equality between a single second derivative and two first derivatives of the form $-\langle D_\rho^2\cdot,\cdot\rangle = \langle D_\rho\cdot, D_\rho\cdot\rangle$ must be answered. For this problem we show one of our main results in this paper.

Proposition 4.1. *Consider a Riemannian metric form (monotonicity not assumed) on matrix spaces written in terms of a pair of operator functions $(\varphi(\rho),\chi(\rho))$ i.e.*

$$-\mathrm{Tr}\Big(D_\rho^2\varphi(\rho)\Big)\chi(\rho)(A,B) = \mathrm{Tr}D_\rho\varphi(\rho)D_\rho\chi(\rho)(A,B). \tag{19}$$

For the validity of this expression of the metric in terms of second Fr'echet derivative on the left side, it is necessary and sufficient that the pair $(\varphi(u),\chi(u))$ is a dual pair $\varphi(u) = u^p$ and $\chi(u) = u^{1-p}$ with some real p (any real number without the monotonicity of the metric).

Proof. It is based on the following identity with second Fréchet derivative for the metric $D^2_\rho\varphi(\rho)(A,B) \equiv D_\rho\Big((D_\rho)\varphi(\rho)(A)\Big)(B)$. Namely,

$$\begin{aligned}\mathrm{Tr}\Big(D^2_\rho\varphi(\rho)\Big)\chi(\rho)(A,B) &= \mathrm{Tr}\Big(\varphi''(\rho)\chi(\rho)A^cB^c + \frac{1}{2}[[\varphi(\rho),\Delta_A],\Delta_B] \\ &\qquad -[[\rho,\Delta_A],\Delta_B]\varphi'(\rho)\Big) \\ &\chi(\rho) + (\mathrm{term}A \leftrightarrow B), \textit{where } \varphi' \textit{ is the ordinary derivative i.e.} \\ \varphi'(\rho) &\equiv \left.\frac{d\varphi(z)}{dz}\right|_{z=\rho}.\end{aligned}$$

(The derivation of this identity by means of contour integration $\varphi(\rho) = \frac{1}{2\pi i}\int_C \frac{\varphi(z)dz}{z-\rho}$, analogous to the one employed in [8], will be presented elsewhere.)

necessity that (φ,χ) is a dual pair. It can be seen that in order for eq.(18) to hold, it is necessary that the term $-\mathrm{Tr}[[\rho,\Delta_A],\Delta_B]\varphi'(\rho)\chi(\rho) = \mathrm{Tr}[\rho,\Delta_A][(\varphi'(\rho)\chi(\rho),\Delta_B]$ identically vanishes for arbitrary operator functions $\chi(\rho)$ i.e.$[\varphi'(\rho)\chi(\rho),\Delta_B] = 0$, hence

$$\varphi'(\rho)\chi(\rho) = \textit{constant} \times I.$$

It should be applicable to the quasi-entropy for a single dual pair, which must be of the form $S_\varphi(\rho,\sigma) = c\mathrm{Tr}\rho(1-\varphi(\sigma)\varphi(\rho)^{-1}) = \mathrm{Tr}(c\rho - \varphi(\sigma)\chi(\rho))$ so that

$$\chi(\rho) \equiv c\varphi(\rho)^{-1}\rho \quad \text{(by virtue of property (a) of quasi-entropy)}$$

Combining eqs.(19) and (20), we see that this is precisely identical to the differential equation discussed by Gibilisco and Isola[7], as

$$\begin{aligned}&\frac{\varphi'(\rho)}{\varphi(\rho)} = \mathrm{const.}\frac{1}{\rho} \quad \text{and setting const} = p(\text{real}) \text{ identifies a dual pair i.e.} \\ &\varphi(\rho) = a\rho^p \quad \text{and} \quad \chi(\rho) = b\rho^{1-p} \quad (p = \frac{1+\alpha}{2} \text{identifies the previous}\,\alpha) \\ &\text{with } c = ab; \text{ and, } a = 1 \text{ by requirement } \varphi(1) = 1;\ b = c = \frac{1}{p(1-p)}.\end{aligned}$$

sufficiency that (φ,χ) is a dual pair. The above analysis provides that the noncommutative part of the metric $\mathrm{Tr}[\varphi(\rho),\Delta_A][\chi(\rho),\Delta_B] = -\mathrm{Tr}[[\varphi(\rho),\Delta_A],\Delta_B]\chi(\rho)$is satisfied by the dual pair $(\varphi(\rho) = \rho^p, \chi(\rho) = \rho^{1-p})$. That this equality is satisfied in the commutative part also by the same dual pair can be seen easily from

$$\begin{aligned}&(\varphi(\rho) = \rho^p \quad \chi(\rho) = \rho^{1-p}), \\ \text{hence} \quad &-\varphi''(\rho)\chi(\rho) = \varphi'(\rho)\chi'(\rho) = p(1-p)\rho^{-1}.\end{aligned}$$

For the sake of completeness of proving Proposition 4.1, we show equality (18) for the Umegaki entropy with dual pair $(\log u, u)$. Namely,

$$\begin{aligned}\textbf{right side} &= \mathrm{Tr} D_\sigma \log \sigma(A) D_\rho \rho(B)\Big|_{\sigma=\rho} \\ &= \mathrm{Tr}(\sigma^{-1}A^c + [\log\sigma, \Delta_A])(B^c + [\rho, \Delta_B])\Big|_{\sigma=\rho} \\ &= \mathrm{Tr}(A^c\rho^{-1}B^c + [\log\rho, \Delta_A][\rho, \Delta_B])\end{aligned} \tag{20}$$

$$\begin{aligned}\textbf{left side} &= \mathrm{Tr}(-D_\sigma^2 \log\sigma(A,B)\chi(\rho))\Big|_{\sigma=\rho} \\ &= \mathrm{Tr}(\rho^{-2}A^cB^c\chi(\rho) - [[\log\rho, \Delta_A], \Delta_B]\chi(\rho)) \\ &= \mathrm{Tr}(A^c\rho^{-1}B^c + [\log\rho, \Delta_A][\rho, \Delta_B])\text{hence} = \textbf{right side}. \quad \square\end{aligned} \tag{21}$$

Remark 4.1 on the validity of eq.(18) for general metrics $D_\sigma^2 S_F(\rho,\sigma)|_{\sigma=\rho} = -D_\sigma D_\rho S_F(\rho,\sigma)$. Proposition 4.1 does not imply its applicability only to convex operator functions but to trace of any paired functions and linear combinations(not necessarily, convex) of such pairs. Therefore, when applied to power-series expansion (15) of S^{ν}_{power}, we can conclude that eq.(18) is valid for the quasi-entropy of power metrics.

Remark 4.2 on the uniqueness for general non-selfdual quasi-entropies In order to substantiate the uniqueness theorem without the pairing restriction, we present
conjecture: $\mathcal{F}_{WYD}$ in Theorem 3.3 is the set of all extreme points of the convex, compactifiable set formed by those Petz functions derived from the non-selfdual class.

5. Parametrization of the matrix manifold

Proposition 5.1. *Let $T_\rho\mathcal{M} \mapsto \mathbf{R}^m$ $(m \le n)$ denote the derivative map of the tangent space $T_\rho\mathcal{M}$ by $\phi \circ \varphi$. There exists a set of linear-independent matrix vectors $\{\Delta_{A_i}\}i = 1,..,m \in T_\rho\phi\circ\varphi \subset T_\rho\mathcal{M}$ by which a tangent vector of $\varphi(\rho)$ at a fixed ρ in the sense of classical geometry can be given by a surjective derivative map of projection of $T_\rho\mathcal{M}$ onto $T_\rho\mathcal{A}_\rho$:*

$$\partial_i\varphi(\rho(\theta)) \equiv D_{\theta^i}(\varphi(\rho(\theta)))(A_i) = \frac{\partial^c\varphi}{\partial\theta^i} + [\varphi(\rho), \Delta_{A_i}]; \quad by \quad Proj(D_{\theta^i})\ T_\rho\mathcal{M} \mapsto T_\rho\mathcal{A}_\rho \tag{22}$$

which generalizes a classically defined tangent vector in the sense that, when the tangent space $T_\rho\mathcal{M}$ is identical to $T_\rho\mathcal{A}_\rho$, $\partial_i\varphi(\rho(\theta)$ reduces to the first *term* $\frac{\partial^c \varphi}{\partial \theta^i}$.

Proposition 5.2. from non-parametric to parametric version of $\mathbf{I}' \to \mathbf{I}$: *Let* $\langle \cdot, \cdot \rangle$ *denote the Hilbert-Schmidt inner product. Then*

$$\langle \partial_i, \partial_j \rangle = \langle \partial_i, \partial_j \rangle^c + \langle \partial_i, \partial_j \rangle^\perp, \qquad where$$

$$\textbf{right side} = (g^F + g^{WYD})_{ij}(\theta), \quad if\ \partial_i \equiv \frac{\partial \varphi}{\partial \theta^i};\ \partial_j \equiv \frac{\partial \chi}{\partial \theta^j}, \quad and$$

$$\textbf{left side} = \partial_i \partial_j \langle -\varphi(\rho(\theta)), \chi(\rho(\theta')) \rangle \Big|_{\theta'=\theta}$$

$$replaces\ D^2_\rho \langle -\varphi(\rho), \chi(\rho') \rangle \Big|_{\rho'=\rho}. \tag{23}$$

Proposition 5.3. *In the parametrized tangent space at a fixed* ρ, *there exist* m *linearly independent tangent vectors* $\partial_i = \partial_i^c + [\cdot, \Delta_{A_i}]$ *that form a natural basis of a vector field defined by* $\rho \mapsto X_\rho \equiv X^i \partial_i$ *in terms of* $\mathcal{A}_\rho$ *functions* $X_i(\rho) \in \mathcal{A}_\rho$, *for all* i *which satisfy*[3]

$$[X_i, X_j] = 0, \quad [\partial_i, \partial_j] = 0 \quad and\ [\partial_i, X_j] = \frac{\partial X_j}{\partial \theta^i} \in \mathcal{A}_\rho for\ all\ i, j \quad and$$

$$for\ a\ pair\ X = X^i \partial_i \quad Y = Y^j \partial_j, \quad [X, Y] = (X^j \partial_j Y^i - Y^j \partial_j X^i) \partial_i. \tag{24}$$

Proposition 5.4. *Submanifold* $\mathcal{T}_\rho \mathcal{A}_\rho$ *of the tangent space* $\mathcal{T}_\rho \mathcal{M}$ *is an autoparallel commutative submanifold* $\mathcal{P}(\chi), \chi$ *finite, in the sense that all differential calculi with commutativity can be made closedly within* $\mathcal{T}_\rho \mathcal{A}_\rho$

$$\nabla_X Y \in \mathcal{T}_\rho \mathcal{A}_\rho\ for\ all\ X, Y \in \mathcal{T}_\rho \mathcal{A}_\rho.$$

Application to deriving $|\alpha| = 3 + 2\frac{F'''(1)}{F''(1)} \leq 3$.
($F'''(1) \leq 0$ by virtue of the complete monotonicity of $-F(u)$)

$$D^3_{ijk} \mathrm{Tr}(-\varphi(\rho)\chi(\sigma)\Big|_{\sigma=\rho} = F''(1)(D_i g_{jk}^{WYD(\alpha)}(\rho) + \Gamma^{(\alpha)}_{ijk}) \quad \text{to yield}$$

$$F'''(1)\rho^{-2} = -F''(1)(1 + \frac{1-\alpha}{2})\rho^{-2} \quad \text{leading to } \alpha = 3 + 2\frac{F'''(1)}{F''(1)}.$$

6. A Concluding Remark–on Tsallis statistics

The present result of uniqueness for information geometry on finite quantum states may contribute to Tsallis statistics (a framework in terms of *Tsallis entropy* $S_q^T = -\frac{1}{q-1}(\mathrm{Tr}\rho^q - 1)$), where the associated relative entropy for a pair of density matrices is found to coincide with the quantum

α-divergence: see, S. Abe's articles [13], if the parameter q is identified with $q = \frac{1+\alpha}{2}$: thus, the present uniqueness theorem should also provide the uniqueness of the "Tsallis relative entropy".

References

1. N.N. Chentsov, *Statistical Decision Rules and Optional Inference* Translations of Mathematical Monograph **53**, AMS, Providence, R.I. (1982), pp159 Theorem 11.1; original paper in Russia 1972.
2. E.A. Morozova and N.N.Chentsov, *Markov invariant geometry on manifolds of states*, Itogi Naukini Teckhniki, **36** (1990) 69-102;English translate,J. Soviet Mathematics **56** (1991), 2648-2669.
3. S. Amari and H. Nagaoka, *Methods of Information Geometry*, AMS monograph **191** (2000); cf. Lecture Notes in Statistics**28**, Springer, Berlin, 1985.
4. D. Petz, *Quasi-entropies for finite quantum systems*, Rep. Math. Phys. **23** (1986) 57-65.
5. D. Petz, *Monotone metrics on matrix spaces*, Linear Alg. Appl. **244** (1996) 81-96.
6. D. Petz, *Covariance and Fisher information in quantum mechanics*, J.Phys.**A** math.
general **35** (2002) 929-939.
7. P. Gibilisco and T. Isola, *On the characterization of paired monotone metrics*, Annals of the Institute of Statistical Mathematics **56**(2004), 369-381.
8. H. Hasegawa, *α-divergence of the non-commutative information geometry*, Rep.Math. Phys.**33**(1993)87-93.
9. H. Hasegawa, *Dual Geometry of the Wigner-Yanase-Dyson InformationContents*, Infinite Dimensional Analysis, Quantum Probability and related topics (IDAQP) **6** (2003)413-430. See also H. Hasegawa and D.Petz, Proc. Quantum Communications and Measurement, ed.O. Hirota et al, Plenum Press, N.Y. (1997).
10. A. Lesniewski and M.B.Ruskai, *Monotone Riemannian metrics and relative entropy on noncommutative probability spaces*, J. Math. Phys. **40** (1999), 5702-5723.
11. E.P. Wigner and M.M. Yanase, *Information content of distributions*, Proc. Nat. Acad. Sci. USA **49** (1963) 910-918.
12. E.H. Lieb, *Convex trace functions and the Wigner-Yanase-Dyson conjecture*, Advances in Math. **11** (1973) 267-288.
13. S. Abe, *Nonadditive generalization of the quantum Kullback-Leibler divergence for measuring the degree of purification*, Phys. Rev. A **68** (2003), 032302-1, and further references therein.

NONCANONICAL REPRESENTATIONS OF A MULTI-DIMENSIONAL BROWNIAN MOTION

YUJI HIBINO
Faculty of Science and Engineering, Saga University,
840-8502, Saga, JAPAN
E-mail: hibinoy@cc.saga-u.ac.jp

We shall construct a noncanonical representation of a d-dimensional Brownian motion by using a method similar to that in [2]. This is also considered as a generalization of a result obtained in [6] .

1. Introduction

P. Lévy [5] pointed out that

$$\widetilde{X}_\alpha(t) = \int_0^t \left(\frac{\alpha+1}{\alpha} - \frac{2\alpha+1}{\alpha} \frac{u^\alpha}{t^\alpha} \right) dB(u), \quad t \geq 0, \tag{1}$$

was a Brownian motion for $-\frac{1}{2} < \alpha (\neq 0)$. This satisfies the property

$$H_t(B) = H_t(\widetilde{X}_\alpha) \oplus LS\left\{ \int_0^t u^\alpha dB(u) \right\}$$

for each $t > 0$, where $H_t(B)$ is the closed linear hull spanned by $\{B(s); s \leq t\}$ and $LS\{\dots\}$ means the linear span of $\{\dots\}$.

In general, the Gaussian process $X = \{X(t); t \geq 0\}$ represented as

$$X(t) = \int_0^t F(t,u) dB(u) \tag{2}$$

satisfies $H_t(B) \supseteq H_t(X)$ for each $t > 0$. If it satisfies $H_t(B) = H_t(X)$, then the representation (2) is said to be *canonical.* (Refer to [3], [4]) It goes without saying that (1) is a *non*canonical representation of a Brownian motion.

In the joint work [2] , we have found that, for any N functions $g_1, g_2 \dots, g_N$, which belong to $L^2[0,t]$ and are linearly independent in $[0,t]$ for any $t > 0$, the Gaussian process defined by

$$\widetilde{B}_{\mathfrak{d}}(t) = B(t) - \int_0^t \int_0^s \mathfrak{d}(s) \gamma^{-1}(s) {}^T\mathfrak{d}(u) dB(u) ds, \quad t \geq 0, \tag{3}$$

is a Brownian motion satisfying

$$H_t(B) = H_t(\widetilde{B}_{\mathfrak{d}}) \oplus LS\left\{\int_0^t g_i(u)dB(u), i = 1, 2, \ldots, N\right\}$$

for each $t > 0$. Here

$$\mathfrak{d}(s) = (g_1(s), \ldots, g_N(s)),$$
$$\gamma(s) = \int_0^s {}^{\tau}\mathfrak{d}(u)\mathfrak{d}(u)du = \left(\int_0^s g_i(u)g_j(u)du\right), \quad \text{(Grammian matrix).}$$

The inverse of $\gamma(s)$ exists for any $s > 0$, since g_i's are linearly independent in $[0, s]$ for any $s > 0$.

Especially, in the case where $N = 1$ and $g_1 \equiv 1$, the representation (3) becomes

$$\widetilde{B}(t) = B(t) - \int_0^t \frac{B(s)}{s} ds, \tag{4}$$

which satisfies

$$H_t(B) = H_t(\widetilde{B}) \oplus LS\{B(t)\}$$

for each $t > 0$. This corresponds to the case of $\alpha = 0$ in (1). Many studies about this representation have been made by M. Yor and his collaborators e.g. [7], [1], [6], etc.

In this article, we extend the above results for a d-dimensional Brownian motion. The statement obtained there is also regarded as a generalization of the result obtained in [6].

2. *d*-dimensional Brownian motions

Let $\mathbb{B}(t) = {}^{\tau}(B_1(t), \ldots, B_d(t))$ be a d-dimensional Brownian motion, each component of which is a mutually independent one-dimensional Brownian motion.

We prepare the following notation:

$$\boldsymbol{g}_j(t) = {}^{\tau}(g_{1j}(t), \ldots, g_{dj}(t)), \quad j = 1, 2, \ldots, N,$$
$$G(t) = \left(\, \boldsymbol{g}_1(t) \cdots \boldsymbol{g}_N(t)\,\right),$$
$$(\boldsymbol{g}_i, \boldsymbol{g}_j)_t = \int_0^t {}^{\tau}\boldsymbol{g}_i(u)\boldsymbol{g}_j(u)du = \int_0^t \sum_{p=1}^d g_{pi}(u)g_{pj}(u)du, \quad \text{(inner product)}$$
$$\Gamma(t) = \int_0^t {}^{\tau}G(u)G(u)du = \left((\boldsymbol{g}_i, \boldsymbol{g}_j)_t\right) = (\Gamma_{ij}(t)), \quad \text{(Grammian matrix)}$$
$$F(t) = G(t)\Gamma(t)^{-1} = (f_{pi}(t)),$$

here we note that $\Gamma(t)$ is invertible when $\boldsymbol{g}_i$'s are linearly independent.

We can concretely construct a noncanonical representation of a d-dimensional Brownian motion having the N-dimensional orthogonal complement analogously to (3).

Theorem 2.1. *Let $\boldsymbol{g}_1, \ldots, \boldsymbol{g}_N$ be N linearly independent d-dimensional vector valued $L^2[0,t]$ functions for any $t > 0$. Then the process $\widetilde{\mathbb{B}}(t) = {}^{\tau}(\widetilde{B}_1(t), \ldots, \widetilde{B}_d(t))$ defined by*

$$\widetilde{\mathbb{B}}(t) = \mathbb{B}(t) - \int_0^t \int_0^s G(s)\Gamma^{-1}(s)^{\tau}G(u)d\mathbb{B}(u)ds \tag{5}$$

is a d-dimensional Brownian motion satisfying

$$H_t(\mathbb{B}) = H_t(\widetilde{\mathbb{B}}) \oplus LS\left\{\int_0^t {}^{\tau}\boldsymbol{g}_j(u)d\mathbb{B}(u), j = 1, 2, \ldots, N\right\},$$

that is to say,

$$H_t(B_1, \ldots, B_d) = H_t(\widetilde{B}_1, \ldots, \widetilde{B}_d) \oplus LS\left\{\sum_{p=1}^{d}\int_0^t g_{pj}(u)dB_p(u), j = 1, 2, \ldots, N\right\}.$$

Proof. First, we note that each component $\widetilde{B}_p$ of $\widetilde{\mathbb{B}}$ is represented as

$$\widetilde{B}_p(t) = B_p(t) - \int_0^t \sum_{k=1}^{N} f_{pk}(s) \int_0^s \sum_{q=1}^{d} g_{qk}(u)dB_q(u)ds, \quad p = 1, 2, \ldots, d.$$

For any p and j,

$$\begin{aligned}
&E\left[\widetilde{B}_p(t) \int_0^{t'} \sum_{r=1}^{d} g_{rj}(u)dB_r(u)\right] \\
&= \int_0^t g_{pj}(u)du - \int_0^t \sum_{k=1}^{N} f_{pk}(s) \int_0^s \sum_{q=1}^{d} g_{qk}(u)g_{qj}(u)duds \\
&= \int_0^t g_{pj}(u)du - \int_0^t \sum_{k=1}^{N} f_{pk}(s)\Gamma_{kj}(s)ds \\
&= \int_0^t g_{pj}(u)du - \int_0^t g_{pj}(s)ds = 0
\end{aligned}$$

holds for any $0 < t < t'$.

Thanks to the famous innovation theorem (refer to [4] and so on), this guarantees the desired result. □

Remark 2.1. Naturally enough, in the case of $d = 1$, (5) coincides with (3).

In order to emphasize the interest of the above theorem, in the rest of this article we shall show some corollaries:

Corollary 2.1.

$$\widetilde{B}_p(t) = B_p(t) - \int_0^t \frac{B_p(s)}{s} ds, \quad p = 1, 2, \dots, d,$$

are d independent Brownian motions which satisfy

$$H_t(B_1, \dots, B_d) = H_t(\widetilde{B}_1, \dots, \widetilde{B}_d) \oplus LS\{B_1(t), \dots, B_d(t)\}.$$

Proof. This is the case where $N = d$ and $g_{pj}(t) \equiv \delta_{pj}$. □

The above corollary is straightforward due to (4).

Next, we consider the case of $N = 1$ and that g_{p1}'s are constant, say $g_{p1}(t) \equiv \alpha_p$, where α_p's are the constant numbers. Without loss of generality, we may assume $\sum_{p=1}^d \alpha_p^2 = 1$.

Corollary 2.2.

$$\widetilde{\mathbb{B}}(t) = \mathbb{B}(t) - \boldsymbol{\alpha}^{\tau}\boldsymbol{\alpha} \int_0^s \frac{1}{s} \mathbb{B}(s) ds$$

is a d-dimensional Brownian motion which satisfies

$$H_t(\mathbb{B}) = H_t(\widetilde{\mathbb{B}}) \oplus LS\left\{\sum_{p=1}^d \alpha_p B_p(t)\right\},$$

where $\boldsymbol{\alpha} = {}^{\tau}(\alpha_1, \dots, \alpha_d)$.

This is a noncanonical representation of a d-dimensional Brownian motion having the one-dimensional orthogonal complement.

In reality, Corollary 2.2 is the same as [6] . This fact shows that our theorem may be regarded as an extension of [6] .

Let us consider, moreover, the case of $d = 2$ and $(g_{11}(t), g_{21}(t)) \equiv (\sqrt{\lambda}, \sqrt{1-\lambda})$ for any fixed $0 < \lambda < 1$.

Corollary 2.3.

$$\widetilde{B}_1(t) = B_1(t) - \int_0^t \left\{\frac{\lambda}{s} B_1(s) + \frac{\sqrt{\lambda - \lambda^2}}{s} B_2(s)\right\} ds$$

$$\widetilde{B}_2(t) = B_2(t) - \int_0^t \left\{\frac{\sqrt{\lambda - \lambda^2}}{s} B_1(s) + \frac{1-\lambda}{s} B_2(s)\right\} ds$$

are two independent Brownian motions satisfying

$$H_t(B_1, B_2) = H_t(\widetilde{B}_1, \widetilde{B}_2) \oplus LS\left\{\sqrt{\lambda}B_1(t) + \sqrt{1-\lambda}B_2(t)\right\}.$$

Such representations are also mentioned in [6] . However, our method can produce them systematically.

In [6], the authors have reached these representations, motivated by the form

$$dX(t) = dB_1(t) + \{f(t)B_2(t) + g(t)B_1(t)\}dt.$$

According to [1], the form

$$dX(t) = dB_1(t) + \frac{B_2(t) - X(t)}{1-t}dt$$

is connected with the insider trading (for *one* insider). Our theorem may be applied to the model of $d-1$ insiders.

References

1. H. Föllmer, C-T. Wu and M. Yor; *Canonical decomposition of linear transformations of two independent Brownian motions motivated by models of insider trading.* Stoch. Proc. Appl. **84** (1999), 137–164.
2. Y. Hibino, M. Hitsuda and H. Muraoka; *Construction of noncanonical representations of a Brownian motion.* Hiroshima Math. J. **27** (1997), 439–448.
3. T. Hida; *Canonical representations of Gaussian processes and their applications.* Mem. Coll. Sci. Univ. Kyoto **33** (1960), 109–155.
4. T. Hida and M. Hitsuda; Gaussian Processes, Representation and Applications, Amer. Math. Soc. (1993).
5. P.Lévy; *Fonctions aléatoires à corrélation linéaire*, Illinois J. Math. **1** (1957), 217–258.
6. C-T. Wu and M. Yor; *Linear transformations of two independent Brownian motions and orthogonal decompositions of Brownian filtrations.* Publ. Mat. **46** (2002), 237–256.
7. M. Yor; Some aspects of Brownian motion. Part I: Some special functionals, Birkhäuser Verlag, Basel, (1992).

SOME OF FUTURE DIRECTIONS OF WHITE NOISE THEORY

TAKEYUKI HIDA

Meijo University, Nagoya 468-8502, Japan
E-mail: thida@ccmfs.meijo-u.ac.jp

Having had a very short review of the white noise analysis, we consider some of its future directions with special emphasis on innovation theory.

1. Introduction

We shall discuss *evolutional random complex systems* such as stochastic processes $X(t)$ and random fields $X(C)$. They are parameterized by the time variable t, or a space-time variable, or a manifold C having special geometric structure in a space.

Our method of the study is the so-called *innovation approach.* From this viewpoint, we shall first make a brief review on the present stage of the white noise analysis and illustrate our idea on how to carry on the innovation approach. Then, we shall propose *ourselves* some of future directions which would be significant in both probability theory and other related fields.

2. White noise analysis

It is well known that an intuitive representation of white noise may be obtained by taking the time derivative of a Brownian motion $B(t)$. Hence, we denote the white noise by $\dot{B}(t)$. Its significance may be understood in many ways. In fact, it is a generalized stochastic process with independent values at every point and is elemental among (generalized) stochastic processes, Gaussian in distribution. In application, it stands for the fluctuation in living organs, the noise in the communication theory and so forth. It can be realized as the probability measure on a certain space of generalized functions on R. A white noise thus defined contains maximum information in under the restriction of a limited power, and is taken to be a good information source in communication theory. While it may be an obstacle when the noise disturbs the message.

It is now clear that functionals of white noise have been well investigated theoretically and in applications.

Our mathematical setup is as follows. Start with a characteristic functional:

$$C(\xi) = \exp[-\frac{1}{2}\|\xi\|^2],$$

where ξ is a member of a suitable nuclear space E dense in $L^2(R^d), d \geq 1$. It defines a probability measure space (E^*, μ), where E^* is the dual space of E. Now, each x in E^* with μ is viewed as a sample function of $\dot{B}(t)$. The Hilbert space $(L^2) = L^2(E^*, \mu)$ is therefore a realization of the space of functionals of a white noise with finite variance.

The second quantization method by using the operator $A = -\Delta+|u|^2+1$ leads us to a Gel'fand triple

$$(S) \subset (L^2) \subset (S)^*.$$

It is an adavantage of our analysis to have $(S)^*$ of generalized white noise functionals that is wide enough to carry on the analysis. If necessary, depending on the problem in question, we may introduce more wider space than $(S)^*$.

Another important tool of our analysis is the infinite dimensional rotation group $O(E)$ consisting of continuous linear operates acting on E and being orthogonal. The collection g^*'s which are adjoints of g's in $O(E)$ forms a group denoted by $O^*(E^*)$. Each g^* keeps the white noise measure μ invariant:

$$g^*\mu = \mu.$$

Finally the so-called S-transform is introduced to have a good representation of a member φ in $(S)^*$:

$$(S\varphi)(\xi) = C(\xi) \int \exp[< x, \xi >]\varphi(x) d\mu(x).$$

The S-transform leads us to appeal to the classical functional analysis.

Another important noise is a Poisson noise denoted by $\dot{P}(t)$, which is the time derivative of a standard Poisson process $P(t)$. This is also an elemental noise that appears as innovation. It has similar probabilistic properties to white noise, and at the same dissimilarity is also interesting.

3. New directions. (Our plans.)

In this report we are going to discuss the following topics which are expected to be developed.

1) Innovation theory.

2) White noise approach to path integral.

3) Quantum information.

4) Application to noncommutative geometry.

The topics listed above will be in order. They are, of course, far from satisfactory stage, but we are interested in working those topics. It is, however, noted that advantages of white noise analysis will be seen explicitly in many places.

In this paper we focus our attention mainly on the topic 1). Some others will be touched upon briefly.

4. Innovation

4.1. The innovation approach is, in a sense, one of the traditional methods of stochastic analysis. It should be reminded that P. Lévy discussed in his monograph [10]. Then, he proposed a following guiding formula for the analysis of stochastic processes in [11].

$$\delta X(t) = \Phi(X(s), s \le t, Y(t), t, dt),$$

where $Y(t)$ is the innovation. This formula is called a *stochastic infinitesimal equation.*

N. Wiener had a similar idea in electrical communication theory, where his idea has appeared explicitly in terms of engineering.

In both approaches L^2-theory does not play any essential role. Namely, sample functions are more important. Actually, the analysis depends heavily on the sample function properties. We have therefore introduced a class (**P**) of random variables where the *convergence in probability* topology is introduced.

4.2. Gaussian processes.

For a Gaussian process, the canonical representation theory has been established, and the innovation theory follows immediately.

A symbolical formula of the representation of a Gaussian process $X(t)$ is

$$X(t) = \int_0^t F(t,u)\dot{B}(u)du,$$

where $F(t,u)$ is a non-random kernel of Volterra type. If $F(t,u)$ is the canonical kernel, i.e. $\mathbf{B}_t(X) = \mathbf{B}_t(\dot{B})$ holds for every $t \geq 0$, then $\dot{B}(t)$ is the innovation.

4.3. Gaussian random fields.

There are various Leitmotive to discuss random fields, among others in quantum field theory. where fluctuation with multi-dimensional parameter plays a key role. The basic concept for the study is again the innovation. The innovation theory is also suggested by the stochastic variational equation which is a generalization of Lévy's stochastic infinitesimal equation.

To fix the idea we consider a Gaussian random field $X(C)$ parameterized by a smooth convex contour in R^2 that runs through a certain class $\mathbf{C}$ which is topologized by the usual method by using the Euclidean metric. Denote by $W = W(u), u \in R^2$, the two-dimensional parameter white noise. The probability distribution of W can be introduced in the space of generalized functions on R^2 in a similar manner to the R^1-parameter case.

Assume that a Gaussian random field $X(C)$ is expressed as a stochastic integral of the form

$$X(C) = \int_{(C)} F(C,u)W(u)du,$$

where the kernel function $F(C,u)$ is locally square integrable in u, and where (C) denotes the domain enclosed by C. For convenience, $F(C,u)$ is assumed to be smooth in (C,u). The stochastic integral is called *causal* representation of $X(C)$ in terms of white noise W. The canonical property of such a causal representation can be defined as in the case of a Gaussian process $X(t)$.

The standard type of the stochastic variational equation for such a field $X(C)$ is of the form

$$\delta X(C) = \int_C F(C,s)\delta n(s)W(s)ds + \int_{(C)} \delta F(C,u)W(u)du.$$

The method of getting the innovation $\{W(s), s \in C\}$ is similar but somewhat complicated compared to the case of $X(t)$. Below, we shall show a typical case.

Theorem 1 Given a stochastic variational equation

$$\delta X(C) = -X(C)\int_C k\delta n(s)ds + X_0\int_C v(s)\partial_s^*\delta n(s)ds, \ C \in \mathbf{C},$$

where $\mathbf{C}$ is taken to be a class of concentric circles, v is a given continuous function and ∂_s^*, the creation operator, is the adjoint operator of ∂_s. Then, the solution $X(C)$, with a trivial initial data, exists and is expressed in the form

$$X(C) = X_0\int_{(C)} \exp[-k\rho(C,u)]\partial_u^* v(u)du,$$

where ρ is the Euclidean metric.

The solution $X(C)$ just obtained is called a Gaussian random field of *Langevin type.*

There are various generalizations of this field (or stochastic variational equation). Further suggestions may be found in the literature [5].

4.4. Functionals of Poisson noise and linear processes.

Having been suggested by the decomposition of a Lévy process, we claim that a Poisson process $P(t)$ or Poisson noise $\dot{P}(t)$ is elemental. The $\dot{P}(t)$ is a stationary generalized stochastic process with independent values at every point. Let the parameter t run through the whole real line. The characteristic functional $C_P(\xi)$ of such a Poisson noise is of the form

$$C_P(\xi) = \exp[\lambda\int_{-\infty}^{\infty}(e^{i\xi(t)} - 1)dt],$$

where $\xi \in E$ and $\lambda(> 0)$ is the intensity. It defines a probability measure μ_P on E^*, and the Hilbert space $L^2(E^*, \mu_P) = (L^2)_P$ is defined.

There have been many attempts towards the analysis on $(L^2)_P$ and a lot of results have been reported, where we see much similarity to the Gaussian case. They are of course very interesting.

Now let us turn our eyes on the dissimilarity to Gaussian case; namely we are keen to find profound properties of Poisson noise which are of different character from Gaussian case.

We now focus our attention upon some properties which seem to be strange. Take a simple, but suggestible example called a *linear process.*

Let a stochastic process $X(t)$ be given by the integral

$$X(t) = \int_0^t F(t,u)\dot{P}(u)du.$$

If $\dot{P}(u)du$ is viewed as a random measure, then the argument is similar to the Wiener integral. There is another understanding of the above integral. Take a sample function of $\dot{P}(u)$. It is a generalized function, so that if the kernel can play the role of a test function (assuming smoothness), then $X(t)$ is a (stochastic) bilinear form. We now prove

Proposition 1. Suppose $F(t,u)$ is smooth and $F(t,t) \neq 0$. Then, we have

$$\mathbf{B}_t(X) = \mathbf{B}_t(P).$$

Proof. By assumption it is easy to see that $X(t)$ and $P(t)$ share the jump points; this fact means the information (and hence sigma-field) is fully transformed from $P(t)$ to $X(t)$ keeping the causality. This proves the equality.

From the above argument, we are led to introduce a space $(\mathbf{P})$ of random variables that come from stochastic processes for which existence of variance is not necessarily required. This may sound to be a vague statement, however we can rigorously define. (cf. [12], [17].)

We are now ready to define a *linear process.* Let $Z(t), t \in T$, be a Lévy process. Define $X(t), t \in T$, by

$$X(t) = \int^t F(t,u)dZ(u),$$

where the integral is defined sample function-wise in the space $(\mathbf{P})$. Such a process $X(t)$ is called a linear process. More general class of linear processes is introduced in [12].

To avoid complex computations, we take a simple linear process. Namely,

$$X_0(t) = \int_0^t F(t,u)\dot{B}(t)dt + \int_0^t G(t,u)\dot{P}(t)dt, \ t \geq 0,$$

where $B(t)$ is a Brownian motion and $P(t)$ is a Poisson process with unit intensity. To compute the characteristic functional of $X_0(t)$, a notation is introduced.

$$F * \xi(u) = \int F(t,u)\xi(t)dt.$$

Proposition 2. The characteristic functional $C_0(\xi)$ of the linear process $X_0(t)$ is expressed in the form

$$C_0(\xi) = C_1(\xi) \cdot C_2(\xi),$$

where

$$C_1(\xi) = \exp[-\frac{1}{2}\|F * \xi\|^2],$$

$$C_2(\xi) = \exp[\int (e^{iG*\xi(u)} - 1)du].$$

Assume that $F(t,u)$ and $G(t,u)$ are canonical kernel defined for representation of Gaussian processes.

Theorem 1. Given a characteristic functional $C_0(\xi)$ of a linear process given above with canonical kernels. Then, the kernels are obtained with the help of the functional derivatives of the second order.

Proof. Set $c(\xi) = \log C_0(\xi)$, which is expressed as $c_1(\xi) + c_2(\xi)$. Take the second functional derivative $\frac{\delta^2}{\delta\xi\delta\eta}c(\xi)$. Then, it is easy to have

$$\int F(t,u)F(s,u)du + \int e^{i(G*\xi)(u)}G(t,u)G(s,u)du.$$

The two terms are discriminated since $(G * \xi)(u), \xi \in E$, spans a dense subset of $L^2([0,\infty))$ because of the canonical property of G. Hence we have two integrals $\int F(t,u)F(s,u)du$ and $\int G(t,u)G(s,u)du$, separately. Since F and G are canonical kernels, the factorisation problem can be solved by the canonical representation theory for Gaussian processes. Thus, F and G are obtained.

Innovation problem is now obvious.

A new framework of stochastic analysis is now proposed, having been suggested by the example in the last subsection. It is clear that one should think of functionals, the variable of which is a path of a stochastic process, and should discuss operations acting on the paths themselves. There is often required to calculate those functionals outside of L^2-space; for one thing, the process in question may not have finite variance, and for another thing, $\dot{B}(t)$ or $\dot{P}(t)$ is not smeared since the time development is expressed explicitly.

Thus, the basic space is to be generated by innovation; a Brownian motion and/or a compound Poisson processes. The topology is, as was explained before, either convergence almost surely or convergence in probability. We are now sure that the suitable space is the $(\mathbf{P})$ introduced before.

Both the space (L^2) and $(\mathbf{P})$ as well as their generalizations are the ground where our game is played.

4.6. Innovations with multi-dimensional parameter.

Whenever the variational calculus for random fields is discussed, we meet innovation with multi-dimensional parameter. The Gaussian case has already been discussed before, so we focus our attention on the Poisson noise with multi-dimensional parameter; let it be denoted by $\{V(u), u \in R^d\}$. The characteristic functional is given by

$$C_P(\xi) = \exp[\lambda \int_{R^d} (e^{i\xi(t)} - 1) dt^d].$$

It determines the probability measure μ_P introduced on the space E^* of generalized functions on R^d.

It can be shown that the stochastic bilinear form $< x, f >$ with $x \in E^*, \xi \in E$ can be defined a.e. μ_P. In particular, if f is taken to be the indicator function I_D of a domain D, then the characteristic functional shows that $< x, I_D >$ is a random variable on the probability space (E^*, μ_P) and is subject to the Poisson distribution with intensity $\lambda|D|$, where $|D|$ denotes the volume of D.

Set $X = X(x) = \langle x, I_D \rangle$, where I_D is the indicator function of a simply connected domain D.

Theorem 2. 1) The random variable $X(x)$ denotes the number of the singular points, with each of which a delta function is associated as a sample function.

2) Under the condition that $X(x)$ involves delta functions as many as n, those points are equally distributed over D.

Proof. The characteristic function $\varphi(x)$ of $X(x)$ is given by

$$\varphi(z) = \exp[\lambda|D|(e^{iz} - 1)].$$

This proves the theorem.

Remark. More precise meanings of the assertion 2) will be given in the forthcoming paper by the author.

We then come to a study of linear functionals of a multi-parameter Poisson noise. There are various motivations that come from applications; for instance, in the quantum optics where photon's behaviour is related to Poisson noise and there is some hope that it will be a background of quantum computing, and in biology. Naturally, there is much interest in a random field $X(C)$ parameterized by a contour C:

$$X(C) = \int_{(C)} G(C, u) V(u) du,$$

where the kernel $G(C,u)$ is continuous in (C,u).

Theorem 3. Let $X(C)$ be given above, and assume that $G(C,s), s \in C$, does not vanish almost everywhere for every C. Then, $V(u)$ is an innovation.

Proof. The variation $\delta X(C)$ exists and it involves the term

$$\int_C G(C,s)\delta n(s)V(s)ds,$$

where $\{\delta n(s)\}$ determines the infinitesimal deformation δC of C. Note that this term can be taken separately from $\delta X(C)$. Here is used the same technique as in the case of [9], so that the values $V(s), s \in C$, can be obtained. It is easy to see that $V(s)$ is the innovation.

References

1. L. Accardi, T. Hida and Si Si, Innovation of some stochastic processes. Volterra Center Notes. No.537. 2002.
2. P.A.M. Dirac, The Lagrangian in quantum mechanics. Phys. Zeitsch. der Sowjetunion. 3, (1933), 64-72..
3. T. Hida, h.-H. Kuo, J. Potthoff and L. Streit, White Noise. An infinite dimensional calculus. Kluwer Academic Press. 1993.
4. T. Hida, Wihte noise analysis: Part I. Theory in Progress. Taiwanese J. Math. 7 (2003), 541-556.
5. T. Hida and Si Si, An innovation approach to random fields. Application of white noise theory. World Scientific Pub. Co. Ltd. 2004.
6. T. Hida and Si Si, Lévy field as a generator of elementary noises. Proc. Levico Conf. 2003.
7. H.-H. Kuo, White Noise Distribution Theory. CRC Press. 1996.
8. R. Léandre and H. Ouerdiane, Connes-Hida calculus and Bismut-Quillen super connections. Les prépub. de l'Institut Élie Cartan. 2003. no.17.
9. P. Leukert and J. Schäfer, A rigorous construction of Abelian Chern-Simons path integrals. Review in Math. Phys. 8 (1996), 445-456.
10. P. Lévy, Théorie de l'addition des variables aléatoires. Gauthier-Villars, 1937.
11. P. Lévy, Random functions: general theory with special reference to Laplacian random functions. Univ. of Calif. Pub. in Stat, vol.1, 1953, 331-390.
12. P. Lévy, Fonctions aléatoires à corrélation linéaire. Illinois J. Math. 1 (1957), 217-258.
13. P. Massani and N. Wiener, Non-linear prediction. Probability and Statistics. Wiley, 1959. 190-212.
14. N. Wiener, Generalized harmonic analysis, Acta Math. 55 (1930), 117-258.
15. N. Wiener, Extrapolation, interpolation and smoothing of stationary time series. The MIT Press. 1949.
16. N. Wiener, Nonlinear problems in random theory. The MIT Press. 1958.
17. Win Win Htay, Note on linear processes. Proc. 2003 Meijo Winter School.

INFORMATION, INNOVATION AND ELEMENTAL RANDOM FIELD

TAKEYUKI HIDA

Meijo University, Nagoya 468-8502, Japan
E-mail: thida@ccmfs.meijo-u.ac.jp

Innovation of a stochastic process or a random field plays one of the most important roles in the stochastic analysis. This fact is illustrated from the view point of information theory.

1. Introduction

We shall first review the analysis of *evolutional random complex systems* such as stochastic processes $X(t)$ and random fields $X(C)$ which are functionals of white noise. Our method of the study is the so-called *innovation approach* by using elemental random fields, where information theoretical viewpoint is always taken into account.

From this line, we shall first make a brief review of white noise analysis and illustrate our idea on how to carry on the innovation approach. Then, we shall propose some of future plans which would be significant in both probability theory and other related fields.

For the review on what has been obtained on white noise analysis we should like to refer to the literatures [5], [6] and [10].

2. White noise analysis

Our mathematical setup is as follows. Start with a characteristic functional of a white noise $\{\dot{B}(t)\}$:

$$C(\xi) = \exp[-\frac{1}{2}\|\xi\|^2],$$

where ξ is a member of a suitable nuclear space E dense in $L^2(R^d), d \geq 1$. It defines a probability measure space (E^*, μ), where E^* is the dual space of E. Now, each x in E^* with μ is viewed as a sample function of $\dot{B}(t)$.

The Hilbert space $(L^2) = L^2(E^*, \mu)$ is therefore a realization of the space of functionals of a white noise with finite variance.

The second quantization method by using the operator $A = -\Delta + |u|^2 + 1$ leads us to a Gel'fant triple

$$(S) \subset (L^2) \subset (S)^*,$$

which defines a space $(S)^*$ of generalized functionals of white noise.

It is one of the adavantages of white noise theory to have a space $(S)^*$ which is reasonably wide enough to carry on the calculus to be required. If necessary, depending on the problem in question, we can introduce much wider space.

Another important tool of our analysis is the infinite dimensional rotation group $O(E)$ consisting of contimuous linear operatoes acting on E and being orhtogonal. The collection g^*'s toperators of $g \in O(E)$ forms a group denoted by $O^*(E^*)$. Each g^* keeps the white noise measure μ invariant:

$$g^*\mu = \mu.$$

Significant role is played by the particular subgroup of $O(E)$, called the *conformal group* involving *whiskers* which are one-parameter subgroups coming from the diffeomorhisms of the parameter space see [6].

Finally the so-called S-transform is introduced in order to have a good analytic representation of a member φ in $(S)^*$:

$$(S\varphi)(\xi) = C(\xi) \int \exp[< x, \xi >]\varphi(x) d\mu(x).$$

With the help of the S-transform we can deal with white noise functionals by appealing to the classical functional analysis.

We now pause to remind that the important, in fact standard class of innovations involves a (Gaussian) white noise and a compound Poisson noise which is a superposition of Poissonj noises with different jumps. Thus, a Gaussian white noise and an individual Poisson noises are *elemental.* Actually, this comes from the Lévy decomposition of a Lévy process (see [13]).

We therefore come to a Poisson noise denoted by $\dot{P}(t)$, which is the time derivative of a standard Poisson process. It has similar probabilistic properties to white noise, and at the same time significant properties different from those of white noise: Among others, method of analysis, optimality of randomness, role of the rotation group, and so forth.

3. Our plans

In this report we are going to discuss the following topics which are expected to be fruitful areas that would be investigated.

1) Innovation theory.
2) White noise approach to path integrals; advanced study.
3) Quantum information with quantum white noise.
4) Application to noncommuitative geometry.

The topics listed above are, of course, far from develped state and we are interested in working those topics. It is, however, noted that various advantages of white noise analysis will be seen explicitly in many places, although the appearance is different.

Because of the reasons mentioned in the last section, our analysis is based on not only white noise (Gaussian noise), but also Poisson noise, as well as compound Poisson noises.

In this note we shall focus our attention on the topic 1) and 2). Some others will be touched upon briefly, and they will be reported in the forthcoming reports.

4. Innovation

4.1. It is reminded that P. Lévy discussed inniovation in his monograph [13]. Then, he proposed to a guiding formula for the analysis of stochastic processes in [14]. The formula is as follows:

$$\delta X(t) = \Phi(X(s), s \leq t, Y(t), t, dt).$$

The formula is called a *stochastic infiniresimal equation.* It gives us how to investigte the probabilistic structure of the given

N. Wiener had a similar idea in electrical communication theory. His idea has been explicily in terme of engineering. Typically in his books and a paper with Masani. Let them be listed: Time series. Prediction, Nonlinear problems in random theory, coding and decoding

In those approaches the L^2-theory does not play any essential role. Namely, sample functions are more important. Actually, the analysis depends heavily on the sample function properties. We have therefore introduced a class $(\mathbf{P})$ of random variables where the *convergence in probabilty* topology is introduced.

4.2. Known cases.

1) Gaussian processes.

For a gaussian process, the canonical representation theory has been established, and the innovation theory follows immidiately.

2) Gaussian random fields.

There are various Leitmotive to discuss random fields, among others in quantum field theory. where fluctuation with multi-dimensional parameter plays an important role. The basic concept for the study is again the innovation. The innovation theory is also suggested by the Lévy's stochastic variational equation which is a generalization of the well known stochastic infinitesimal equation.

To fix the idea we consider a Gaussian random field $X(C)$ parameterized by a smooth convex contour in R^2 that runs through a certain class $\mathbf{C}$ which is topologized by the usual method by using the Euclidean metric. Denote by $W = W(u), u \in R^2$, the two-dimensional parameter white noise.

Assume that a Gaussian random field $X(C)$ is expressed as a stochastic integral of the form

$$X(C) = \int_{(C)} F(C,u)W(u)du,$$

where the kernel function $F(C,u)$ is locally squre integrable in u, and where (C) denotes the domain enclosed by C. For convenience, $F(C,u)$ is assumed to be smooth in (C,u). The stochastic integral may be said to be a *causal* representation of $X(C)$ in terms of white noise W. The canonical property of such a causal representation can be defined .

The standard type of a stochastic variational equation for $X(C)$ is

$$\delta X(C) = \int_C F(C,s)\delta n(s)W(s)ds + \int_{(C)} \delta F(C,u)W(u)du.$$

The method of getting the innovation $\{W(s), s \in C\}$ is easy.

Theorem 4.1. *Given a stochastic variational equation*

$$\delta X(C) = -X(C)\int_C k\delta n(s)ds + X_0\int_C v(s)\partial_s^*\delta n(s)ds, \ C \in \mathbf{C},$$

where $\mathbf{C}$ *is taken to be a class of concentric circles,* v *is a given continuous function and* ∂_s^**, the creation operator, is the adjoint operator of* ∂_s*. Then, the solution* $X(C)$*, with a trivial initial data, exists and is expressed in the form*

$$X(C) = X_0\int_{(C)} \exp[-k\rho(C,u)]\partial_u^* v(u)du,$$

where ρ is the Euclidean metric.

It is east to obtain a relationship with innovation.

Useful suggestions may be found in the literature [6]. The motivation for path integrals is seen in [3] and we can discuss furhter development.

3) Linear processes.

A Poisson noise $\dot{P}(t)$ is a stationary generalized stochastic process with independent values at every point. Let the parameter t run through R^1. The characteristic functional $C_P(\xi)$ of the Poisson noise is

$$C_P(\xi) = \exp[\lambda \int_{-\infty}^{\infty} (e^{i\xi(t)} - 1)dt].$$

It defines a probability measure μ_P on E^*, and the Hilbert space $L^2(E^*, \mu_P) = (L^2)_P$ is defined.

A linear process $X(t)$ is given by the integral of the form

$$X(t) = \int_0^t F(t,u)\dot{B}(u)du + \int_0^t G(t,u)\dot{P}(u)du.$$

As for the second term there are two ways to undrstand the integral with respect to $\dot{P}(u)du$. One comes from the understanding that $\dot{P}(u)du$ is a random measure, and the other is the view point that a sample function of $\dot{P}(u)$ is a generalized function, so that the integral is a bilinear form. They are actually different. With this remark in mind, we can discuss a method to identify two kernels F and G by using the second variation of the characteristic functional of $X(t)$.

For more general linear processes, we can establish a general theory (see [7]).

5. Path integrals

Path integral should not be understood as an application of an integration theory over a function space, but it suggests us another direction of stochastic analysis. We are inspired by the pioneering work in Lagrangian dynamics by Dirac [3] to come to our proposal of a path integral. Also the Feynman's paper [4] has suggested us to consider a measure on the space of functions which are quantum mechanical trajectories to define the integral that he had in mind, and it has given us a chance to use our generalized white noise functionals (called Brownina functionals in early days of white noise theory).

With such a history we proposed an approach to Feynman integral in [17]. Our idea was to take the following $y(s), s \in [0,1]$, to be a possible quantum mechanical trajectory:

$$y(s) = x(s) + \sqrt{\frac{\hbar}{m}} B_b(s), \;\; s \in [0,1],$$

where $x(s)$ is the classical path uniquely determined by the Lagarangian and where $B_b(s)$ is a Brownian bridge on the unit time interval.

Let $L(x, \dot{x})$ be the Lagrangian. Then, the propagator (or transformation function) is given by

$$E\{N \exp[\frac{i}{\hbar} \int_0^t L(y, \dot{y}) d\tau \cdot \exp[\frac{1}{2} \int_0^t \dot{B}_b(\tau)^2 d\tau]\},$$

We may replace a Brownian bridge with the ordinary Brownian motion $B(s)$ and put a delta function $\delta(B(t))$. The second factor makes the white noise measur to be a flat measure. The factor does not exist in the classical sense, but now it is understoos ad a generalized white noise functional.

A characterization of a Brownian bridge

The action principle (see Dirac's text book, Chapt. V) suggests the fluctuation is a Markov process. It is expected to be reversible. It is tacitly assumed to be Gaussian and satisfies some additional conditions. To fix the idea, let us consider the time parameter of the process in question runs through $[0, 1]$. We now claim

Theorem 5.1. *Let $X(t), t \in [0,1]$, be a non-trivial, meaqn continuous Gaussian process satisfying*
i) Markov,
ii) bridged over $[0, 1]$,
iii) Its normalized process $Y(t)$ enjoys the projective invariance.
iv) Local continuity at $t = 0$ of sample functions.

Then, $X(t)$ is a Brownina bridge up to constant.

Proof. [Outline of the proof] Since $X(t)$ is a Gaussian Markov process, it has the canonical representation of the form

$$X(t) = f(t) \int_0^t g(u) \dot{B}(u) du.$$

The covariance function $\Gamma_X(t,s)$ is expressed in the form

$$\Gamma_X(t,s) = f(t)f(s)G(s), \;\; s \le t,$$

where $G(s) = \int_0^s g(u)^2 du$. With some note, it is shown that the covariance $\Gamma_Y(t,s)$ of the normalized process is

$$\Gamma_Y(t,s) = \sqrt{\frac{G(s)}{G(t)}}.$$

It is, by assumption iii), expressed in the form by a suitable function f, for $s \le t$,

$$\Gamma_Y(t,s) = f((0,1;s,t)) = f(\frac{s(1-t)}{t(1-s)}),$$

where $(0,1;s,t)$ is the anharmonic ration of $0,1,s,t$. Hence, we may write

$$\frac{G(s)}{G(t)} = h(\frac{s/(1-s)}{t/(1-t)}),$$

with a suitable function h. The above functional equation asserts that h is linear, so that we complete the proof. □

Hence, we can write, up to a constant,

$$X(t) = B_0(t) = (1-t)\int_0^t \frac{1}{1-u}\dot{B}(u)du.$$

Going back to the Feynman's path integral, it is noted that the functional $\int_0^t \dot{B}(u)^2 du$ as well as its exponentials play an important role. Those functionals belong to the space of *generalized white noise functionals* for which rigorous definition is given and the analysis has been established, e.g. in connection with the Lévy Laplacian.

A generalization

There is a hope that the above discussion would be generalized to the case of a field parameterized by a curve (or a surface) C. As a first step we have stablished the following result (see [6]).

Let $x(u), u \in R^2$, be a two-dimensional parameter white noise, and let $\{C_r, r > 0\}$ be a family of concentric circles. Denote by (C) the disc enclosed by C. Now we define a Gaussian random field $X(C_r)$ by

$$X(C_r) = f(r)\int_{(C_r)-(C_0)} g(u)x(u)du, \;\; C_0 = C_{r_0},$$

where

$$f(r) = \sqrt{\frac{(r_1^2 - r_0^2)(r_1^2 - r^2)}{r^2 - r_0^2}}$$

and

$$g(u) = (r_1^2 - |u|^2)^{-1}.$$

It is easy to prove that the covariance function of $X(C_r)$ is given by

$$\pi\sqrt{(s^2, t^2; r_0^2, r_1^2)}.$$

Theorem 5.2. *Apply a conformal transformation to $x(u)$:*

$$x(u) \longrightarrow x(\frac{r_0 r_1}{|u|^2} u)\frac{r_0 r_1}{|u|^2}.$$

Then, we have the same Gaussian random field.

This property may be called the conformalinvariance of the field $X(C)$.

Remark 5.1. Let $X(C_r)$ be the same as above. Take a diffeomorphism g which is in the infinite dimensional roattion group $O(E)$. Then, the same conformal invariance hols for the field $\{X(gC_r)\}$.

It is noted that L. Accardi and I. Volovich [1] have given another rigorous definition of the Feynman functional integral by using the Feynman causal $i\epsilon$-prescription and white noise calculus.

6. Concluding remark

Under the same technique that was emplyed in the last section, we can discuss the Chern-Simon-Witten action integral. There we shall use a generalization of the Brownian Bridge whose value is higher (actually two) dimensional parameter.

The action has arosen from quantum dynamics and non commutative geometry, and it has now developed to be an interesting subject of stochastic analysis. For details, see [11]. [12] and references listed there.

References

1. L. Accardi and I.V. Volovich, Feynman functional integrals and the stochastic limit. Volterra Center Notes N.317, 1998.

2. L. Accardi, T. Hida and Si Si, Innovation of some stochastic processes. Volterra Center Notes. No.537. 2002.
3. P.A.M. Dirac, The Lagrangian in quantum mechanics. Phys. Zeitsch. der Sowjetunion. 3, (1933), 64-72..
4. R. Feynman, Space-time approach to non-relativistic quantum mechanics. Review of Modern Phys. 20 (1948), 367-387.
5. T. Hida, h.-H. Kuo, J. Potthoff and L. Streit, White Noise. An infinite dimensional calculus. Kluwer Academic Press. 1993.
6. T. Hida and Si Si, An innovation approach to random fields. Application of white nois theory. World Scientific Pub. Co. Ltd. 2004.
7. T. Hida and Si Si, Lévy field as a generator of elementar noises. Proc. Levico Conf. 2003.
8. T. Hida, Some of future directions of white nnoise theory. presented at International Conference on Quantum Information held at Tokyo Univ. of Sci. Nov. 2003.
9. T. Kuna, L. Streit and W. Westerkamp, Feynman integrals for a class of exponentially growing potentials. J/ Math. Phys. 39 (1998), 4476-4491.
10. H.-H. Kuo, White Noise Distribution Theory. CRC Press. 1996.
11. R. Léandre and H. Ouerdiane, Connes-Hida calculus and Bismut-Quillen superconnections.
12. L. Streit and W. Westerkamp, Feynman integrals for a class of exponentially growing potentials. J/ Math. Phys. 39 (1998), 4476-4491. Les prépub. de l'Institut Élie Cartan. 2003. no.17.
13. P.Lévy, Théorie de l'addition des variables aléatoires. Gauthier-Villars, 1937.
14. P. Lévy, Random functions: general theory with special reference to Laplacian random functions. Univ. of Calif. Pub. in Stat, vol.1, 1953, 331-390.
15. P. Lévy, Fonctions aléatoires à corrélation linéaire. Illinois J. Math. 1 (1957), 217-258.
16. P. Massani and N. Wiener, Non-linear prediction. Probability and Statistics. H. Cramér volume. ed. U. Grenander. John Wiley Sons, 1959. 190-212.
17. L. Streit and T. Hida, Generalized Brownian functionals anf the Feynman integral. Stochastic Processes and their Applications. 16 (1983), 55-69.
18. N. Wiener, Exprapolation, interpolation and smoothing of stationary time series. The MIT Press. 1949.
19. N. Wiener, Nonlinear problems in random theory. The MIT Press. 1958.
20. Win Win Htay, Note on linear processes. Proc. Nagoya Winter School. Jan. 2003. to appear.

GENERALIZED QUANTUM TURING MACHINE AND ITS APPLICATION TO THE SAT CHAOS ALGORITHM

SATOSHI IRIYAMA, MASANORI OHYA

Department of Information Sciences
Science University of Tokyo
Chiba 278-8510 Japan
E-mail: ohya@is.noda.tus.ac.jp, iriyama@is.noda.tus.ac.jp

IGOR VOLOVICH

Steklow Mathematical Institute
Gubkin St. 8, 117966, GSP-1 Moskow, Russia
E-mail: volovich@mi.ras.ru

Ohya and Volovich have proposed a new quantum algorithm with chaotic amplification to solve the SAT problem, which went beyond usual quantum algorithm. In this paper, we generalize quantum Turing machine in terms of general channel, not always unitary, transformation. Moreover, some computational classes in generalized quantum Turing machine are introduced, and it is shown that the Ohya-Volovich (OV) SAT algorithm can be described in this generalized Turing machine.

1. Introduction

The problem whether NP-complete problems can be P problem has been considered as one of the most important problems in theory of computational complexity. Various studies have been done for many years [1]. Ohya and Volovich [2,3] proposed a new quantum algorithm with chaotic amplification to solve the SAT problem, which went beyond usual quantum algorithm. This quantum chaos algorithm enabled to solve the SAT problem in a polynomial time [2,3,4].

In this paper we generalize quantum Turing machine so that it enables to describe non-unitary evolution of states. This study is based on mathematical studies of quantum communication channels [5,6]. It is discussed in this generalized quantum Turing machine (GQTM) that we can treat the OV SAT algorithm.

In Section 2, we generalize QTM by rewriting usual QTM in terms of channel transformation so that it contains both dissipative and unitary dynamics. In Section 3, the SAT problem is reviewed and fundamental quantum unitary gates are presented. In Section 4, based on the papers [4,8], we concretely construct the fundamental gates needed for computation of the SAT problem. In Section 5, we rewrite the total process including a measurement process and amplifier process with chaotic dynamics by GQTM.

2. Generalized Quantum Turing Machine

Classical Turing machine(TM or CTM) M_{cl} is defined by a triplet (Q, Σ, δ), where Σ is a finite alphabets with an identified blank symbol $\#$, Q is a finite set of states (with an initial state q_0 and a set of final states q_f) and $\delta : Q \times \Sigma \to Q \times \Sigma \times \{-1, 0, 1\}$ is a transition function. Note that $\{-1, 0, 1\}$ indicates moving direction of the tape head of TM. The deterministic TM has a deterministic transition function $\delta : Q \times \Sigma \to 2^Q \times \Sigma \times \{-1, 0, 1\}$, that is, δ is a non-branching map, in other words, the range of δ for each $(q, a) \in Q \times \Sigma$ is unique. A TM M is called non-deterministic if it is not deterministic.

Quantum Turing machine (QTM) was introduced by Deutsch [9] and was studied by Bernstein and Vazirani [10]. In this section, we introduce a generalized quantum Turing machine (GQTM) which contains QTM as a special case.

The Hilbert space $\mathcal{H}$ of QTM consists from complex functions defined on the space of classical configurations.

Definition 2.1. Usual Quantum Turing machine M_q is defined by a quadruplet $M_q = (Q, \Sigma, \mathcal{H}, U)$, where $\mathcal{H}$ is a Hilbert space described below in (1)and U is a unitary operator on the space $\mathcal{H}$ of the special form described below in (2).

Let $\mathcal{C} = Q \times \Sigma \times \mathbb{Z}$ be the set of all configurations of the Turing machine, where $\mathbb{Z}$ is the set of all integers. It is a countable set and one has

$$\mathcal{H} = \left\{ \varphi \mid \varphi : \mathcal{C} \to \mathbb{C}; \sum_{C \in \mathcal{C}} |\varphi(C)|^2 < \infty \right\}. \tag{1}$$

Since the configuration $C \in \mathcal{C}$ can be written as $C = (q, A, i)$ one can say that the set of functions $\{\mid q, A, i >\}$ is a basis in the Hilbert space $\mathcal{H}$. Here $q \in Q$, $i \in \mathbb{Z}$ and A is a function $A : \mathbb{Z} \to \not{k}$. We will call this basis the *computational basis.*

By using the computational basis we now state the conditions to the unitary operator U. We denote the set $\Gamma \equiv \{-1,0,1\}$. One requires that there is a function $\delta : Q\times\Sigma\times Q\times\Sigma\times\Gamma \to \widetilde{\mathbb{C}}$ which takes values in the field of computable numbers $\widetilde{\mathbb{C}}$ and such that the following relation is satisfied:

$$U\left|q,A,i\right\rangle = \sum_{p,b,\sigma} \delta(q,A(i),p,b,\sigma)\left|p,A_i^b,i+\sigma\right\rangle . \tag{2}$$

Here the sum runs over the states $p \in Q$, the symbols $b \in \Sigma$ and the elements $\sigma \in \Gamma$. Actually this is a finite sum. The function $A_i^b : \mathbb{Z} \to \not{A}$ is defined as

$$A_i^b(j) = \begin{cases} b \text{ if } j = i, \\ A(j) \text{ if } j \neq i. \end{cases}$$

The restriction to the computable number field $\tilde{\mathbb{C}}$ instead of all the complex number $\mathbb{C}$ is required since otherwise we can not construct or design a Turing Machine.

Note that if, for some integer $t \in \mathbb{N} \equiv \{1,2,...\}$, the quantum state $U^t\left|q_0,A,0\right\rangle$ is a final quantum state, i.e., $\|E_Q(q_F)U^s\left|q_0,A,0\right\rangle\| = 1$ and for any $s<t$, $s\in\mathbb{N}$ one has $\|E_Q(q_F)U^s\left|q_0,A,0\right\rangle\| = 0$, then one says that the quantum Turing machine *halts with running time* t on input A.

Now we define the generalized quantum Turing machine (GQTM) by using of a channel Λ (see below) instead of a unitary operator U.

Definition 2.2. Generalized Quantum Turing machine M_{gq} (GQTM) is defined by a quadruplet $M_{gq} = (Q,\Sigma,\mathcal{H},\Lambda)$, where Q, Σ and $\mathcal{H}$ are the same as above and Λ is a channel on the space of states on $\mathcal{H}$ as described below.

Let us explain GQTM in more detailed. GQTM M_{gq} is defined by quadruplet $(Q,\Sigma,\mathcal{H},\Lambda)$, where Q is a processor configuration, Σ is a set of alphabet including a blank symbol and Λ is a quantum transition function sending a quantum state to a quantum state. Q and Σ are represented by a density operator on Hilbert space $\mathcal{H}_Q$ and $\mathcal{H}_\Sigma$,which are spanned by canonical basis $\{\left|q\right\rangle; q\in Q\}$ and $\{\left|a\right\rangle; a\in\Sigma\}$, respectively. A tape configuration A is a sequence of elements of Σ represented by a density operator on Hilbert space $\mathcal{H}_\Sigma$ spanned by a canonical basis $\{\left|A\right\rangle; A\in\Sigma^*\}$, where Σ^* is the set of sequences of alphabets in Σ. A position of tape head is represented by a density operator on Hilbert space $\mathcal{H}_Z$ spanned by a canonical basis $\{|i\rangle ; i \in \mathbf{Z}\}$. Then a configuration ρ of GQTM M_{gq} is described by a density operator in $\mathcal{H} \equiv \mathcal{H}_Q \otimes \mathcal{H}_\Sigma \otimes \mathcal{H}_Z$. Let $\mathfrak{S}(\mathcal{H})$ be the set of all density operators in Hilbert space $\mathcal{H}$. A quantum transition function is given by a completely positive (CP) channel

$$\Lambda : \mathfrak{S}(\mathcal{H}) \to \mathfrak{S}(\mathcal{H}),$$

whose explicit form is discussed below. For instance, given a configuration $\rho \equiv \sum_k \lambda_k |\psi_k\rangle \langle\psi_k|$, where $\sum \lambda_k = 1, \lambda_k \geq 0$ and $\psi_k = |q_k\rangle \otimes |A_k\rangle \otimes |i_k\rangle$ $(q_k \in Q, A_k \in \Sigma^*, i_k \in \mathbf{Z})$ is a vector in a basis of $\mathcal{H}$. This configuration changes to a new configuration ρ' by one step transition as $\rho' = \Lambda(\rho) = \sum_k \mu_k |\psi_k\rangle \langle\psi_k|$ with $\sum \mu_k = 1, \mu_k \geq 0$.

One requirement on GQTM $M_{gq} = (Q, \Sigma, \mathcal{H}, \Lambda)$ is the correspondence with QTM. If the channel Λ in GQTM is a unitary operator U, then GQTM $M_{gq} = (Q, \Sigma, \mathcal{H}, \Lambda = U \cdot U^*)$ reduces to QTM $M_q = (Q, \Sigma, \mathcal{H}, U)$.

Several studies have been done on QTM whose transition function is represented by unitary operator, in which various theorems and computational classes in QTM were discussed in [10,11].

Let us explain how to construct a QTM. Let δ be a function

$$\delta : Q \times \Sigma \times Q \times \Sigma \times \{-1, 0, 1\} \to \mathbb{C}.$$

For any $q \in Q, a \in \Sigma$, it holds

$$\sum_{p \in Q, b \in \Sigma, d \in \{-1,0,1\}} |\delta(q, a, p, b, d)|^2 = 1.$$

For any $q \in Q, a \in \Sigma$, $q' (\neq q) \in Q, a' (\neq a) \in \Sigma$, it holds

$$\sum_{p \in Q, b \in \Sigma, d \in \{-1,0,1\}} \delta(q', a', p, b, d)^* \delta(q, a, p, b, d) = 0.$$

Given QTM M_q and its configuration $\rho = |\varphi\rangle \langle\varphi|$ with $|\varphi\rangle = |q, A, i\rangle$, after one step, this configuration is changed by the transition function δ as

$$\Lambda(|q, A, i\rangle \langle i, A, q|) = \sum_{p, b, \sigma, p', b', \sigma'} \delta(q, A(i), p, b, \sigma) \delta^*(q, A(i), p', b', \sigma') \left|p, A_i^b, i + \sigma\right\rangle \langle i + \sigma', A_i^{b'}, p'|$$

Remark 2.1. For any $q, p \in Q, a, b \in \Sigma, d \in \{-1, 0, 1\}$, let $\delta(q, a, p, b, d) = \{0, 1\}$, then QTM is a reversal TM.

A transition of GQTM is regarded as a transition of amplitude of each configuration vector. We categorize GQTMs by a property of CP channel Λ as below.

Definition 2.3. A GQTM M_{gq} is called unitary QTM (UQTM, i.e., usual QTM), if all of quantum transition function Λ in M_{gq} are unitary CP channel.

For all configuration $\rho = \sum_n \lambda_n \rho_n$ $(\Sigma_n \lambda_n = 1, \lambda_n \geq 0)$, a GQTM M_{gq} is called LQTM M_{lq} if Λ is affine ; $\Lambda\left(\sum_n \lambda_n \rho_n\right) = \sum_n \lambda_n \Lambda\left(\rho_n\right)$. Since a measurement defined by $\Lambda_M \rho = \sum\limits_k P_k \rho P_k$ with a PVM $\{P_k\}$ on $\mathcal{H}$ is a linear CP channel, LQTM may include the measurement process of the above type.

For a more general channel the state change is expressed as

$$\Lambda(|q, A\,(i)\,, i\rangle\,\langle q, A\,(i)\,, i|) = \sum_{p,b,\sigma,p',b',\sigma'} \delta(q, A(i), p, b, \sigma, p', b', \sigma') \left|p, A_i^b, i+\sigma\right\rangle \langle p, A_i^{b'}, i+\sigma|$$

with some function $\delta(q, A(i), p, b, \sigma, p', b', \sigma')$ such that the RHS of this relation is a state. The quantum transition function Λ is one of GQTM, which is defined through a transition function $\delta : Q \times \Sigma \times Q \times \Sigma \times \Gamma \times Q \times \Sigma \times \Gamma \rightarrow \widetilde{\mathbb{C}}$.

Thus we define two more classes of GQTM for non-unitary CP channels.

Definition 2.4. A GQTM M_{gq} is called a linear QTM(LQTM) if its quantum transition function Λ is a linear quantum channel.

Unitary operator is linear, hence UQTM is a sub-class of LQTM. Moreover, classical TM is a special class of LQTM.

Definition 2.5. A GQTM M_{gq} is called non-linear QTM ($NLQTM$) if its quantum transition function Λ contains non-linear CP channel.

A chaos amplifier used in [2,3] is a non-linear CP channel, the details of this channel and its application to the SAT problem will be discussed in the sequel.

2.1. *Computational class for GQTM*

Let us review some language classes which classical Turing machine recognizes.

Definition 2.6. The class of languages is in P if its language is recognized by a deterministic Turing machine in polynomial time of input size.

Definition 2.7. The class of languages is in NP if there is a non-deterministic Turing machine, called the *verifier, which recognize languages with some informations in polynomial time* of input size. Besides, if a language L_1 ∈NP and L_1 reduces to L_2 ∈NP in polynomial time, a language L_1 is NP-complete.

Definition 2.8. If languages are accepted by non-deterministic Turing machine in polynomial time of input size with a certain probability, then this class of languages is called the class of bounded probability polynomial time(BPP).

A NP-complete language is the most difficult one in NP. If there is a polynomial time algorithm to solve it in the above sense, it implies P=NP. The existence of such a algorithm is demonstrated in [2,3] in an extended quantum domain, as is reviewed in the next section. We will show that this OV algorithm can be written by GQTM in the sequel section.

Given a GQTM $M_{gq} = (Q, \Sigma, \delta)$ and an input configuration $\rho_0 = |v_{in}\rangle \langle v_{in}|$, $(|v_{in}\rangle = |q_0\rangle \otimes |T\rangle \otimes |0\rangle)$, a computation process is described as the following product of several different types of channels

$$\Lambda_1 \circ \cdots \circ \Lambda_t (\rho_0) = \rho_f \equiv |v_f\rangle \langle v_f|$$

where $\Lambda_1, \cdots, \Lambda_t$ are CP channels. Applying the CP channels to an initial state, we obtain a final state ρ_f and we measure this state by a projection (or PVM)

$$P_f = |q_f\rangle \langle q_f| \otimes I_\Sigma \otimes I_Z,$$

where I_Σ, I_Z are identity operators on $\mathcal{H}_\Sigma, \mathcal{H}_Z$, respectively. Let $p \geq 0$ be a halting probability such that

$$tr_{\mathcal{H}_\Sigma \otimes \mathcal{H}_Z} (P_f \rho_f) = p\, |q_f\rangle \langle q_f|\,.$$

Then, we define the *acceptance (rejection)* of GQTM and some classes of languages.

Definition 2.9. Given GQTM M_{gq} and a language L, if there exists N steps when we obtain the configuration of acceptance (*or rejection*)by the probability p, we say that the GQTM M_{gq} accepts (*or rejects*)L by the probability p and its computational complexity is N.

Definition 2.10. A language L is bounded quantum probability polynomial time GQTM(BGQPP) if there is a polynomial time GQTM M_{gq} which accepts L with probability $p \geq \frac{1}{2}$.

Similarly, we can define the class of languages BUQPP(= BQPP), BLQPP, BNLQPP corresponding to UQTM, LQTM and NLQTM, respectively.

In Section 2, it is pointed out that LQTM includes classical TM, which implies

$$BPP \subseteq BLQPPL \subseteq BNLQPP \subseteq BGQPP.$$

Moreover, if NLQTM accepts the SAT OV algorithm in polynomial time with probability $p \geq \frac{1}{2}$, then we may have the inclusion

$$NP \subseteq BGQPP$$

We will discuss this inclusion in Sec. 4 by constructing GQTM which accepts the SAT OV algorithm.

3. SAT Problem

Let $X \equiv \{x_1, \ldots, x_n\}, n \in \mathbf{N}$ be a set. x_k and its negation $\overline{x}_k$ $(k = 1, \ldots, n)$ are called literals. Let $\overline{X} \equiv \{\bar{x}_1, \ldots, \bar{x}_n\}$ be a set, then the set of all literals is denoted by $X' \equiv X \cup \bar{X} = \{x_1, \ldots, x_n, \bar{x}_1, \ldots, \bar{x}_n\}$. The set of all subsets of X and $\bar{X}$ is denoted by $F(X)$ and $F(\bar{X})$, respectively. An element $C \in F(X) \cup F(\bar{X})$ is called a clause if $(C \cap \mathcal{F}(X)) \cap (C \cap \mathcal{F}(\bar{X})) = \emptyset$. We take a truth assignment to all variables x_k. If we can assign the truth value to at least one element of C, then C is called satisfiable. When C is satisfiable, the truth value $t(C)$ of C is regarded as true, otherwise, that of C is false. Take the truth values as "true $\leftrightarrow$1, false $\leftrightarrow$0". Then $C$is satisfiable iff $t(C) = 1$.

Let $L = \{0, 1\}$ be a Boolean lattice with usual join $\vee$ and meet $\wedge$, and $t(x)$ be the truth value of a literal x in X. Then the truth value of a clause C is written as $t(C) \equiv \vee_{x \in C} t(x)$.

Moreover the set $\mathcal{C}$ of all clauses C_j $(j = 1, 2, \cdots, m)$ is called satisfiable iff the meet of all truth values of C_j is 1; $t(\mathcal{C}) \equiv \wedge_{j=1}^{m} t(C_j) = 1$. Thus the SAT problem is written as follows:

Definition 3.1. SAT Problem: Given a Boolean set $X \equiv \{x_1, \cdots, x_n\}$ and a set $\mathcal{C} = \{C_1, \cdots, C_m\}$ of clauses, determine whether $\mathcal{C}$ is satisfiable or not.

That is, this problem is to ask whether there exists a truth assignment to make C satisfiable. It is known in usual algorithm that it is polynomial time to check the satisfiability only when a specific truth assignment is given, but we can not determine the satisfiability in polynomial time when an assignment is not specified.

In [4] we discussed the quantum algorithm of the SAT problem, which was rewritten in [8] with showing that the OM SAT-algorithm is combinatorial. In [2,3] it is shown that the chaotic quantum algorithm can solve the SAT problem in polynomial time.

Ohya and Masuda pointed out [4] that the SAT problem, hence all other NP problems, can be solved in polynomial time by quantum computer if the superposition of two orthogonal vectors $|0\rangle$ and $|1\rangle$ is physically detected. However this detection is considered not to be possible in the present technology. The problem to be overcome is how to distinguish the pure vector $|0\rangle$ from the superposed one $\alpha|0\rangle+\beta|1\rangle$, obtained by the OM SAT-quantum algorithm, if β is not zero but very small. If such a distinction is possible, then we can solve the NPC problem in the polynomial time. In [2,3] it is shown that it can be possible by combining nonlinear chaos amplifier with the quantum algorithm, which implies the existence of a mathematical algorithm solving NP=P. The algorithm of Ohya and Volovich is not known to be in the framework of quantum Turing algorithm or not. This aspect is studied in this paper.

3.1. *Quantum computation*

In this subsection, we review fundamentals of quantum computation (see, for instance, [3,12]) for the SAT problem. Let $\mathbb{C}$ be the set of all complex numbers, and $|0\rangle$ and $|1\rangle$ be the two unit vectors $\binom{1}{0}$ and $\binom{0}{1}$, respectively. Then, for any two complex numbers α and β satisfying $|\alpha|^2+|\beta|^2=1$, $\alpha|0\rangle+\beta|1\rangle$ is called a qubit. For any positive integer N, let $\mathcal{H}$ be the tensor product Hilbert space defined as $\left(\mathbb{C}^2\right)^{\otimes N}$ and let $\left\{|e_i\rangle\,;0\leq i\leq 2^{N-1}\right\}$ be the basis whose elements are denoted as

$$\begin{aligned}
|e_0\rangle &= |0\rangle \otimes |0\rangle \cdots \otimes |0\rangle \equiv |0,0,\cdots,0\rangle\,,\\
|e_1\rangle &= |1\rangle \otimes |0\rangle \cdots \otimes |0\rangle \equiv |1,0,\cdots,0\rangle\,,\\
|e_2\rangle &= |0\rangle \otimes |1\rangle \cdots \otimes |0\rangle \equiv |0,1,\cdots,0\rangle\,,\\
&\vdots\\
|e_{2^N-1}\rangle &= |1\rangle \otimes |1\rangle \cdots \otimes |1\rangle \equiv |1,1,\cdots,1\rangle\,.
\end{aligned}$$

For any two qubits $|x\rangle$ and $|y\rangle$, $|x,y\rangle$ and $\left|x^N\right\rangle$ is written as $|x\rangle \otimes |y\rangle$ and $\underbrace{|x\rangle \otimes \cdots \otimes |x\rangle}_{N \text{ times}}$, respectively.

The usual (unitary) quantum computation can be formulated mathematically as the multiplication by unitary operators. Let U_{NOT},U_{CN} and U_{CCN} be the three unitary operators defined as

$$\begin{aligned}
U_{NOT} &\equiv |1\rangle\langle 0| + |0\rangle\langle 1|\,,\\
U_{CN} &\equiv |0\rangle\langle 0| \otimes I + |1\rangle\langle 1| \otimes U_{NOT},\\
U_{CCN} &\equiv |0\rangle\langle 0| \otimes I \otimes I + |1\rangle\langle 1| \otimes |0\rangle\langle 0| \otimes I + |1\rangle\langle 1| \otimes |1\rangle\langle 1| \otimes U_{NOT}.
\end{aligned}$$

U_{NOT},U_{CN} and U_{CCN} represent the NOT-gate, the Controlled-NOT gate and the Controlled-Controlled-NOT gate, respectively. Moreover, Hadamard transformation H is defined as the transformation on $\mathbb{C}^2$ such as

$$H|0\rangle = \frac{1}{\sqrt{2}}(|0\rangle + |1\rangle)\,,\ H|1\rangle = \frac{1}{\sqrt{2}}(|0\rangle - |1\rangle)\,.$$

These four operators U_{NOT}, U_{CN}, U_{CCN} and H are called the elementary gates here. For any $k \in \mathbb{N}$, $U_H^{(N)}(k)$ denotes the k-tuple Hadamard transformation on $\left(\mathbb{C}^2\right)^{\otimes N}$ defined as

$$U_H^{(N)}(k)\left|0^N\right\rangle = \frac{1}{2^{k/2}}(|0\rangle + |1\rangle)^{\otimes k}\left|0^{N-k}\right\rangle = \frac{1}{2^{k/2}}\sum_{i=0}^{2^{k-1}}|e_i\rangle \otimes \left|0^{N-k}\right\rangle.$$

The above unitary operators can be extended to the unitary operators on $\left(\mathbb{C}^2\right)^{\otimes N}$:

$$U_{NOT}^{(N)}(u) \equiv I^{\otimes u-1} \otimes (|0\rangle\langle 1| + |1\rangle\langle 0|)\, I^{\otimes N-u-1}$$

$$U_{CN}^{(N)}(u,v) \equiv I^{\otimes u-1} \otimes |0\rangle\langle 0| \otimes I^{\otimes N-u-1} + I^{\otimes u-1} \otimes |1\rangle\langle 1| \otimes I^{\otimes v-u-1} \otimes U_{NOT} \otimes I^{\otimes N-v-1}$$

$$\begin{aligned} U_{CCN}^{(N)}(u,v,w) = {} & I^{\otimes u-1} \otimes |0\rangle\langle 0| \otimes I^{\otimes N-u-1} + I^{\otimes u-1} \otimes |1\rangle\langle 1| \\ & \otimes I^{\otimes v-u-1} \otimes |0\rangle\langle 0| \otimes I^{\otimes N-v-1} \\ & + I^{\otimes u-1} \otimes |1\rangle\langle 1| \otimes I^{\otimes v-u-1} \otimes |1\rangle\langle 1| \otimes \\ & I^{\otimes w-t-1} \otimes U_{NOT} \otimes I^{\otimes N-w-1}, \end{aligned}$$

where u, v and w be a positive integers satisfying $1 \leq u < v < w \leq N$.

4. SAT Algorithm

In this section, we explain the algorithm of the SAT problem which was introduced by Ohya-Masuda [4] and developed by Accardi-Sabbadini [8]. The computation of the truth value can be done by by a combination of the unitary operators on a Hilbert space $\mathcal{H}$, so that the computation is described by the unitary quantum algorithm. The detail of this section is given in the papers [4,3,8,?], so we will discuss just the essence of the OM algorithm. Throughout this section, let n be the total number of Boolean variables used in the SAT problem. Let $\mathcal{C}$ be a set of clauses whose cardinality is equal to m. Let $\mathcal{H} = \left(\mathbb{C}^2\right)^{\otimes n+\mu+1}$ be a Hilbert space and $|v_0\rangle$ be the initial state $|v_0\rangle = |0^n, 0^\mu, 0\rangle$, where μ is the number of dust qubits which is determined by the following proposition. Let $U_{\mathcal{C}}^{(n)}$ be a unitary operator for the computation of the SAT:

$$U_{\mathcal{C}}^{(n)} |v_0\rangle = \frac{1}{\sqrt{2^n}} \sum_{i=0}^{2^n-1} |e_i, x^\mu, t_{e_i}(\mathcal{C})\rangle \equiv |v_f\rangle$$

where x^μ denotes the μ strings in the dust bits and $t_{e_i}(\mathcal{C})$ is the truth value of $\mathcal{C}$ with e_i. In [4,8], $U_{\mathcal{C}}^{(n)}$ was constructed.

Let $\{s_k; k = 1, \ldots, m\}$ be the sequence defined as

$$\begin{aligned} s_1 &= n+1, \\ s_2 &= s_1 + card(C_1) + \delta_{1,card(C_1)} - 1, \\ s_i &= s_{i-1} + card(C_{i-1}) + \delta_{1,card(C_{i-1})}, \quad 3 \leq i \leq m, \end{aligned}$$

where $card(C_i)$ means the cardinality of a clause C_i. And let define s_f as

$$s_f = s_m - 1 + card(C_m) + \delta_{1,card(C_m)}.$$

Note that the number m of the clause is at most $2n$. Then we have the following proposition and theorem.

Proposition 4.1. *For $m \geq 2$, the total number of dust qubits μ is*

$$\begin{aligned}\mu &= s_f - 1 - n \\ &= \sum_{k=1}^{m} card\left(C_k\right) + \delta_{1,card(C_k)} - 2.\end{aligned}$$

Determining μ and the work spaces for computing $t\left(C_k\right)$, we can construct $U_C^{(n)}$ concretely. We use the following unitary operators for this concrete expression:

$$U_A^{(n)}\left(k\right) = \begin{cases} U_{CCN}^{(n)}\left(s_{k+1}-1, s_{k+2}-2, s_{k+2}-1\right), & \left(1 \leq k \leq m-2\right) \\ U_{CCN}^{(n)}\left(s_m - 1, s_f - 1, s_f\right), & \left(k = m-1\right) \end{cases},$$

Here, we consider an decomposition of the set of indices $J\left(C_k\right)$ of elements in C_k such as $J\left(C_k\right) = j\left(C_k\right) \cup \bar{j}\left(C_k\right)$, where $j\left(C_k\right) = \left\{i; x_i \in C_k \cap \mathcal{F}\left(X\right)\right\}$ and $\bar{j}\left(C_k\right) = \left\{i; x_i \in C_k \cap \mathcal{F}\left(\bar{X}\right)\right\}$. Since $\left(C \cap \mathcal{F}\left(X\right)\right) \cap \left(C \cap \mathcal{F}\left(\bar{X}\right)\right) = \emptyset$, this decomposition is unique. Then we construct the followins unitary gates to calculate $t\left(C_k\right)$:

$$U_O^{(n)}\left(k\right) = U_N^{(n)}\left(C_k\right) U_{O1}^{(n)}\left(C_k\right) U_N^{(n)}\left(C_k\right),$$

where

$$\begin{aligned}U_N^{(n)}\left(C_k\right) &= \prod_{i \in \bar{j}(C_k)} U_{NOT}^{(n)}\left(i\right), \\ U_{O1}^{(n)}\left(C_k\right) &= U_{O2}^{(n)}\left(j_{card(C_k)}, s_k - card\left(C_k\right) - 1, s_k - card\left(C_k\right) - 2\right) \cdot \cdots \cdot \\ &\quad \cdot U_{O2}^{(n)}\left(j_3, s_k, s_k + 1\right) \cdot U_{O2}^{(n)}\left(j_1, j_2, s_k\right), \\ U_{O2}^{(n)}\left(u, v, w\right) &= U_{CCN}^{(N)}\left(u, v, w\right) \cdot U_{CN}^{(N)}\left(v, w\right) \cdot U_{CN}^{(N)}\left(u, w\right)\end{aligned}$$

and where $j_1, j_2, j_3, \cdots, j_{card(C_k)} \in J\left(C_k\right)$.

Theorem 4.1. *The unitary operator $U_C^{(n)}$, is represented as*

$$\begin{aligned}U_C^{(n)} &= U_A^{(n+\mu+1)}\left(m-1\right) \cdot U_A^{(n+\mu+1)}\left(m-2\right) \cdot \cdots \cdot U_A^{(n+\mu+1)}\left(1\right) \\ &\quad \cdot U_O^{(n+\mu+1)}\left(m\right) \cdot U_O^{(n+\mu+1)}\left(m-1\right) \cdot \cdots \cdot U_O^{(n+\mu+1)}\left(1\right) \cdot U_H^{(n+\mu+1)}\left(n\right).\end{aligned}$$

4.1. *The resulting state in the SAT algorithm*

Applying the above unitary operator to the initial state, we obtain the final state ρ.The result of the computation is registered as $|t(\mathcal{C})\rangle$ in the last section of the final vector, which will be taken out by a projection $P_{n+\mu,1} \equiv I^{\otimes n+\mu} \otimes |1\rangle\langle 1|$ onto the subspace of $\mathcal{H}$ spanned by the vectors $|\varepsilon^n, \varepsilon^\mu, 1\rangle$.

The following theorem is easily seen.

Theorem 4.2. *$\mathcal{C}$ is SAT if and only if*

$$P_{n+\mu,1} U_{\mathcal{C}}^{(n)} |v_0\rangle \neq 0$$

According to the standard theory of quantum measurement, after a measurement of the event $P_{n+\mu,1}$, the state $\rho = |v_f >< v_f|$ becomes

$$\rho \to \frac{P_{n+\mu,1} \rho P_{n+\mu,1}}{Tr \rho P_{n+\mu,1}} =: \overline{\rho}$$

Thus the solvability of the SAT problem is reduced to check that $\overline{\rho} \neq 0$. The difficulty is that the probability

$$Tr \overline{\rho} P_{n+\mu,1} = \| P_{n+\mu,1} |v_f\rangle \|^2 = \frac{|T(\mathcal{C}_0)|}{2^n}$$

is very small in some cases, where $|T(\mathcal{C}_0)|$ is the cardinality of the set $T(\mathcal{C}_0)$, of all the truth functions t such that $t(\mathcal{C}_0) = 1$.

We put $q \equiv \sqrt{\frac{r}{2^n}}$ with $r \equiv |T(\mathcal{C}_0)|$. Then if r is suitably large to detect it, then the SAT problem is solved in polynomial time. However, for small r, the probability is very small so that we in fact do not get an information about the existence of the solution of the equation $t(\mathcal{C}_0) = 1$, hence in such a case we need further deliberation.

Let go back to the SAT algorithm. After the quantum computation, the quantum computer will be in the state

$$|v_f\rangle = \sqrt{1-q^2}\, |\varphi_0\rangle \otimes |0\rangle + q\, |\varphi_1\rangle \otimes |1\rangle$$

where $|\varphi_1\rangle$ and $|\varphi_0\rangle$ are normalized n ($=n+\mu$) qubit states and $q = \sqrt{r/2^n}$. Effectively our problem is reduced to the following 1 qubit problem: The above state $|v_f\rangle$ is reduced to the state

$$|\psi\rangle = \sqrt{1-q^2}\, |0\rangle + q\, |1\rangle\,,$$

and we want to distinguish between the cases $q = 0$ and $q > 0$(small positive number). Let us denote the correspondence from $\rho_0 \equiv |v_0\rangle\langle v_0|$ with $\rho \equiv |\psi\rangle\langle\psi|$ by a channel Λ_I;

$$\rho = \Lambda_I \rho_0.$$

It is argued in [14] that quantum computer can speed up **NP** problems quadratically but not exponentially. The no-go theorem states that if the inner product of two quantum states is close to 1, then the probability that a measurement distinguishes which one of the two is exponentially small. And one may claim that amplification of this distinguishability is not possible in usual quantum algorithm. At this point we emphasized [3] that we do not propose to make a measurement which will be overwhelmingly likely to fail. What we did is a proposal to use the output $|\psi\rangle$ of the quantum computer as an input for another device which uses chaotic dynamics. *The amplification would be not possible if we use the standard model of quantum computations with a unitary evolution. However the idea of the paper* [2,3] *is different. In* [2,3] *it is proposed to combine quantum computer with a chaotic dynamics amplifier. Such a quantum chaos computer is a new model of computations and we demonstrate that the amplification is possible in the polynomial time.*

One could object that we do not suggest a practical realization of the new model of computations. But at the moment nobody knows of how to make a practically useful implementation of the standard model of quantum computing ever. It seems to us that the quantum chaos computer considered in [3] deserves an investigation and has a potential to be realizable.

4.2. *Chaotic dynamics*

Various aspects of classical and quantum chaos have been the subject of numerous studies ([5,12] and ref's therein). Here we will briefly review how chaos can play a constructive role in computation (see [2,3] for the details).

Chaotic behavior in a classical system usually is considered as an exponential sensitivity to initial conditions. It is this sensitivity we would like to use to distinguish between the cases $q = 0$ and $q > 0$ discussed in the previous subsection.

Consider the so called logistic map which is given by the equation

$$x_{n+1} = ax_n(1 - x_n) \equiv g(x), \quad x_n \in [0, 1] .$$

The properties of the map depend on the parameter a. If we take, for example, $a = 3.71$, then the Lyapunov exponent is positive, the trajectory is very sensitive to the initial value and one has the chaotic behavior [3]. It is important to notice that if the initial value $x_0 = 0$, then $x_n = 0$ for all n.

The state $|\psi\rangle$ of the previous subsection is transformed into the density matrix of the form

$$\overline{\rho} = q^2 P_1 + \left(1 - q^2\right) P_0$$

where P_1 and P_0 are projectors to the state vectors $|1\rangle$ and $|0\rangle$. One has to notice that P_1 and P_0 generate an Abelian algebra which can be considered as a classical system. The density matrix $\overline{\rho}$ above is interpreted as the initial data, and we apply the channel $\Lambda \equiv \Lambda_{CA}$ due to the logistic map as

$$\Lambda_{CA}\left(\overline{\rho}\right) = \frac{\left(I + g\left(\overline{\rho}\right)\sigma_3\right)}{2},$$

where I is the identity matrix and σ_3 is the z-component of Pauli matrices.

$$\overline{\rho}_k = \Lambda_{CA}^k\left(\overline{\rho}\right)$$

To find a proper value k we finally measure the value of σ_3 in the state ρ_k such that

$$M_k \equiv tr\overline{\rho}_k\sigma_3.$$

We obtain [3]

Theorem 4.3.

$$\overline{\rho}_k = \frac{(I + g^k(q^2)\sigma_3)}{2}, \ \textit{and}\ M_k = g^k(q^2).$$

Thus the question is whether we can find such a number k in polynomial steps of n satisfying the inequality $M_k \geq \frac{1}{2}$ for very small but non-zero q^2. Here we have to remark that if one has $q = 0$ then $\overline{\rho} = P_0$ and we obtain $M_k = 0$ for all k. If $q \neq 0$, the chaotic dynamics leads to the amplification of the small magnitude q in such a way that it can be detected. The transition from $\overline{\rho}$ to $\overline{\rho}_k$ is nonlinear and can be considered as a classical evolution because our algebra generated by P_0 and P_1 is abelian. The amplification can be done within at most 2n steps due to the following propositions. Since $g^k(q^2)$ is x_k of the logistic map $x_{k+1} = g(x_k)$ with $x_0 = q^2$, we use the notation x_k in the logistic map for simplicity.

Theorem 4.4. *For the logistic map $x_{n+1} = ax_n\left(1 - x_n\right)$ with $a \in [0,4]$ and $x_0 \in [0,1]$, let x_0 be $\frac{1}{2^n}$ and a set J be $\{0,1,2,\ldots,n,\ldots,2n\}$. If a is 3.71, then there exists an integer k in J satisfying $x_k > \frac{1}{2}$.*

Theorem 4.5. *Let a and n be the same in above theorem. If there exists k in J such that $x_k > \frac{1}{2}$, then $k > \frac{n-1}{\log_2 3.71 - 1}$.*

Corollary 4.1. *If* $x_0 \equiv \frac{r}{2^n}$ *with* $r \equiv |T(\mathcal{C})|$ *and there exists* k *in* J *such that* $x_k > \frac{1}{2}$, *then there exists* k *satisfying the following inequality if* $\mathcal{C}$ *is SAT.*

$$\left[\frac{n-1-\log_2 r}{\log_2 3.71 - 1}\right] \leq k \leq \left[\frac{5}{4}(n-1)\right].$$

From these theorems, for all k, it holds

$$M_k \begin{cases} = 0 & \text{iff } \mathcal{C} \text{ is not SAT} \\ > 0 & \text{iff } \mathcal{C} \text{ is SAT} \end{cases}$$

5. SAT algorithm in GQTM

In this section, we construct a GQTM for the OV SAT algorithm. The GQTM with the chaos amplifier belongs to NLQTM because the chaos amplifier is described by a non-linear CP channel. The OV algorithm runs from an initial state $\rho_0 \equiv |v_0\rangle \langle v_0|$ to $\overline{\rho}_k$ through $\rho \equiv |v_f\rangle \langle v_f|$. The computation from $\rho_0 \equiv |v_0\rangle \langle v_0|$ to $\rho \equiv |v_f\rangle \langle v_f|$ is due to unitary channel $\Lambda_C \equiv U_C \bullet U_C$, and that from $\rho \equiv |v_f\rangle \langle v_f|$ to $\overline{\rho}_k$ is due to a non-unitary channel $\Lambda_{CA}^k \circ \Lambda_I$, so that all computation can be done by $\Lambda_{CA}^k \circ \Lambda_I \circ \Lambda_C$, which is a completely positive, so the whole computation process is deterministic. It is a multi-track (actually 4 tracks) GQTM that represents this whole computation process.

A multi-track GQTM has some workspaces for calculation, whose tracks are independent each other. This independence means that the TM can operate only one track at one step and all tracks do not affect each other. Let us explain our computation by a multi-track GQTM. The first track stores the input data and the second track stores the value of literals. The third track is used for the computation of $t(C_i), (i = 1, \cdots, m)$ described by unitary operators. The fourth track is used for the computation of $t(\mathcal{C})$ denoting the result. The work of GQTM is represented by the following 8 steps:

- Step 1 : Store the counter $c = 0$ in Track 1. Calculate $\left[\frac{5}{4}(n-1)\right] + 1$, we take this value as the maximum value of the counter. Then, store it in Track 4.
- Step 2 : Calculate $c + 1$ and store it in Track 4.
- Step 3 : Apply the Hadamard transform to Track 2.
- Step 4 : Calculate $t(C_1), \cdots t(C_m)$ and store them in Track 3.
- Step 5 : Calculate $t(\mathcal{C})$ by using the value of the third track, and store $t(\mathcal{C})$ in Track 4.

- Step 6 : Empty the first, second and third Tracks.
- Step 7 : Apply the chaos amplifier to the result state obtained up to the step 6.
- Step 8 : If $c = \left[\frac{5}{4}(n-1)\right]+1$ or GQTM is in the final state, GQTM halts. If GQTM is not in the final state, GQTM runs the step 2 to the step 8 again.

Let us explain the above steps for unitary computation (OM algorithm; i.e., up to the steps 6 above) by an example. Let the number of literals be n and that of clauses be m. Then the language is represented by the following strings

$$0^n X \prod_{i=1}^{m} C_S G(C_i) C_E,$$

where

$$G(C_i) = \varepsilon_1 \varepsilon_2 \ldots \varepsilon_n Y \overline{\varepsilon_1 \varepsilon_2} \ldots \overline{\varepsilon_n}$$

$$\varepsilon_k = \begin{cases} 0 & k \notin I_i \\ 1 & k \in I_i \end{cases}$$

$$\overline{\varepsilon_k} = \begin{cases} 0 & k \notin I_i' \\ 1 & k \in I_i' \end{cases}$$

and X, C_S, Y, C_E are used as particular symbols of clauses. For example, given $X = \{1,2,3\}$, $\mathcal{C} = \{C_1, C_2, C_3\}$, $C_1 = (\{1,2\},\{3\})$, $C_2 = (\{3\},\{2\})$, $C_3 = (\{1\},\{2,3\})$, the input tape will be

$$000XC_S110Y001C_EC_S001Y010C_EC_S100Y011C_E$$

First, our GQTM applies the DFT (discrete Fourier transform) to a part of literals on the track 2. The transition function for DFT is written by the following table. Put the vector in $\mathcal{H}_Q$ by $q.$ instead of $|q.\rangle$ and denote the direction moving the tape head by R for the right and L to the left (Note that O is the starting position).

	#	0	1	X
q_0		$q_a, 0, R$	$q_a, 1, R$	
q_a		$q_a, 0, R$	$q_a, 1, R$	q_b, X, L
q_b	$q_f, \#, R$	$\frac{1}{\sqrt{2}} q_b, 0, L + \frac{1}{\sqrt{2}} \varphi_b, 1, L$	$\frac{1}{\sqrt{2}} q_b, 0, L - \frac{1}{\sqrt{2}} q_b, 1, L$	

The tape head moves to the right until it reads a symbol C_S. When the tape head reads C_S, GQTM increases a program counter by one, while moves to the right until it reads 1. Then GQTM stops increasing the counter and the tape head moves to the top of the tape. According to the program counter, the tape head moves to the right as reducing the counter by one. When the counter becomes zero, GQTM reads the data and calculates OR with the data in the track 2, then GQTM writes the result in the track 3. GQTM goes back to the top of the track 1 and repeats the above processes until it reads Y.

When GQTM reads Y, it calculates OR with the negation and repeats the processes as above. When it reads C_E, it writes down f_{C_k} in the track 3 and clean the workspace for the next calculation. Then GQTM reads the blank symbol #, and it begins to calculate AND. The calculation of AND is done on the track 4. GQTM calculates them as moving to the left because the position of the tape head is at the end of the track 3 when the OR calculation is finished. Then the result of the calculation is showed on the top of the track 4.

The transition function of OR calculation is described, similar as classical TM, by the following three tables:

	0	1	C_S
q_0	$q_a, 0, R$		q_b, C_S, R
q_a	$q_a, 0, R$	$q_a, 1, R$	q_b, C_S, R
q_b	$q_{b,1}, 0, R$	$q_{c,1}, 0, L$	
$q_{b,1}$	$q_{b,2}, 0, R$	$q_{c,1}, 0, L$	
$\vdots$			
$q_{b,k}$	$q_{b,k+1}, 0, R$	$q_{c,k+1}, 0, L$	
$\vdots$			
$q_{b,n-1}$	$q_{b,n}, 0, R$	$q_{c,n}, 0, L$	
$q_{c,1}$	$q_{c,1}, 0, L$	$q_{c,1}, 1, L$	$q_{c,1}, X, L$
$\vdots$			
$q_{c,n}$	$q_{c,n}, 0, L$	$q_{c,n}, 1, L$	$q_{c,n}, C_S, L$
$q_{d,1}$	$q_{t2,0}, 0, N$	$q_{t2,1}, 1, N$	
$q_{d,2}$	$q_{d,1}, 0, R$	$q_{d,1}, 1, R$	
$\vdots$			
$q_{d,k}$	$q_{d,k-1}, 0, R$	$q_{d,k-1}, 1, R$	
$\vdots$			
$q_{d,n}$	$q_{d,n-1}, 0, R$	$q_{d,n-1}, 0, R$	

	X	Y	$\#$
q_0		$q_{\bar{O}}, Y, R$	
q_a	q_a, X, R	$q_{\bar{O}}, Y, R$	$q_A, \#, N$
$q_{b,n}$		$q_{\bar{O}}, Y, R$	
$q_{c,1}$	$q_{c,1}, X, L$		$q_{d,1}, \#, R$
$\vdots$			
$q_{c,n}$	$q_{c,n}, X, L$		$q_{d,n}, \#, R$

	0	1	Y	C_S
$q_{\bar{O}}$	$q_{g,1,}, 0, R$	$q_{h,1}, 0, L$		
q_e	$q_e, 0, R$		q_g, Y, R	q_e, C_S, R
q_g	$q_{g,1}, 0, R$	$q_{h,1}, 0, L$		
$q_{g,1}$	$q_{g,2}, 0, R$	$q_{h,2}, 0, L$		
$\vdots$				
$q_{g,k}$	$q_{g,k+1}, 0, R$	$q_{h,k+1}, 0, L$		
$\vdots$				
$q_{g,n-1}$	$q_{g,n}, 0, R$	$q_{h,n}, 0, L$		
$q_{g,n}$				
$q_{h,1}$	$q_{h,1}, 0, L$	$q_{h,1}, 1, L$	$q_{h,1}, Y, L$	$q_{h,1}, C_S, L$
$\vdots$				
$q_{h,n}$	$q_{h,n}, 0, L$	$q_{h,n}, 1, L$	$q_{h,n}, Y, L$	$q_{h,n}, C_S, L$
$q_{i,1}$	$q_{t2,1}, 0, N$	$q_{t2,0}, 1, N$		
$q_{i,2}$	$q_{i,1}, 0, R$	$q_{i,1}, 1, R$		
$\vdots$				
$q_{i,k}$	$q_{i,k-1}, 0, R$	$q_{i,k-1}, 1, R$		
$\vdots$				
$q_{i,n}$	$q_{i,n-1}, 0, R$	$q_{i,n-1}, 0, R$		
q_j	$q_j, 0, L$		$q_j, 0, L$	$q_{t2,a}, 0, N$

	C_E	X	#
$q_{g,n}$	$q_j, 0, L$		
$q_{h,1}$		$q_{h,1}, X, L$	$q_{i,1}, \#, R$
$\vdots$			
$q_{h,n}$		$q_{h,n}, X, L$	$q_{i,n}, \#, R$

	0	1	#
$q_{t3,0}$	$q_{t3,0}, 0, R$	$q_{t3,1}, 1, R$	$q_a, 0, N$
$q_{t3,1}$	$q_{t3,1}, 0, R$	$q_{t3,1}, 1, R$	$q_a, 1, N$
$q_{t3,a}$	$q_{t4,0}, 0, N$	$q_{t4,1}, 1, N$	
$q_{t3,b}$	$q_{t3,b}, \#, L$	$q_{t3,b}, \#, L$	$q_{t3,c}, \#, R$
$q_{t3,c}$			$q_a, 0, N$

The transition function of AND calculation is described by the following table:

	0	1	#
$q_{t4,0}$			$q_{t3,b}, 0, R$
$q_{t4,1}$			$q_{t3,b}, 1, R$
q_A			$q_{t4,a}, \#, L$
$q_{t4,a}$	$q_{t4,a}, \#, L$	$q_{t4,b}, \#, L$	$q_{t4,c}, \#, R$
$q_{t4,b}$	$q_{t4,a}, \#, L$	$q_{t4,b}, \#, L$	$q_{t4,d}, \#, R$
$q_{t4,c}$			$q_5, 0, L$
$q_{t4,d}$			$q_5, 1, R$

Let q_6 be the processor state of GQTM after the step 6 and $T_i, i = 1, \ldots, 4$ be the strings of the i-th track. Then the OM algorithm showed that the computation of the SAT problem of the above example gives us the resulting state ρ_6 expressed as

$$\rho_6 = q^2 |q_6\rangle \langle q_6| \otimes |T_1, T_2, T_3, T_4(1)\rangle \langle T_1, T_2, T_3, T_4(1)| \otimes |O\rangle \langle O| \\ + (1 - q^2) |q_6\rangle \langle q_6| \otimes |T_1, T_2, T_3, T_4(0)\rangle \langle T_1, T_2, T_3, T_4(0)| \otimes |O\rangle \langle O|,$$

where $T_4(1)$ (resp. $T_4(0)$) indicates that the value in the track 4 is 1 (resp. 0).

Next step, as the three tracks (1,2,3) can be empty, we can apply the chaos amplifier to the above ρ_6 in the following manner: We put

$$P_0 = |q_7\rangle \langle q_7| \otimes |T_1, T_2, T_3, T_4(0)\rangle \langle T_1, T_2, T_3, T_4(0)| \otimes |O\rangle \langle O|$$
$$P_1 = |q_f\rangle \langle q_f| \otimes |T_1, T_2, T_3, T_4(1)\rangle \langle T_1, T_2, T_3, T_4(1)| \otimes |O\rangle \langle O|$$

then $P_0 \perp P_1$ and P_0 and P_1 are Aberian. For any state in the form $\rho = (1 - q^2) P_0 + q^2 P_1$ where $q^2 \in \left\{0, \frac{1}{2^n}, \frac{2}{2^n}, \cdots, \frac{2^n - 1}{2^n}, 1\right\}$, define a map

$$\delta : \mathfrak{S}(\mathcal{H}) \times \mathfrak{S}(\mathcal{H}) \to [0, 1]$$

as

$$\delta(\rho, P_0) = 1 - g(q^2)$$
$$\delta(\rho, P_1) = g(q^2)$$

Using this δ, the transition function of the step 7 denoted by the chaos amplifier is formally written as

$$\Lambda_{CA}(\rho_6) = \delta(\rho_6, P_1) P_1 + \delta(\rho_6, P_0) P_0 \\ = g(q^2) P_1 + (1 - g(q^2)) P_0$$

$$(\Lambda_{CA})^k (\rho_6) = g^k(q^2) P_1 + (1 - g^k(q^2)) P_0$$

where g is the logistic map explained in Section 4.2. According to 4.1, GQTM halts in at most $\left[\frac{5}{4}(n-1)\right]$ steps with the probability $p \geq \frac{1}{2}$, by which we can claim that $\mathcal{C}$ is SAT.

5.1. *Computational complexity of the SAT algorithm*

We define the computational complexity of the OV SAT algorithm as the product of $T_Q\left(U_{\mathcal{C}}^{(n)}\right)$ and $T_{CA}(n)$,where $T_Q\left(U_{\mathcal{C}}^{(n)}\right)$ is the complexity of unitary computation and $T_{CA}(n)$ is that of chaos amplification.

The following theorem is essentially discussed in [13,3,4].

Theorem 5.1. *For a set of clauses $\mathcal{C}$ and n Boolean variables, the computational complexity of the OV SAT algorithm including the chaos amplifier, denoted by $T(\mathcal{C}, n)$, is obtained as follows.*

$$T_{GQTM}(\mathcal{C}, n) = T_Q\left(U_{\mathcal{C}}^{(n)}\right) T_{CA}(n) = \mathcal{O}(poly(n)),$$

where $poly(n)$ denotes a polynomial of n.

The computational complexity of quantum computer is determined by the total number of logical quantum gates. This inequality implies that the computational complexity of SAT algorithm is bounded by $\mathcal{O}(n)$ for the size of input n while a classical algorithm is bounded by $\mathcal{O}(2^n)$.

Acknowledgment

One (MO) of the authors thanks IIAS for finatial supports.

References

1. J.Gu, P.W.Purdom, J.Franco, and B.W.Wah, "Algorithms for the Satisfiability (SAT) Problem: a Survey," Preliminary version, 1996. http://citeseer.nj.nec.com/56722.html
2. M.Ohya and I.V.Volovich, *Quantum computing and chaotic amplification*, J. opt. B, **5**,No.6 639-642, 2003.
3. M.Ohya and I.V.Volovich, *New quantum algorithm for studying NP-complete problems*, Rep.Math.Phys., **52**, No.1,25-33 2003.
4. M.Ohya and N.Masuda, *NP problem in Quantum Algorithm,* Open Systems and Information Dynamics, 7 No.1 33-39, 2000.
5. M.Ohya, *Complexities and Their Applications to Characterization of Chaos,* Int. Journ. of Theoret. Physics, **37** 495, 1998.
6. L.Accardi and M.Ohya, Compound channels, transition expectations, and liftings, Appl. Math. Optim., Vol.39, 33-59, 1999.

7. L.Accardi and M.Ohya (2004) A Stochastic Limit Approach to the SAT Problem,Open Systems and Information dynamics, 11,1-16.
8. L.Accardi and R.Sabbadini, On the Ohya–Masuda quantum SAT Algorithm, Preprint Volterra, N. **432**, 2000.
9. D.Deutsch, Quantum Theory, the Church-Turing Principle and the Universal Quantum Computer, Proc. Roy. Soc, A, **400** 97-117, 1985.
10. E.Bernstein and U.Vazirani, *Quantum Complexity Theory*, In Proc. 25th ACM Symp. on Theory of Computation, 11–20, 1993.
11. H. Nishimura and M. Ozawa: Computational Complexity of Uniform Quantum Circuit Families and Quantum Turing machines, quant-ph/9906095, 2000.
12. M.Ohya and I.V.Volovich, Mathematical Foundation of *Quantum Information and Quantum Computation,* to appear.
13. S.Iriyama and S.Akashi, Estimation of Complexity for the Ohya-Masuda-Volovich SAT Algorithm (to appear).
14. C.H.Bennett, E.Bernstein, G.Brassard, U.Vazirani, *Strengths and Weaknesses of Quantum Computing*, SICOMP Vol. 26 Number 5 pp. 1510-1523. 1997.

A STROBOSCOPIC APPROACH TO QUANTUM TOMOGRAPHY

A. JAMIOŁKOWSKI

Institute of Physics
Nicolaus Copernicus University
Grudziadzka 5, 87–100 Toruń, Poland

In this paper we discuss some problems of stroboscopic characterization of quantum states. In particular, the minimal number η of observables $Q_1, \ldots, Q_\eta$, whose expectation values at some time instants $t_1, \ldots, t_r$ determine the trajectory of a d-level quantum system ("qudit") governed by a semigroup of linear transformations are analyzed. We use the unitary and the Gaussian evolutions as examples. We assume that the macroscopic information about the system in question is given by the mean values ${}_j(Q_i) = \mathrm{Tr}\,(Q_i\rho(t_j))$ of n selfadjoint operators $Q_1, \ldots, Q_n$ at some time instants $t_1 < t_2 < \ldots < t_r$, where $n < d^2 - 1$ and $r \leq \deg \mu(\lambda, \mathbb{L})$. Here $\mu(\lambda, \mathbb{L})$ stands for the minimal polynomial of the generator

$$\mathbb{L}\rho = -i[H,\rho] \mathrm{or} \mathbb{L}\rho = -\frac{1}{2}[H,[H,\rho]]$$

of the unitary and Gaussian flows, respectively.

1. Introduction

The information contained in the state of a physical system allows us to predict its future evolution and its effect on other systems. It appears that the possibility to create and manipulate (control) quantum states is becoming increasingly important in physical research with implications for others areas of science such as: cryptography, quantum communication and computing.

According to one of the basic assumptions of quantum mechanics, the achievable information about the state of a physical system is encoded in the density matrix ρ, which allows one to evaluate all possible expectation values of observables trough the formula

$$\langle Q \rangle = \mathrm{Tr}\,(\rho Q)\,, \tag{1}$$

where Q is a self-adjoint operator representing a particular physical quantity.

Thus, in order to have full information about a given quantum system we need to know its density matrix ρ. In particular, a very useful tool in this regard is quantum state tomography (QST) which provides means for the complete reconstruction of the density matrix for a qudit (or a set of qubits). Two fundamental restrictions limit the possibility of introducing state reconstruction methods. On the one hand, we cannot recover the quantum state from measurements on a single copy of a system, on the other hand, the "no-cloning" theorem implies that it is not possible to make exact copies of an unknown state of a quantum system. Hence, the only general method relies on the possibility to reproduce a large number of identical (although unknown) states and to perform a series of measurements on complementary aspects of the state within an ensemble.

Suppose that we can prepare a quantum system repeatedly in the same state and make a series of experiments such that we can measure the expectation values

$$\mathcal{E}_t(Q) = \mathrm{Tr}(Q\rho(t)) \tag{2}$$

of some observables $Q_1, \ldots, Q_n$ at different time moments $t_1 < t_2 < \ldots < t_r$. The fundamental question arises:

Can we find the average value of any desired operator from the set of measured outcomes of a given set $Q_1, \ldots, Q_n$

$$\begin{bmatrix} \mathcal{E}_{t_1}(Q_1) & \cdots & \mathcal{E}_{t_r}(Q_1) \\ \vdots & \cdots & \vdots \\ \mathcal{E}_{t_1}(Q_n) & \cdots & \mathcal{E}_{t_r}(Q_n) \end{bmatrix}, \tag{3}$$

where $0 \leq t_1 < \ldots < t_r \leq T$, for an interval $[0, T]$?

Among the existing tomographic techniques for quantum systems, the so-called homodyne tomography has received much attention in the literature [1,2,3]. In the phase-space formulation of quantum mechanics there is a one-to-one relation between a quantum state and the so-called Wigner function. Its marginals are accessible experimentally, and an inverse (Radon) transformation allows one to reconstruct the phase-space distribution associated with the unknown quantum state.

The question of how to reconstruct states of finite dimensional systems (qudits) is also natural. In this case various methods have been proposed to determine ρ [4,5,6]. If the problem under consideration is static, then the state of a d-level open quantum system (a qudit) can be uniquely determined only if $n = d^2 - 1$ expectation values of linearly independent observables are at our disposal. However, if we assume that we know the dynamics of our system, i.e. we know the generator of the time evolution, then we can use the stroboscopic approach based on (3). In general, the term "tomography" will be used collectively to denote any kind of state-reconstruction method.

With reference to the terminology used in the system theory, we assume the following definition:

Definition 1.1. *A d-level open quantum system $\mathcal{S}$ is said to be $(Q_1, \ldots, Q_n)$-reconstructible on an interval $[0, T]$, if there exists at least one set of time instants $0 \leq t_1 < \ldots < t_r \leq T$ such that the state trajectory can be uniquely determined by the correspondence*

$$[0, T] \ni t_j \longmapsto \mathcal{E}_{t_j}(Q_i) = \operatorname{Tr}(\rho(t_j) Q_i) \tag{4}$$

for $i = 1, \ldots, n$, $j = 1, \ldots, r$.

The above definition is equivalent to the following one

Definition 1.2. *A d-level open quantum system $\mathcal{S}$ is said to be $(Q_1, \ldots, Q_n)$-reconstructible on an interval $[0, T]$, if for every two trajectories with distinct initial states there exists at least one $\hat{t} \in [0, T]$ and at least one operator $Q_k \in \{Q_1, \ldots, Q_n\}$ such that*

$$\operatorname{Tr}(Q_k \rho_1(\hat{t})) \neq \operatorname{Tr}(Q_k \rho_2(\hat{t})). \tag{5}$$

Remark 1.1. *In the above definitions we assume that the time evolution of the system is given in terms of a completely positive semigroup of operators. Arguments in favour of completely positive semigroups as the foundation of non-Hamiltonian dynamics as well as the discussion of properties of such semigroups can be found in papers of Kraus [7], Lindblad [8], and Gorini et al. [9].*

In particular, in Lindblad's paper [8] the general form of the generator of an arbitrary completely positive semigroup was derived. A linear operator $\mathbb{L}$ on a set of linear operators $B(\mathcal{H}) := M(\mathbb{C}^d)$, where $\mathcal{H} \simeq \mathbb{C}^d$ is a d-dimensional Hilbert space and M denotes the set of matricies with complex entries, proves to be the generator of a completely positive semigroup if and only if it can be represented in the form

$$\mathbb{L}\rho = -i[H, \rho] + \frac{1}{2} \sum_{j=1}^{\kappa} \left([V_j \rho, V_j^*] + [V_j, \rho V_j^*] \right), \tag{6}$$

where $V_j \in B(\mathcal{H})$ for $j = 1, \ldots, \kappa$, and H is a self-adjoint operator also belonging to $B(\mathcal{H})$ (cf. also [10]).

Remark 1.2. *It is important that for the number r of time instants $t_1, \ldots, t_r$ we do not formulate any restriction (except that it is finite).*

Remark 1.3. *The question of obvious physical interest is to find the minimal number of observables $Q_1, \ldots, Q_\eta$ for which the d-level quantum system $\mathcal{S}$ with the generator $\mathbb{L}$ can be $(Q_1, \ldots, Q_\eta)$-reconstructible. It can be shown that if the time evolution of the system $\mathcal{S}$ is described by the master equation,*

$$\frac{d}{dt}\rho(t) = \mathbb{L}\rho(t) , \tag{7}$$

then there exists [4] a set of observables $Q_1, \ldots, Q_\eta$, where

$$\eta = \max_{\lambda \in \sigma(\mathbb{L})} \left\{ dim\, Ker\, (\lambda \mathbb{1} - \mathbb{L}) \right\} \tag{8}$$

such that the system $\mathcal{S}$ is $(Q_1, \ldots, Q_\eta)$-reconstructible. Moreover, if we have another set of observables $\widetilde{Q}_1, \ldots, \widetilde{Q}_{\tilde{\eta}}$ with this property, then $\tilde{\eta} \geq \eta$. The number η given by (8) we will call the index of cyclicity of the system $\mathcal{S}$.

2. *Polynomial Representation of Flows*

The main idea of the stroboscopic approach to quantum tomography is based on the polynomial representation of the flow defined by the general master equation. Namely, we have

$$\Phi(t) = \exp(\mathbb{L}t) = \sum_{k=0}^{m-1} \alpha_k(t)\mathbb{L}^k , \tag{9}$$

where by Cauchy's theorem

$$\alpha_k(t) := \frac{1}{2\pi i} \oint_{\partial D} \frac{\mu_k(z)}{\mu(z, \mathbb{L})} \exp(tz)\, dz . \tag{10}$$

In the above expression ∂D is any simple closed contour enclosing the spectrum of the operator $\mathbb{L}$ in the complex plane and

$$\mu(z, \mathbb{L}) = \sum_{k=0}^{m-1} d_k z^k \tag{11}$$

denotes the minimal polynomial of the generator $\mathbb{L}$. The symbol $\mu_k(z)$ denote auxialiary polynomials constructed from $\mu(z, \mathbb{L})$ (c.f. [12]). It is interesting that there are ways to compute the functions $\alpha_k(t)$ in (9) without summing the exponential series or without knowing the Jordan canonical form of $\mathbb{L}$. Namely, differentiating (9) with respect to t and using the minimal polynomial of $\mathbb{L}$ one finds that the functions $\alpha_k(t)$ for $k = 0, \ldots, m-1$ satisfy the system of equations

$$
\begin{aligned}
\frac{d\alpha_0(t)}{dt} &= d_0 \alpha_{m-1}(t)\,, \\
\frac{d\alpha_1(t)}{dt} &= \alpha_0(t) + d_1 \alpha_{m-1}(t)\,, \\
\cdots \quad & \cdots \quad \cdots \quad \cdots \\
\frac{d\alpha_{m-1}(t)}{dt} &= \alpha_{m-2}(t) + d_{m-1}\alpha_{m-1}(t)\,,
\end{aligned}
\tag{12}
$$

with initial conditions $\alpha_k^{(i)}(0) = \delta_{ik}$, where the coefficients d_i are defined by (11). It can be shown that functions $\alpha_k(t)$ are mutually linearly independent, therefore for a given T there exists at least one set of time instants $t_1, \ldots, t_m$ ($m = deg\, \mu(\lambda, \mathbb{L}^*)$) such that $0 \leq t_1 < \ldots < t_m \leq T$ and $det\,[\alpha_k(t_j)] \neq 0$.

Taking into account the above conditions one finds that the state $\rho(0)$ can be determined uniquely (and the trajectory $\Phi(t)\rho(0)$ can be reconstructed) if and only if operators of the form $(\mathbb{L}^*)^k Q_i$ for $i = 1, \ldots, n$ and $k = 0, 1, \ldots, m-1$ span the space $B(H)$.

If the dynamical semigroup is completely positive, then the general form of the generator $\mathbb{L}$ is given by (6). In this case the criterion for reconstructibility of a d-level quantum system can be formulated using the operators H and V_j. In particular, if we consider an isolated quantum system characterized by a Hamiltonian H_0 and $V_j = 0$ for $j = 1, \ldots, \kappa$, then the minimal number of observables $Q_1, \ldots, Q_\eta$ for which the system is $(Q_1, \ldots, Q_\eta)$-reconstructible is given by

$$
\eta :=: n_1^2 + n_2^2 + \cdots + n_r^2\,,
$$

where $n_i = \dim \mathrm{Ker}\,(\lambda_1 I - H_0)$ for all $\lambda_i \in \sigma(H_0)$, $i = 1, \ldots, r$ (for details cf. [5]).

3. *The Minimal Number of Observables for Qudits Governed by Gaussian Semigroups*

Let us assume that the time evolution of a d-level quantum system S is described by the generator $\mathbb{L}$ given by

$$\mathbb{L}\rho = \frac{1}{2}\{[H\rho, H] + [H, \rho H]\} = -\frac{1}{2}[H,[H,\rho]] \tag{13}$$

that is, the semigroup $\Phi(t)$ has the form (cf. e.g. [11])

$$\Phi(t)\rho = \frac{1}{\sqrt{2\pi t}} \int_{-\infty}^{\infty} ds\, e^{-s^2/(2t)} e^{-iHs} \rho e^{iHs}. \tag{14}$$

The symbol H in (13) and (14) denotes a self-adjoint operator with the spectrum

$$\sigma(H) = \{\lambda_1, \ldots, \lambda_m\}. \tag{15}$$

In the sequel n_i stands for the multiplicity of the eigenvalue λ_i for $i = 1, \ldots, m$. One can assume that the elements of the spectrum of H are numbered in such a way that the inequalities $\lambda_1 < \lambda_2 < \ldots < \lambda_m$ are fulfilled. The following theorem holds:

Theorem 3.1. *The index of cyclicity of the Gaussian semigroup with a generator $\mathbb{L}$ given by (13) is expressed by the formula*

$$\eta = \max\{\kappa, \gamma_1, \ldots, \gamma_r\}, \tag{16}$$

where $r = (m-1)/2$ if m is odd or $r = (m-2)/2$ if m is even, and

$$\kappa := n_1^2 + n_2^2 + \ldots + n_m^2, \tag{17}$$

$$\gamma_k := 2\sum_{i=1}^{m-k} n_i\, n_{i+k}. \tag{18}$$

Proof. In order to determine the value of η for the generator $\mathbb{L}$ defined by (13) we must find the number of nontrivial invariant factors of the operator $\mathbb{L}$. Let us observe that if $\sigma(H) = \{\lambda_1, \ldots, \lambda_m\}$ then the spectrum of the operator $\mathbb{L}$ is given by

$$\sigma(\mathbb{L}) = \left\{\nu_{ij} \in \mathbb{R};\ \nu_{ij} = (\lambda_i - \lambda_j)^2,\ i,j = 1, \ldots, m\right\}. \tag{19}$$

The above statement follows from the fact that the operator $\mathbb{L}$ can also be represented as

$$\mathbb{L} = H^2 \otimes \mathbb{1} + \mathbb{1} \otimes H^2 - 2H \otimes H, \tag{20}$$

where $\mathbb{1}$ denotes the identity in the space $B(H)$. Since H is self-adjoint therefore the algebraic multiplicity of λ_i, i.e. the multiplicity of λ_i as the root of the characteristic polynomial of H, is equal to the geometric multiplicity of λ_i, $n_i = dim\, Ker\,(\lambda_i \mathbb{1} - H)$. Of course, we have $n_1 + \ldots + n_m = dim\, H$.

From (20) we can see that the multiplicities of the eigenvalues of the operator $\mathbb{L}$ are not determined uniquely by the multiplicities of $\lambda_i \in \sigma(H)$. But if we assume that $\lambda_1 < \ldots < \lambda_m$ and $\lambda_k = (k-1)c + \lambda_1$, where $k = 1, \ldots, m$, and $c = const > 0$, then the multiplicities of all eigenvalues of $\mathbb{L}$ are given by

$$\gamma_{|i-j|} = dim\, Ker\,[(\lambda_i - \lambda_j)^2 \mathbb{1} - \mathbb{L}] \tag{21}$$

for $i \neq j$ and

$$dim\, Ker\,(\mathbb{L}) = n_1^2 + \ldots + n_m^2 = \kappa \tag{22}$$

when $i = j$. Now, as we know, the minimal number of observables $Q_1, \ldots, Q_\eta$ for which the qudit S can be $(Q_1, \ldots, Q_\eta)$-reconstructible is given by (8), so in our case

$$\eta = \max_{i,j=1,\ldots,m} \left\{ dim\, Ker\,[(\lambda_i - \lambda_j)^2 \mathbb{1} - \mathbb{L}] \right\}, \tag{23}$$

where $\lambda_i \in \sigma(H)$. Using the above formulae and the inequality $\gamma_k < \kappa$ for $k > r$, where r is given by $(m-1)/2$ if m is odd and $(m-2)/2$ if m is even, we can observe that also without the assumption $\lambda_k = (k-1)c + \lambda_1$ one obtains

$$\eta = \max\{\kappa, \gamma_1, \ldots, \gamma_r\}. \tag{24}$$

This completes the proof. □

According to Theorem 3.1 if the quantum system governed by a Gaussian semigroup is $(Q_1, \ldots, Q_n)$-reconstructible then the number n of observables must satisfy the inequality $n \geq \eta$. In this case there exists a set of time instants $t_1 < t_2 < \ldots < t_m$ ($m = deg\, \mu(\lambda, \mathbb{L})$) such that the knowledge of the expectation values $E_j(Q_i) = \mathrm{Tr}\,(\rho(t_j) Q_i)$ for $i = 1, \ldots, n$ and $j = 1, \ldots, m$ uniquely determines the trajectory of the system.

The natural question arises: what are the criteria governing the choice of time instants $t_1, \ldots, t_m$? The following theorem holds:

Theorem 3.2. *Let us assume that* $0 \leq t_1 < t_2 < \ldots < t_m \leq T$. *Suppose that the mutual distribution of time instants* $t_1, \ldots, t_m$ *is fixed, i.e. a set of* *nonnegative numbers* $c_1 < \ldots < c_m$ *is given and* $t_j := c_j t$ *for* $j = 1, \ldots, m$, *and* $t \in \mathbb{R}_+$. *Then for* $T > 0$ *the set*

$$\tau(T) := \left\{(t_1, \ldots, t_m) : \quad t_j = c_j t, \quad 0 \leq t \leq \frac{T}{c_m}\right\}$$

contains almost all sequences of time instants $t_1, \ldots, t_m$, *i.e. all of them except a finite number.*

Proof. As one can check, the expectation values $E_{t_j}(Q_i)$ and the operators $(\mathbb{L}^*)^k Q_i$ are related by the equality

$$\mathcal{E}_{t_j}(Q_i) \;=\; \sum_{k=0}^{m-1} \alpha_k(c_j t)\Big((\mathbb{L}^*)^k Q_i, \rho_0\Big)\,, \tag{25}$$

where we assume that $t_j = c_j t$ and the bracket $(\cdot, \cdot)$ denotes the Hilbert-Schmidt product in $B(H)$. One can determine ρ_0 from (25) for all those values $t \in \mathbb{R}_+$ for which the determinant $\alpha(t)$ is different from zero, i.e.

$$\alpha(t) \;:=\; det\left[\alpha_k(c_j t)\right] \;\neq\; 0\,. \tag{26}$$

One can prove that the range of the parameter $t \in \mathbb{R}_+$ for which $\alpha(t) = 0$ consists only of isolated points on the semiaxis $\mathbb{R}_+$, i.e. does not possess any accumulation points on $\mathbb{R}_+$. To this end let us note that since the functions $t \to \alpha_k(t)$ for $k = 0, 1, \ldots, m-1$, are analytic on $\mathbb{R}$, the determinant $\alpha(t)$ defined by (26) is also an analytic function of $t \in \mathbb{R}$. If $\alpha(t)$ can be proved to be nonvanishing identically on $\mathbb{R}$, then, making use of its analyticity, we shall be in position to conclude that the values of t, for which $\alpha(t) = 0$, are isolated points on the axis $\mathbb{R}$.

It is easy to check that for $k = m(m-1)/2$

$$\left.\frac{d^k \alpha(t)}{dt^k}\right|_{t=0} = \prod_{1 \leq j < i \leq m} (c_i - c_j)\,. \tag{27}$$

According to the assumption $c_1 < c_2 < \ldots < c_m$, we have $\alpha^{(k)}(0) \neq 0$ if $k = m(m-1)/2$. This means that the analytic function $t \to \alpha(t)$ does not vanish identically on $\mathbb{R}$ and the set of values of t for which $\alpha(t) = 0$ cannot contain accumulation points. In other words, if we limit ourselves to an arbitrary finite interval $[0, T]$, then $\alpha(t)$ can vanish only on a finite number of points belonging to $[0, T]$. This completes the proof. □

Acknowledgment

The author is grateful to Prof. M. Ohya and N. Watanabe for invitation to the conferences in Tokyo and Kyoto and for their kind hospitality.

References

1. *K. Vogel and H. Risken, Phys. Rev. A* ***40****, 2847 (1989).*
2. *M. G. Raymer, Contemp. Physics* ***38****, 343 (1997).*
3. *U. Leonhardt, Measuring the quantum states of light, Cambridge Univ. Press, Cambridge, 1997.*
4. *A. Jamiołkowski, Int. J. Theor. Phys.* ***22****, 369 (1983); A. Jamiołkowski, Rep. Math. Phys.* ***21****, 101 (1985).*
5. *G. Cassinelli, G. M. D'Ariano, E. De Vito, J. Math. Phys.* ***41****, 7940 (2000).*
6. *S Weigert, Phys. Rev. A* ***53****, 2078 (1996).*
7. *K. Kraus, Ann. Phys.* ***64****, 119, (1971).*
8. *G. Lindblad, Comm. Math. Phys.* ***48****, 119 (1976).*
9. *V. Gorini, A. Kossakowski, E. C. G. Sudarshan, J. Math. Phys.* ***17****, 149 (1976).*
10. *R. Alicki, K. Lendi, Quantum Dynamical Semigroups and Applications, Springer, Berlin, 1987.*
11. *H. Spohn, Rev. Mod. Phys.* ***50****, 569 (1980).*
12. *A. Jamiołkowski, Rep. Math. Phys.* ***46****, 469 (2000).*

POSITIVE MAPS AND SEPARABLE STATES IN MATRIX ALGEBRAS

A. KOSSAKOWSKI

Institute of Physics
Nicolaus Copernicus University
Grudziadzka 5, 87–100 Toruń, Poland
E-mail: kossak@phys.uni.torun.pl

A class of linear positive, trace preserving maps in M_n is given in terms of affine maps in $\mathbb{R}^{n^2-1}$ which map the close unit ball into itself. The above class contains atomic maps considered by Tanahashi and Tomiyama.

1. Introduction

A linear mapping from a C^*-algebra A into a C^*-algebra B is called *positive* if it carries positive elements of A into positive elements of B. Such a map is said to be *normalised* if the image of the unity of A coincides with the unity of B.

Positive maps have become a common interest to mathematicians and physicist, c.f. [1 – 31] and [32, 33]. There are several reasons for this:

(i) The notion of positive map generalizes that of state, $*$-homomorphism, Jordan homomorphism, and conditional expectation.

(ii) By duality, a normalised positive map defines an affine mapping between sets of states of C^*-algebras. If these algebras coincide with the C^*-algebras of observables assigned to a physical system, then this affine mapping corresponds to some operation which can be performed on the system.

(iii) It has been recently shown that there exists a strong connection between the classification of the entanglement of quantum states and the structure of positive maps.

Let $M_k(A)$ be the algebra of $k \times k$ matrices with elements from A and $M_k^+(A)$ be the positive cone in $M_k(A)$. For $k \in \mathbb{N}$ and $\varphi : A \to B$

define maps $\varphi_k, \overline{\varphi}_k : M_k(A) \to M_k(B)$, where $\varphi_k([a_{ij}]) = [\varphi(a_{ij})]$ and $\overline{\varphi}_k([a_{ij}]) = [\varphi(a_{ji})]$. The map φ is said to be *k-positive* (*k-copositive*) if the map φ_k ($\overline{\varphi}_k$) is positive, respectively.

A map φ is said to be *k-decomposable* if φ_k is positive whenever $[a_{ij}]$ and $[a_{ji}]$ belong to $M_k^+(A)$. A map $\varphi : A \to B$ is called *completely positive*, *completely copositive* and *decomposable* if it is k-positive, k-copositive and and k-decomposable for all $k \in \mathbb{N}$, respectively. According to the Størmer theorem [10], every decomposable map $\varphi : A \to B(\mathcal{H})$ ($B(\mathcal{H})$ is the set of bounded linear operators on a complex Hilbert space $\mathcal{H}$) has the form $\varphi(a) = v^* j(a) v$, where $j : A \to B(\mathcal{K})$ is a Jordan homomorphism and v is a bounded operator from $\mathcal{H}$ into $\mathcal{K}$. Since every Jordan homomorphism is the sum of a $*$-homomorphism and a $*$-antihomomorphism [3] every decomposable map $\varphi : A \to B(\mathcal{H})$ has the form $\varphi = \varphi_1 + \varphi_2$, where φ_1 is completely positive and φ_2 is completely copositive. A positive map which is not decomposable is called an *indecomposable* one.

In the case $A = B(\mathcal{H}) = M_n(\mathbb{C})$ every decomposable map $\varphi : M_n(\mathbb{C}) \to M_n(\mathbb{C})$ can be written in the form

$$\varphi(a) \;=\; \sum_k a_k^* a a_k + \sum_k b_k^* a^T b_k \,, \tag{1.1}$$

where $a_1, a_2, \ldots , b_1, b_2, \ldots \in M_n(\mathbb{C})$ and ^{T} is the transpose map in M_n. It is known [2, 12, 16] that every positive map $\varphi : M_2(\mathbb{C}) \to M_2(\mathbb{C})$ is decomposable. The first example of an indecomposable map in $M_3(\mathbb{C})$ was given by Choi [7, 8] and its generalizations can be found in [17 – 31].

There are two known types of indecomposable maps in M_n for $n \geq 3$:

(i) A series of maps $\varphi_k : M_n(\mathbb{C}) \to M_n(\mathbb{C})$, $k = 1, \ldots, n-2$; c.f. [21, 31]. Let ϵ be the projection of norm one in $M_n(\mathbb{C})$ to the diagonal part with respect to a given basis in $\mathbb{C}^n$ and s be the unitary shift in $M_n(\mathbb{C})$ such that $s = [\delta_{i,j+1}]$, where the indices are understood to be mod n. The maps φ_k are defined as follows

$$\varphi_k(a) \;=\; (n-k)\epsilon(a) + \sum_{i=1}^{k} \epsilon(s^i a s^{*i}) - a \,. \tag{1.2}$$

Since $\varphi_k(\mathbb{1}_n) \;=\; (n-1)\mathbb{1}_n$ and $\mathrm{Tr}\varphi_k(a) = (n-1)\mathrm{Tr}\, a$, the maps

$$\tau_k(a) \;=\; \frac{1}{n-1}\varphi_k(a)\,, \qquad k = 1, \ldots, n-2\,, \tag{1.3}$$

are bistochastic ones.

(ii) A class of maps $\varphi(p_0, p_1, \ldots, p_n) : M_n \to M_n$, c.f. [28, 31] of the form $\varphi(p_0, p_1, \ldots, p_n)(a) = a'$ with

$$\begin{aligned} a'_{11} &= p_0 a_{11} + p_n a_{nn} \\ a'_{22} &= p_0 a_{22} + p_1 a_{11} \\ \ldots & \quad \ldots \\ a'_{nn} &= p_0 a_{nn} + p_{n-1} a_{n-1,n-1} \\ a'_{ij} &= -a_{ij} , \end{aligned} \tag{1.4}$$

where

$$p_0, p_1, \ldots, p_n > 0 , \quad n-1 > p_0 \geq n-2 , \quad p_1 \cdot \ldots \cdot p_n \geq (n-1-p_0)^n , \quad n \geq 3 . \tag{1.5}$$

The above maps are atomic, i.e., they cannot be decomposed into the sum of a 2-positive and a 2-copositive maps [23, 31].

A state ω on $M_n \otimes M_n$, i.e., a positive normalized linear functional on $M_n \otimes M_n$, is said to be *separable* if it can be written in the form

$$\omega = \sum_\alpha p_\alpha \rho_\alpha \otimes \sigma_\alpha , \tag{1.6}$$

where $p_\alpha \geq 0$, $\sum_\alpha p_\alpha = 1$, and $\rho_\alpha, \sigma_\alpha$, $\alpha = 1, 2, \ldots$ are states on M_n.

Let $P[M_n, M_n]$ be the set of all linear, positive and trace preserving maps of M_n into M_n. According to [32], a state ω on $M_n \otimes M_n$ is separable iff $(\mathbb{1}_n \otimes \varphi)\omega$ is a state on $M_n \otimes M_n$ for all $\varphi \in P[M_n, M_n]$, where $\mathbb{1}_n$ is the identity map.

2. A Class of Positive Maps in M_n

Let φ be a positive, trace preserving map in $M_n(\mathbb{C})$ i.e., a dynamical map [34]. Since φ is self-adjoint, i.e., $\varphi(a^*) = \varphi(a)^*$ it is enough to consider it in the space

$$H_n = \{a \in M_n : a = a^*\} . \tag{2.1}$$

The linear space H_n is the real Hilbert space with the scalar product $(a, b) = \mathrm{Tr}\,(ab)$ and the norm $\|a\|^2 = (a, a)$.

If $a \in H_n$ and $\mathrm{Tr}\, a = p > 0$, the necessary condition for positivity of a is $\mathrm{Tr}\, a^2 \leq (\mathrm{Tr}\, a)^2 = p^2$, which can also be written in the form

$$\left\| a - \frac{\mathbb{1}_n}{n} \mathrm{Tr}\, a \right\|^2 \leq \frac{n-1}{n} (\mathrm{Tr}\, a)^2 . \tag{2.2}$$

Let us define the following convex sets in H_n:

$$B_n(p) = \left\{ a \in H_n \,:\, \left\| a - \frac{\mathbb{1}_n}{n} p \right\|^2 \leq \frac{n-1}{n} p^2, \quad \mathrm{Tr}\, a = p \right\}, \tag{2.3}$$

$$S_n(p) = \left\{ a \in B_n(p) \,:\, a \geq 0 \right\}, \tag{2.4}$$

$$B_n^0(p) = \left\{ a \in H_n \,:\, \left\| a - \frac{\mathbb{1}_n}{n} p \right\|^2 \leq \frac{1}{n(n-1)} p^2, \quad \mathrm{Tr}\, a = p \right\}. \tag{2.5}$$

The set $S_n(p = 1)$ is just the set of all density operators on $\mathbb{C}^n$.

Lemma 2.1. *If $a \in B_n^0(p)$ then $a \geq 0$, i.e., the inclusions*

$$B_n^0(p) \subseteq S_n(p) \subseteq B_n(p) \tag{2.6}$$

hold.

Proof. Let us observe that B_n^0 is invariant with respect to the mapping $\rho \to U^* \rho U$, $U \in \mathrm{SU}\,(n)$, and the set

$$b_n^0 = \left\{ p_1, \ldots, p_n \,:\, \sum_{i=1}^{n} \left(p_i - \frac{1}{n} p \right)^2 \leq \frac{p^2}{n(n-1)}, \quad \sum_{i=1}^{n} p_i = p \right\}$$

is the ball inscribed in the set

$$s_n(p) = \left\{ p_1, \ldots, p_n \,:\, p_i \geq 0\,,\ i = 1, \ldots, n\,, \quad \sum_{i=1}^{n} p_i = p > 0 \right\}.$$

Since the eigenvalues of $a \in B_n^0$ lie in b_n^0, they are non-negative. □

It should be pointed out that in the case $n = 2$ one has

$$B_2^0(p) = S_2(p) = B_2(p)\,. \tag{2.7}$$

Let $\varepsilon_p : B_n(p) \to B_n^0(p)$ be defined as follows:

$$\varepsilon_p(a) = \frac{\mathbb{1}_n}{n} p + \frac{1}{n-1} \left(a - \frac{\mathbb{1}_n}{n} p \right) = \frac{1}{n-1} a + \left(1 - \frac{1}{n-1} \right) \frac{\mathbb{1}_n}{n} p\,. \tag{2.8}$$

The map ϵ_p is positive and trace preserving. It is convenient to introduce an orthonormal base in H_n: $f_1, \ldots, f_{n^2}$ such that $(f_\alpha, f_\beta) = \delta_{\alpha\beta}$ and $f_{n^2} = \mathbb{1}_n/\sqrt{n}$, i.e. $\mathrm{Tr}\, f_\alpha = 0$ for $\alpha = 1, \ldots, n^2 - 1$. Every element of $a \in H_n$ can be written in the form

$$a = \sum_{\alpha=1}^{n^2} x_\alpha f_\alpha\,, \qquad x_\alpha = \mathrm{Tr}\,(a f_\alpha) \tag{2.9}$$

or

$$a = \frac{\mathbb{1}_n}{n}\operatorname{Tr} a + \langle f, x\rangle\,, \tag{2.10}$$

where $x = (x_1, \ldots, x_{n^2-1}) \in \mathbb{R}^{n^2-1}$, $f = (f_1, \ldots, f_{n^2-1})$, and

$$\langle f, x\rangle = \sum_{\alpha=1}^{n^2-1} f_\alpha x_\alpha\,. \tag{2.11}$$

Let $B(\mathbb{R}^n, r)$ be the closed ball in $\mathbb{R}^n$ with the radius r, i.e.,

$$B(\mathbb{R}^n, r) = \{x \in \mathbb{R}^n \,:\, \langle x, x\rangle \leq r\}\,. \tag{2.12}$$

Let $\mathcal{D}_n$ denote the set of affine maps $(T, b) : \mathbb{R}^n \to \mathbb{R}^n$, where

$$(T, b)x = Tx + b\,, \qquad T \in M_n(\mathbb{R})\,, \quad b \in \mathbb{R}^n\,, \tag{2.13}$$

which maps the closed unit ball $B(\mathbb{R}^n, 1)$ into itself. Since $\mathcal{D}_n$ is convex, compact and finite-dimensional, each element of $\mathcal{D}_n$ can be written as a finite convex combination of its extreme elements.The extreme elements of $\mathcal{D}_n$ have been classified in [16] in the following manner:

$$\operatorname{Extr}\mathcal{D}_n = \Big\{(T, b) \,:\, T = R_1 \Lambda R_2\,, \quad b = R_1 c\,, \quad R_1, R_2 \in \mathrm{O}(n)\,,$$

$$\Lambda = \|\lambda_i \delta_{ij}\|\,,\ \lambda_1 = \ldots = \lambda_{n-1} = [1 - \delta^2(1-\kappa^2)]^{1/2},\ \lambda_n = \kappa[1-\delta^2(1-\kappa^2)]^{1/2}\,,$$

$$c_1 = \ldots = c_{n-1} = 0\,,\ c_n = \delta(1-\kappa^2)\,, \quad 0 \leq \kappa \leq 1\,, \quad 0 < \delta \leq 1\Big\}\,. \tag{2.14}$$

It is clear that (T, rb) maps the ball $B(\mathbb{R}^n, r)$ into itself provided $(T, y) \in \mathcal{D}_n$.

Let us observe that the subset $\operatorname{Extr}\mathcal{D}_n^0$ of $\operatorname{Extr}\mathcal{D}_n$ corresponding to $\kappa = 1$ has the form

$$\operatorname{Extr}\mathcal{D}_n^0 = \{(T, y) \,:\, T \in \mathrm{O}\,(n)\,,\ y = 0\}\,. \tag{2.15}$$

Taking into account (2.3), (2.9) and (2.10) one finds that an element $a \in B_n(p)$ can be represented in the form

$$a = \frac{\mathbb{1}_n}{n}p + \langle f, x\rangle\,, \quad p = \operatorname{Tr} a\,, \text{where} x \in B\Big(\mathbb{R}^{n^2-1}, p\Big(\frac{n-1}{n}\Big)^{1/2}\Big)\,, \tag{2.16}$$

while the map ε_p, c.f. (2.8) takes the form

$$\varepsilon_p(a) = \frac{\mathbb{1}_n}{n}p + \frac{1}{n-1}\langle f, x\rangle\,. \tag{2.17}$$

Composing the map ε_p with the map $\psi : B_n(p) \to B_n(p)$ given by

$$\psi\Big[T, yp\Big(\frac{n-1}{n}\Big)^{1/2}\Big](a) = \frac{1\!\!1_n}{n}p + \Big\langle f, Tx + yp\Big(\frac{n-1}{n}\Big)^{1/2}\Big\rangle \tag{2.18}$$

one finds that the composed map $\varphi[(T,y)] : B_n(p) \to B_n^0(p) \subseteq B_n(p)$ has the form

$$\varphi[T,y](a) = \frac{1\!\!1_n}{n}p + \frac{1}{n-1}\Big\langle f, Tx + py\Big(\frac{n-1}{n}\Big)^{1/2}\Big\rangle, \tag{2.19}$$

where $(T,y) \in \mathcal{D}_{n^2-1}$, is a positive one by the construction. Since the map (2.19) is homogeneous it can be extended to H_n , and its extension to M_n is given by

$$\varphi[T,y](c+id) = \varphi[T,y](c) + i\varphi[T,y](d) \quad \text{for} \quad c,d \in H_n\,. \tag{2.20}$$

The main result can be summarized in the following

Theorem 2.1. *Every affine map (T,y) which maps the closed unit ball in $\mathbb{R}^{n^2-1}$ into itself induces a positive, trace preserving map $\varphi(T,y) : M_n \to M_n$*

$$\varphi[T,y](a) = \frac{1\!\!1_n}{n}\mathrm{Tr}\,a + \frac{1}{n-1}\Big\langle f, Tx + y\Big(\frac{n-1}{n}\Big)^{1/2}\mathrm{Tr}\,a\Big\rangle, \tag{2.21}$$

where $x = (x_1,\dots,x_{n^2-1}) \in \mathbb{R}^{n^2-1}$, $x_\alpha = \mathrm{Tr}(af_\alpha)$, and $f = (f_1, f_2, \dots, f_{n^2-1})$, $f_\alpha = f_\alpha^*$, $\mathrm{Tr}(f_\alpha f_\beta) = \delta_{\alpha\beta}$, $\mathrm{Tr}\,f_\alpha = 0$, $\alpha,\beta = 1,\dots,n^2-1$.

It should be pointed out that in the case $n = 2$ every positive, trace preserving map has the form (2.21).

A systematic construction of the operators $f_1,\dots,f_{n^2-1} \in H_n$ (generators of SU(n)) is well known, c.f. [35]. They are given by

$$(f_1,\dots,f_{n^2-1}) = (d_\ell, u_{k\ell}, v_{k\ell})\,, \quad \ell = 1,\dots,n-1\,, \quad 1 \le k < \ell \le n\,, \tag{2.22}$$

$$d_\ell = \frac{1}{\sqrt{\ell(\ell+1)}}\sum_{k=1}^{\ell}(e_{kk} - ke_{k+1,k+1})\,, \quad \ell = 1,\dots,n-1 \tag{2.23}$$

$$u_{k\ell} = \frac{1}{\sqrt{2}}(e_{k\ell} - e_{\ell k})\,, \tag{2.24}$$

$$v_{k\ell} = \frac{-i}{\sqrt{2}}(e_{k\ell} - e_{\ell k})\,, \tag{2.25}$$

$$e_{k\ell} = e_k(e_\ell, \cdot\,)\,, \tag{2.26}$$

where $(e_1,\dots,e_n)$ is an orthonormal base in $\mathbb{C}^n$.

As an illustration of the formula (2.21) let us consider the case $n = 3$. Using the notation

$$x_1 = \mathrm{Tr}(ad_1)\,, \qquad x_2 = \mathrm{Tr}(ad_2) \tag{2.27}$$

and

$$x_{k\ell} = \mathrm{Tr}(au_{k\ell})\,, \qquad y_{k\ell} = \mathrm{Tr}(av_{k\ell}) \tag{2.28}$$

let us consider the rotation $R(\alpha) \in \mathrm{SO}(8)$ given by

$$\begin{aligned} x_1' &= x_1 \cos\alpha - x_2 \sin\alpha\,, \\ x_2' &= x_1 \sin\alpha + x_2 \cos\alpha\,, \\ x_{k\ell}' &= -x_{k\ell}\,, \\ y_{k\ell}' &= -y_{k\ell}\,, \qquad 1 \le k < \ell \le n\,. \end{aligned} \tag{2.29}$$

Taking into account the explicit form of d_1, d_2, $u_{k\ell}$, $v_{k\ell}$, $1 \le k < \ell \le 3$, one finds that the corresponding map $\varphi[\alpha] \equiv \varphi[R(\alpha), 0] : M_3(\mathbb{C}) \to M_3(\mathbb{C})$ has the form $a' = \varphi[\alpha](a)$, where

$$\begin{aligned} a_{11}' &= \frac{1}{2}[\lambda(\alpha)a_{11} + \mu(\alpha)a_{22} + \nu(\alpha)a_{33}]\,, \\ a_{22}' &= \frac{1}{2}[\nu(\alpha)a_{11} + \lambda(\alpha)a_{22} + \mu(\alpha)a_{33}]\,, \\ a_{33}' &= \frac{1}{2}[\mu(\alpha)a_{11} + \nu(\alpha)a_{22} + \lambda(\alpha)a_{33}]\,, \\ a_{ij}' &= -\frac{1}{2}a_{ij}\,, \qquad i \neq j\,, \end{aligned} \tag{2.30}$$

and

$$\begin{aligned} \lambda(\alpha) &= \frac{2}{3}(1 + \cos\alpha)\,, \\ \mu(\alpha) &= \frac{2}{3}\Big(1 - \frac{1}{2}\cos\alpha - \frac{\sqrt{3}}{2}\sin\alpha\Big)\,, \\ \nu(\alpha) &= \frac{2}{3}\Big(1 - \frac{1}{2}\cos\alpha + \frac{\sqrt{3}}{2}\sin\alpha\Big)\,, \\ \lambda(\alpha) + \mu(\alpha) &+ \nu(\alpha) = 2\,. \end{aligned} \tag{2.31}$$

The special cases are:

(i) $a' = \varphi[\alpha = \pi/3](a)$

$$\begin{aligned} a'_{11} &= \frac{1}{2}(a_{11} + a_{33}), \\ a'_{22} &= \frac{1}{2}(a_{22} + a_{11}), \\ a'_{33} &= \frac{1}{2}(a_{33} + a_{22}), \\ a'_{ij} &= -\frac{1}{2}a_{ij}. \end{aligned}$$

(ii) $a' = \varphi[\alpha = -\pi/3](a)$

$$\begin{aligned} a'_{11} &= \frac{1}{2}(a_{11} + a_{22}), \\ a'_{22} &= \frac{1}{2}(a_{22} + a_{33}), \\ a'_{33} &= \frac{1}{2}(a_{33} + a_{11}), \\ a'_{ij} &= -\frac{1}{2}a_{ij}. \end{aligned}$$

(iii) $a' = \varphi[\alpha = \pi](a) = 1/2(\mathbb{1}_3 \mathrm{Tr}\, a - a)$

(iv) $a' = \varphi[\alpha = 0](a) = 1/3\varphi[\alpha = \pi](a) + (1 - 1/3)\psi(a)$

$$\begin{aligned} a'' &= \psi(a), \\ a''_{11} &= a_{11}, \quad a''_{22} = a_{22}, \quad a''_{33} = a_{33}, \\ a''_{ij} &= -\frac{1}{2}a_{ij}. \end{aligned}$$

The maps (i) and (ii) are indecomposable Choi's maps, the map (iii) is completely copositive, while (iv) is decomposable.

Let $[x_{ij}] \in M_3(M_3)$ be the matrix

$$[x_{ij}] = \left(\begin{array}{ccc|ccc|ccc} 1 & 0 & 0 & 0 & 1 & 0 & 0 & 0 & 1 \\ 0 & p & 0 & 0 & 0 & 0 & 0 & 0 & 0 \\ 0 & 0 & p^{-1} & 0 & 0 & 0 & 0 & 0 & 0 \\ \hline 0 & 0 & 0 & p^{-1} & 0 & 0 & 0 & 0 & 0 \\ 1 & 0 & 0 & 0 & 1 & 0 & 0 & 0 & 1 \\ 0 & 0 & 0 & 0 & 0 & p & 0 & 0 & 0 \\ \hline 0 & 0 & 0 & 0 & 0 & 0 & p & 0 & 0 \\ 0 & 0 & 0 & 0 & 0 & 0 & 0 & p^{-1} & 0 \\ 1 & 0 & 0 & 0 & 1 & 0 & 0 & 0 & 1 \end{array}\right) \qquad p > 0. \tag{2.32}$$

Then both $[x_{ij}]$ and $[x_{ji}]$ belong to $M_3(M_3)^+$. It is easily seen that the matrix $[\varphi[\alpha](x_{ij})]$ is not positive for $0 < \alpha < \pi$, i.e., the map $\varphi[\alpha]$ is indecomposable for $0 < \alpha < \pi$.

It should be pointed out that the indecomposable maps (1.4) does not belong to the class considered above except the case $p_0 = n - 2$ and $p_1 = \ldots = p_n = 1$.

References

1. W.F. Stinespring, Proc. Amer. Math. Soc. **6**, 211 (1955).
2. E. Størmer, Acta Math. **110**, 233 (1963).
3. E. Størmer, Trans. Amer. Math. Soc. **120**, 438 (1965).
4. E. Størmer, in: Lecture Notes in Physics **29**, Springer Verlag, Berlin, 1974, pp. 85–106.
5. W. Arverson, Acta Math. **123**, 141 (1969).
6. M.D. Choi, Lin. Alg. Appl. **10**, 285 (1975).
7. M.D. Choi, Lin. Alg. Appl. **12**, 95 (1975).
8. M.D. Choi, J. Operator Theory **4**, 271 (1980).
9. E. Størmer, Math. Ann. **247**, 21 (1980).
10. E. Størmer, Proc. Amer. Math. Soc. **86**, 402 (1982).
11. E. Størmer, J. Funct. Anal. **66**, 235 (1985).
12. S.L. Woronowicz, Rep. Math. Phys. **10**, 165 (1976).
13. S.L. Woronowicz, Comm. Math. Phys. **51**, 243 (1976).
14. A. Jamiołkowski, Rep. Math. Phys. **3**, 275 (1972).
15. L.E. Labuschagne, W.A. Majewski, M. Marciniak, to appear.
16. V. Gorini, E.C.G. Sudarshan, Comm. Math. Phys. **46**, 43 (1976).
17. B.M. Terhal, Lin. Alg. Appl. **323**, 61 (2001).
18. A.G. Robertson, Math. Proc. Camb. Phil. Soc. **94**, 291 (1983).
19. A.G. Robertson, Proc. Roy. Soc. Edinburgh **94 A**, 71 (1983).
20. A.G. Robertson, J. Lond. Math. Soc. (2) **32**, 133 (1985).
21. T. Ando, seminar note 1985.
22. J. Tomiyama, Contemporary Math. **62**, 357 (1987).
23. K. Tanahashi, J. Tomiyama, Canadian Math. Bull. **31**, 308 (1988).
24. T. Takasaki, J. Tomiyama, Math. Japonica **1**, 129 (1982).
25. S.H. Kye, in: *Elementary operators and applications*, M. Mathion, ed., World Scientific, 1992.
26. S.J. Cho, Lin. Alg. Appl. **171**, 213 (1992).
27. H. Osaka, Lin. Alg. Appl. **153**, 73 (1991).
28. H. Osaka, Publ. RIMS Kyoto Univ. **28**, 747 (1992).
29. H. Osaka, Lin. Alg. Appl. **186**, 45 (1992).
30. K.C. Ha, Publ. RIMS Univ. **34**, 591 (1998).
31. K.C. Ha, Lin. Alg. Appl. **359**, 277 (2003).
32. M. Horodecki, P. Horodecki, R. Horodecki, Phys. Rev. Lett A **223**, 1 (1996).
33. G. Albert, T. Beth, M. Horodecki, P. Horodecki, R. Horodecki, M. Rötter, H. Weinfurter, R. Werner, A. Zeinlinger, *Quantum Information*, Springer Tracts

in Modern Physics **173**, Springer Verlag, Berlin, 2001.
34. E. C. G. Sudarshan, P. M. Mathews, J. Rau, Phys. Rev. **121**, 920 (1961).
35. F. T. Hioe, J. H. Eberly, Phys. Rev. Lett. **47**, 838 (1981).

SIMULATING OPEN QUANTUM SYSTEMS WITH TRAPPED IONS

SABRINA MANISCALCO

INFM and MIUR, Dipartimento di Scienze Fisiche ed Astronomiche dell'Università di Palermo, via Archirafi 36, 90123 Palermo, Italy
E-mail: sabrina@fisica.unipa.it

This paper focus on the possibility of simulating the open system dynamics of a paradigmatic model, namely the damped harmonic oscillator, with single trapped ions. The key idea consists in using a controllable physical system, i.e. a single trapped ion interacting with an engineered reservoir, to simulate the dynamics of other open systems usually difficult to study. The exact dynamics of the damped harmonic oscillator under very general conditions is firstly derived. Some peculiar characteristic of the system's dynamics are then presented. Finally a way to implement with trapped ion the specific quantum simulator of interest is discussed.

1. Introduction

The dynamics of closed systems may be calculated exactly by solving directly the Schrödinger equation. In realistic physical conditions, however, quantum mechanical systems need to be regarded as open systems due to the fact that, as in classical physics, any realistic system is coupled to an uncontrollable environment which influences it in a non–negligible way. In general, the microscopic description of the total (system plus reservoir) closed system is extremely complicated. Moreover, even if one would be able to solve the microscopic equations of motion of the total closed system, one would be left with an overwhelming amount of information. Indeed, in most of the cases, one is interested only in the description of the dynamics of a small sub-ensemble of the total closed system, which by definition is called 'system'. For these reasons, a huge deal of attention has been devoted to the study of the open systems dynamics [1].

Nowadays the interest in the broad field of open quantum systems has notably increased mainly for two reasons. On the one hand experimental advances in the coherent control of single or few atoms and ions have paved the way to the realization of the first basic elements of quantum computers, c-not [2] and phase quantum gates [3]. Moreover, the first quantum cryptographic [4] and quantum teleportation [5] schemes have been experimentally implemented. These technological applications rely on the persistence of quantum coherence. Thus, understanding decoherence and dissipation arising from the unavoidable interaction between the system and its surrounding is necessary in order to implement real-size quantum computers [6] and other quantum technologies.

On the other hand, one of the most debated aspects of quantum theory, namely the quantum measurement problem, has been recently interpreted in terms of environment induced decoherence [7]. According to this interpretation the emergence of the classical world from the quantum world can be seen as a decoherence process due to the interaction between the system and the environment [8]. For this reason the study of some paradigmatic models of open systems allows to gain new insight in the theory of quantum measurement and in the related fundamental issues of quantum theory.

The models describing the interaction between a system and its surrounding are necessarily phenomenological since, in general, one can only infer the Hamiltonian of the environment and the structure of its coupling with the system. For this reason, in order to study the dynamics of open quantum systems, shedding light on the mechanisms of decoherence and dissipation, it is of great importance that the microscopic modelling of the environment is the most general possible and, at the same time, leads to an exact analytic description in terms of the density matrix of the reduced system.

A paradigmatic model of the theory of open systems is the damped harmonic oscillator or quantum Brownian motion (QBM) model, namely a harmonic oscillator linearly coupled with a reservoir described as an infinite set of non interacting oscillators. This model is central in many physical contexts, e.g., quantum field theory [9], quantum optics [1,10,11], and solid state physics [12]. The importance of the damped harmonic oscillator, is also due to the fact that it is one of the few exactly solvable non-trivial systems. Indeed, an exact master equation for the reduced density matrix can be formulated and exactly solved [13,14,15,16,17,18,19].

This paper focus on the possibility of studying experimentally quantum Brownian motion in a harmonic potential by simulating the system dynamics with single trapped ions coupled to artificial reservoirs. The idea of simulating the dynamics of a given (closed) quantum systems by using other systems, more easily controllable and measurable, was introduced by Feynman in [20]. Following this idea, some experimental schemes for realizing quantum simulators have been proposed in the trapped ion context [21,22]. In this paper we propose to extend the concept of quantum simulators from closed to open quantum systems. The possibility of realizing an open system quantum simulator stems from the recent experimental achievements in the realization of artificial reservoirs with trapped ion systems. In this context, indeed, it is possible not only to engineer experimentally an artificial reservoir but also to synthesize both its spectral density and the coupling with the system oscillator [23,24]. This makes it possible to think of new types of experiments aimed at testing the predictions of fundamental models as the one of quantum Brownian motion (or its high T limit: the famous Caldeira-Leggett model [15]).

The paper is structured as follows. In Sec. 2 the exact master equation for QBM is presented and an analytical method to solve it is discussed. In Sec. 3, the time evolution of the mean energy of the particle is studied by using the exact solution. This analysis brings to light two different regions of the system and reservoir parameter space in correspondence of which the system dynamics changes drastically. In Sec. 4 the experimental conditions for simulating QBM and observing the peculiar dynamics of the mean energy of the systems are studied. Finally, in Sec. 5 conclusions are presented.

2. Quantum Brownian motion: the exact dynamics

2.1. *Master Equation*

The dynamics of a harmonic oscillator linearly coupled with a quantized reservoir, modelled as an infinite chain of quantum harmonic oscillators, may be described exactly by means of a generalized Master Equation of the form [1,16,25,26]

$$\frac{d\rho_S(t)}{dt} = \frac{1}{i\hbar}\mathbf{H}_0^S \rho_S(t) - \Big[\Delta(t)(\mathbf{X}^S)^2 - \Pi(t)\mathbf{X}^S\mathbf{P}^S \\ -\frac{i}{2}r(t)(\mathbf{X^2})^S + i\gamma(t)\mathbf{X}^S\mathbf{P}^\Sigma\Big]\rho_S(t). \tag{1}$$

We indicate with $\mathbf{X}^{S(\Sigma)}$ and $\mathbf{P}^{S(\Sigma)}$ the commutator (anticommutator) position and momentum superoperators respectively and with $\mathbf{H}_0^S$ the commutator superoperator relative to the system Hamiltonian. The time dependent coefficients appearing in the Master Equation can be written, to the second order in the coupling strength, as follows

$$\Delta(t) = \int_0^t \kappa(\tau)\cos(\omega_0\tau)d\tau, \tag{2}$$

$$\gamma(t) = \int_0^t \mu(\tau)\sin(\omega_0\tau)d\tau, \tag{3}$$

$$\Pi(t) = \int_0^t \kappa(\tau)\sin(\omega_0\tau)d\tau, \tag{4}$$

$$r(t) = 2\int_0^t \mu(\tau)\cos(\omega_0\tau)d\tau, \tag{5}$$

where

$$\kappa(\tau) = \alpha^2\langle\{E(\tau), E(0)\}\rangle, \tag{6}$$

and

$$\mu(\tau) = i\alpha^2\langle[E(\tau), E(0)]\rangle, \tag{7}$$

are the noise and dissipation kernels, respectively. In the previous equations we indicate with α the system–reservoir coupling constant, with ω_0 the frequency of the system oscillator and with E the generalized reservoir position operator.

The Master Equation (1) is local in time, even if non-Markovian. This feature is typical of all the generalized Master Equations derived by using the time-convolutionless projection operator technique [1,28] or equivalent approaches such as the superoperatorial one presented in [19,29].

The time dependent coefficients appearing in Eq. (1) contain all the information about the short time system-reservoir correlation. The coefficient $r(t)$ gives rise to a time dependent renormalization of the frequency of the oscillator. The term proportional to $\gamma(t)$ is a classical damping term while the coefficients $\Delta(t)$ and $\Pi(t)$ are diffusive terms.

In what follows we study the time evolution of the heating function $\langle n(t)\rangle$ with n quantum number operator. The dynamics of $\langle n(t)\rangle$ depends only on the diffusion coefficient $\Delta(t)$ and on the classical damping coefficient $\gamma(t)$ [19]. Furthermore, the quantum number operator n belongs to a class of observables not influenced by a secular approximation which takes the Master Equation (1) into the form [19,27]

$$\begin{aligned}\frac{d\rho_S(t)}{dt} = & \frac{\Delta(t)+\gamma(t)}{2}\left[2a\rho_S(t)a^\dagger - a^\dagger a\rho_S(t) - \rho_S(t)a^\dagger a\right] \\ & \frac{\Delta(t)-\gamma(t)}{2}\left[2a^\dagger\rho_S(t)a - aa^\dagger\rho_S(t) - \rho_S(t)aa^\dagger\right].\end{aligned} \tag{8}$$

For this reason, in order to calculate the exact time evolution of the heating function, one can use the solution of the approximated Master Equation (8). In the previous equation we have introduced the bosonic annihilation and creation operators $a = (X + iP)/\sqrt{2}$ and $a^\dagger = (X - iP)/\sqrt{2}$, with X and P dimensionless position and momentum operator. Note that the above Master Equation is of Lindblad type as far as the coefficients $\Delta(t) \pm \gamma(t)$ are positive [30].

2.2. *Time evolution of the Quantum Characteristic Function*

The superoperatorial Master Equation (1) can be exactly solved by using specific algebraic properties of the superoperators [19]. The solution for the density matrix of the system is derived in terms of the Quantum Characteristic Function (QCF) $\chi_t(\xi)$ at time t, defined through the equation [9]

$$\rho_S(t) = \frac{1}{2\pi} \int \chi_t(\xi)\, e^{\left(\xi a^\dagger - \xi^* a\right)} d^2\xi. \tag{9}$$

It is worth noting that one of the advantages of this approach is the easiness in calculating the analytic expression for the mean values of observables of interest by using the relation

$$\langle a^{\dagger m} a^n \rangle = \left(\frac{d}{d\xi} \right)^m \left(-\frac{d}{d\xi^*} \right)^n e^{|\xi|^2/2} \chi(\xi) \bigg|_{\xi=0} . \tag{10}$$

The exact analytic expression for the time evolution of the heating function can be obtained from the solution of Eq. (8). In the secular approximation the QCF takes the form [19]

$$\chi_t(\xi) = e^{-\Delta_\Gamma(t)|\xi|^2} \chi_0 \left[e^{-\Gamma(t)/2} e^{-i\omega_0 t} \xi \right], \tag{11}$$

with χ_0 QCF of the initial state of the system. The quantities $\Delta_\Gamma(t)$ and $\Gamma(t)$ appearing in Eq. (11) are defined in terms of the diffusion and dissipation coefficients $\Delta(t)$ and $\gamma(t)$ respectively as follows

$$\Gamma(t) = 2 \int_0^t \gamma(t_1)\, dt_1, \tag{12}$$

$$\Delta_\Gamma(t) = e^{-\Gamma(t)} \int_0^t e^{\Gamma(t_1)} \Delta(t_1) dt_1. \tag{13}$$

Eq. (11) shows that the QCF is the product of an exponential factor, depending on both the diffusion $\Delta(t)$ and the dissipation $\gamma(t)$ coefficients, and a transformed initial QCF. The exponential term accounts for energy dissipation and is independent of the initial state of the system. Information on the initial state is given by the second term of the product, the transformed initial QCF.

In what follows we focus on the dynamics of the heating function $\langle n \rangle$, related to the mean energy of the system oscillator through the equation $\langle H_0 \rangle = \hbar\omega_0 \left(\langle n \rangle + 1/2 \right)$. Having in mind Eq. (11) and using Eq. (10), one gets the following expression for the heating function

$$\langle n(t) \rangle = e^{-\Gamma(t)} \langle n(0) \rangle + \frac{1}{2} \left(e^{-\Gamma(t)} - 1 \right) + \Delta_\Gamma(t). \tag{14}$$

The asymptotic long time behavior of the heating function is readily obtained by using the Markovian stationary values for $\Delta(t)$ and $\gamma(t)$. For a thermal reservoir, one gets

$$\langle n(t) \rangle = e^{-\Gamma t} \langle n(0) \rangle + n(\omega_0) \left(1 - e^{-\Gamma t} \right). \tag{15}$$

In the next section we will discuss in detail the dynamics of the heating process and we will show the changes in the short time behavior due to the variations of typical reservoir parameters such as its cut-off frequency.

3. Time evolution of the mean energy: Lindblad-type and non-Lindblad-type dynamics

In a previous paper we have presented a theory of heating for a single trapped ion interacting with a natural reservoir able to describe both its short time non-Markovian behavior and the asymptotic thermalization process [31]. Here we focus instead on the case of interaction with engineered reservoirs. In the trapped ion context, it is experimentally possible to engineer artificial reservoirs and couple them to the system in a controlled way. Since the coupling with the natural reservoir is negligible for long time intervals [32,33], this allows to test fundamental models of open system dynamics as the one for QBM we are interested in. By using the analytic solution, one can look for ranges of the relevant parameters of both the reservoir and the system in correspondence of which deviations from Markovian dissipation become experimentally observable.

In the experiments on artificially engineered amplitude reservoirs [24] the high temperature condition $\hbar\omega_0/KT \ll 1$ is always satisfied. For this reason here we concentrate on this regime of the parameters. For an Ohmic reservoir spectral density with Lorentz-Drude cut-off

$$J(\omega) = \frac{2\omega}{\pi} \frac{\omega_c^2}{\omega_c^2 + \omega^2}, \tag{16}$$

the dissipation and damping coefficients $\gamma(t)$ and $\Delta(t)$, appearing in the Master Equation (8), to second order in the coupling constant, take the form

$$\gamma(t)=\frac{\alpha^2\omega_0 r^2}{1+r^2}\left[1-e^{-\omega_c t}\cos(\omega_0 t)-re^{-\omega_c t}\sin(\omega_0 t)\right], \tag{17}$$

and

$$\Delta(t)=2\alpha^2 kT\frac{r^2}{1+r^2}\left\{1-e^{-\omega_c t}\left[\cos(\omega_0 t)\right.\right.$$
$$\left.\left.-(1/r)\sin(\omega_0 t)\right]\right\}, \tag{18}$$

with $r=\omega_c/\omega_0$ and ω_c cut-off frequency. Equation (18) has been derived in the high T limit, while $\gamma(t)$ does not depend on temperature. Comparing Eq. (18) with Eq. (17), one notices immediately that in the high temperature regime, $\Delta(t)\gg\gamma(t)$. Having this in mind it is easy to prove that the heating function, given by Eq. (14), may be written in the following approximated form

$$\langle n(t)\rangle\simeq\Delta_\Gamma(t), \tag{19}$$

where we have assumed that the initial state of the ion is its vibrational ground state, as it is actually the case at the end of the resolved sideband cooling process [34,35]. For times much bigger than the reservoir correlation time $\tau_R=1/\omega_c$ the asymptotic behavior of Eq. (19) is given by Eq. (15). This equation gives evidence for a second characteristic time of the dynamics, namely the thermalization time $\tau_T=1/\Gamma$, with $\Gamma=\alpha^2\omega_0 r^2/(r^2+1)$. The thermalization time depends both on the coupling strength and on the ratio $r=\omega_c/\omega_0$ between the reservoir cut-off frequency and the system oscillator frequency. Usually, when one studies QBM, the condition $r\gg 1$, which corresponds to a natural flat reservoir, is assumed. In this case the thermalization time is simply inversely proportional to the coupling strength. For an "out of resonance" engineered reservoir with $r\ll 1$, τ_T is notably increased and therefore the thermalization process is slowed down.

A further approximation to the heating function of Eq. (19) can be obtained for times $t\ll\tau_T$:

$$\langle n(t)\rangle\simeq\int_0^t\Delta(t_1)dt_1=\frac{2\alpha^2 KT}{\omega_c}\frac{r^2}{(r^2+1)^2}\left\{\omega_c t(r^2+1)\right. \tag{20}$$
$$\left.-(r^2-1)\left[1-e^{-\omega_c t}\cos(\omega_0 t)\right]-re^{-\omega_c t}\sin(\omega_0 t)\right\}. \tag{21}$$

This approximation shows a clear connection between the sign of the diffusion coefficient $\Delta(t)$ and the time evolution of the heating function before thermalization. The diffusion coefficient is indeed the time derivative of the heating function. We remind that, since $\Delta(t) \gg \gamma(t)$ for the case considered here, whenever $\Delta(t) > 0$ the Master Equation (8) is of Lindblad type, whilst the case $\Delta(t) < 0$ corresponds to a non-Lindblad type Master Equation. From Eq. (20) one sees immediately that while for $\Delta(t) > 0$ the heating function grows monotonically, when $\Delta(t)$ assumes negative values it can decrease and present oscillations.

To better understand such a behavior we study in more details the dynamics for three exemplary values of the ratio r between the reservoir cut-off frequency and the system oscillator frequency: $r \gg 1$, $r = 1$ and $r \ll 1$. As we have already noticed the first case corresponds to the assumption commonly done when dealing with natural reservoir while the last case corresponds to an engineered "out of resonance" reservoir.

For $r \gg 1$ the diffusion coefficient, given by Eq. (18), is positive for all t and r since

$$\Delta(t) \propto 1 - e^{-\omega_c t} \cos(\omega_0 t) \geq 0. \tag{22}$$

Therefore the Master Equation is always of Lindblad type and the heating function grows monotonically from its initial null value. Equation (20) shows that, for times $t \ll \tau_R$ and for $r \gg 1$, $\langle n(t) \rangle \simeq (\alpha^2 \omega_c kT) t^2$, that is the initial non-Markovian behavior of the heating function is quadratic in time.

For $r = 1$, a similar behavior is observed since also in this case $\Delta(t)$ is positive at all times.

Finally, in the case $r \ll 1$, Eq. (18) shows that, if r is sufficiently small, $\Delta(t)$ oscillates acquiring also negative values. It is worth noting, however, that the long time asymptotic value of $\Delta(t)$ is always positive. Whenever the diffusion coefficient is negative, the heating function decreases, so the overall heating process is characterized by oscillations of the heating function. The decrease in the population of the ground state of the system oscillator, after an initial increase due to the interaction with the high T reservoir, is due to the emission and subsequent reabsorption of the same quantum of energy. Such an event is possible since the reservoir correlation time $\tau_R = 1/\omega_c$ is now much longer than the period of oscillation $\tau_s = 1/\omega_0$. We underline that, although the Master Equation in this case is not of Lindblad type, it conserves the positivity of the reduced density matrix. This of course does not contradict the Lindblad theorem since the semigroup property is clearly violated for the reduced system dynamics [1].

4. Experiment for simulating QBM with trapped ions

Let me begin discussing the technique used to engineer an artificial reservoir coupled to a single trapped ion. Reference [24] presents recent experimental results showing how to couple a properly engineered reservoir with a quantum oscillator, namely the quantized center of mass (c.m.) motion of the ion. These experiments aim at measuring the decoherence of a quantum superposition of coherent states and Fock states due to the presence of the reservoir. Several types of engineered reservoirs are demonstrated, e.g., thermal amplitude reservoirs, phase reservoirs, and zero temperature reservoirs.

A high T amplitude reservoir is obtained by applying a random electric field $\vec{E}$ whose spectrum is centered on the axial frequency $\omega_z/2\pi = 11.3\text{MHz}$ of oscillation of the ion [24]. The trapped ion motion couples to this field due to the net charge q of the ion: $H_{int} = -q\vec{x}\cdot\vec{E}$, with $\vec{x} = (X, Y, Z)$ displacement of the c.m. of the ion from its equilibrium position. Remembering that $\vec{E} \propto \sum_i \vec{\epsilon}_i(b_i + b_i^\dagger)$, with b_i and $b_i^\dagger$ annihilation and creation operators of the fluctuating field modes, and that $X \propto \left(a + a^\dagger\right)$ one realizes that this coupling is equivalent to the bilinear one assumed to derive Eq. (1).

The random electric field is applied to the endcap electrodes through a network of properly arranged low pass filters limiting the natural environmental noise but allowing deliberately large applied fields to be effective. This type of drive simulates an infinite-bandwidth amplitude reservoir [24]. It is worth stressing that, for the times of duration of the experiment, namely $\Delta t = 20\mu\text{s}$, the heating due to the natural reservoir is definitively negligible [24].

The reservoir considered in our paper is a thermal reservoir with spectral distribution given by

$$\begin{aligned} I(\omega) &= J(\omega)[n(\omega) + 1/2] \\ &= \frac{\omega}{\pi}\frac{\omega_c^2}{\omega_c^2 + \omega^2}\coth\left(\omega/KT\right), \end{aligned} \tag{23}$$

where Eq. (16) has been used. For high T, Eq. (23) becomes

$$I(\omega) = \frac{2KT}{\pi}\frac{\omega_c^2}{\omega_c^2 + \omega^2}. \tag{24}$$

The infinite-bandwidth amplitude reservoir realized in the experiments corresponds to the case $\omega_c \to \infty$ in the previous equation. Therefore, for high

T, the reservoir discussed in the paper can be realized experimentally by filtering the random field, used in the experiments for simulating an infinite-bandwidth reservoir, with a Lorentzian shaped low pass filter at frequency ω_c. The change of the ratio r thus would be accomplished simply by changing the low pass filter.

It is well known that non-Markovian features usually occur in the dynamics for times $t \ll \tau_R = 1/\omega_c$. In general, since $\omega_c \gg \omega_0$ and typically $\omega_0 \simeq 10^7$Hz for trapped ions, this means that deviations from the Markovian dynamics appear for times $t \ll 0.1\mu$s. This is the reason why the initial quadratic behaviour of the heating function is not observed in the experiments, wherein the typical time scales go from 1 to 100μs.

A way to force non-Markovian features to appear is to 'detune' the trap frequency from the reservoir spectral density. This corresponds, for example, to the case in which $r = \omega_c/\omega_0 = 0.1$. In this case the reservoir correlation time is bigger than the period of oscillation of the ion and this leads to the oscillatory behaviour of the heating function predicted by Eq. (20) for $r \ll 1$. Under this condition $\tau_c = 1\mu$s, and therefore the non-Markovian features show up in the time evolution and can be measured. Detuning the trap frequency from the reservoir, however, decreases the effective coupling between the system and the environment and, for this reason, in order to obtain values of the heating function big enough to be measured we need to increase either the coupling constant α^2, which correspond to an increase in the intensity of the voltage applied to the electrodes, or the strength of the fluctuations $\langle V^2 \rangle$, which correspond to an increase in the effective temperature of the reservoir.

Let us look in more detail to Eq. (20). For $r \ll 1$ this equation becomes

$$\langle n(t) \rangle \simeq \frac{2\alpha^2 KT}{\pi \omega_c} r^2 \left\{ \omega_c t + 1 - e^{-\omega_c t} \cos(\omega_0 t) \right. \\ \left. - r e^{-\omega_c t} \sin(\omega_0 t) \right\} . \qquad (25)$$

From a comparison with the experiment of reference [24] one derives the following value for the front factor $2\alpha^2 KT/\pi \simeq 0.84 \cdot 10^7$Hz [36]. Increasing of two orders of magnitude this front factor, and for $r = 0.1$, Eq. (25) predicts the behaviour for the heating function shown in Fig. 1 (a). We believe that for this range of $\langle n \rangle$ and of times the oscillations of the heating function could be experimentally measurable.

Summarizing, in order to observe the virtual exchanges of phonons between the system and the reservoir, leading to the oscillations of the heating function, one needs to increase of two order of magnitude the coefficient

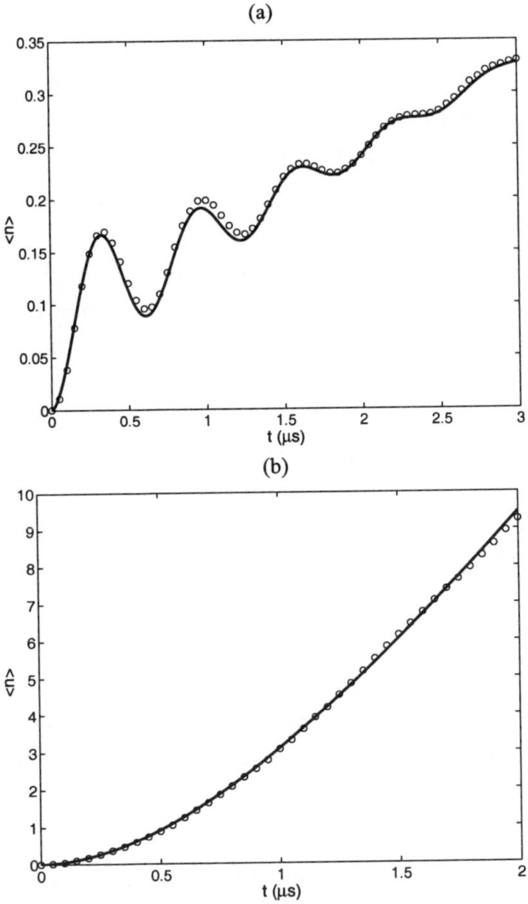

Figure 1. Time evolution of the heating function for (a) $2\alpha^2 KT/\pi = 0.84 \cdot 10^9$Hz, $\omega_c = 1$MHz, $r = 0.1$; and (b) $2\alpha^2 KT/\pi = 0.84 \cdot 10^9$Hz, $\omega_c = 1$MHz, $r = 10$. Solid line is the analytical and circles the simulation result.

$2\alpha^2 KT/\pi$. This can be done either increasing the intensity or increasing the fluctuations of the applied noise, or combining an increase in the intensity with an increase in the fluctuations [36]. Moreover one needs to use a low pass filter for the applied noise having cut off frequency $\omega_c = 0.1\omega_0$.

We now examine briefly the conditions for which the quadratic behaviour of the heating function, can be observed. We remind that this corresponds to the case in which $r \gg 1$ and the time evolution of the density matrix is of Lindblad-type. In view of the considerations done at the beginning of this section, in order to reveal non-Markovian dynamics in a time scale of $1-100\mu s$ we need to have $\tau_R \geq 1\mu s$, that is, for $r = 10$, $\omega_0 \leq 0.1$MHz. This means that one actually needs a 'loose' trap. For example, for $\omega_0 = 100$kHz, with the same applied noise used in [24]), i.e. $2\alpha^2 KT/\pi = 0.84 \cdot 10^7$Hz, the time evolution of the heating function is the one shown in Fig 1 (b).

5. Conclusions

In this paper the dynamics of a single harmonic oscillator coupled to a quantized high temperature reservoir is studied, focussing in particular on the non-Markovian heating dynamics typical of short times. In this regime the system time evolution is influenced by correlations between the system and the reservoir. For certain values of the system and reservoir parameters, virtual exchanges of energy between the system and its environment become dominant. These virtual processes strongly affect the short time dynamics and are responsible for the appearance of oscillations in the heating function (non-Lindblad type dynamics).

Extending the ideas of using trapped ions for simulating quantum optical systems, a QBM quantum simulator with single trapped ions coupled to artificial reservoirs is proposed. I have carefully analyzed the possibility of revealing, by using present technologies, the non-Markovian dynamics of a single trapped ion interacting with an engineered reservoir, underlining the conditions under which non-Markovian features become observable.

Acknowledgements

The author gratefully acknowledge Jyrki Piilo, Francesco Petruccione, Francesco Intravaia and Antonino Messina, which have worked with her on the results reviewed in this paper, and the EU network COCOMO (contract HPRN-CT-1999-00129) for financial support.

References

1. H.-P. Breuer and F. Petruccione, *The Theory of Open Quantum systems* (Oxford University Press, 2002).
2. J.I. Cirac and P. Zoller, Phys. Rev. Lett. **74**, 4091 (1995); C. Monroe *et al.*, Phys. Rev. Lett. **75**, 4714 (1995); B. DeMarco *et al.*, Phys. Rev. Lett. **89**, 267901 (2002); F. Schmidt-Kaler *et al.*, Nature **422**, 408 (2003).
3. G. Falci *et al.*, Nature **407**, 355 (2000); D. Leibfried *et al.*, Nature **422**, 412 (2003), J. Pachos and H. Walther, Phys. Rev. Lett. **89**, 187903 (2002).
4. N. Gisin, G. Ribordy, W. Tittel, and H. Zbinden, Rev. Mod. Phys. **74**, 145 (2002).

5. I. Marcikic *et al.*, Nature **421**, 509 (2003).
6. D. Kielpinski, C. Monroe, and D.J. Wineland, Nature **417**, 709 (2002).
7. W.H. Zurek, Phys. Today **44**, 36 (1991); for comments see also Phys. Today **46**, 13 and 81 (1991); W.H. Zurek, Rev. Mod. Phys. **75**, 715 (2003).
8. E. Joos, H.-D. Zeh, C. Kiefer, D. Giulini, J. Kupsch, and I.-O. Stamatescu, *Decoherence and the Appearence of a Classical World in Quantum Theory, 2nd ed.* (Springer-Verlag, Berlin, 2003).
9. C. Cohen-Tannoudj, J. Dupont-Roc, and G. Grynberg, *Atom-Photon Interactions* (John Wiley and sons, 1992).
10. H.J. Carmichael, *An Open System Approach to Quantum Optics* (Springer-Verlag, Berlin, 1993).
11. V. Bužeck and P.L. Knight in *Progress in Optics XXXIV* (Elsevier, Amsterdam,1995).
12. U. Weiss, *Quantum Dissipative Systems* 2nd ed (World Scientific, Singapore, 1999).
13. F. Haake and R. Reibold, Phys. Rev. A **32**, 2462 (1985).
14. R.P. Feynman and F.L. Vernon, Ann. Phys. **24**, 118 (1963).
15. A.O. Caldeira and A.J. Leggett, Phys. A **121**, 587 (1983).
16. B.L. Hu, P. Paz and Y. Zhang, Phys. Rev. D **45**, 2843 (1992).
17. G.W. Ford and R.F. O'Connell, Phys. Rev. D **64**, 105020 (2001).
18. H. Grabert, P. Schramm, and G.-L. Ingold, Phys. Rep. **168**, 115 (1988).
19. F. Intravaia, S. Maniscalco, and A. Messina, Phys. Rev. A **67**, 042108 (2003).
20. R.P. Feynman, Int. J. Theor. Phys. A **21**, 467 (1982).
21. D.J. Wineland *et al.*, Phys. Scr. **T76**, 147 (1998).
22. D. Leibfried *et al.*, Phys. Rev. Lett. **89**, 247901 (2002).
23. J.F. Poyatos, J.I. Cirac, and P. Zoller, Phys. Rev. Lett. **77**, 4728 (1996).
24. C. J. Myatt *et al.*, Nature **403**, 269 (2000); Q.A. Turchette *et al.*, Phys. Rev. A **62**, 053807 (2000).
25. F. Intravaia, S. Maniscalco, and A. Messina, Eur. Phys. J. B **32**, 97 (2003).
26. H.-P. Breuer, B. Kappler, and F. Petruccione, Ann. Phys. **291**, 36 (2001).
27. R. Karrlein and H. Grabert, Phys. Rev. E **55**, 153 (1997).
28. H. P. Breuer, B. Kappler, and F. Petruccione, Phys. Rev. A **59**, 1633 (1999).
29. A. Royer, Phys. Rev. A **6**, 1741 (1972); A. Royer, Phys. Lett. A **315**, 335 (2003).
30. S. Maniscalco, F. Intravaia, J. Piilo, and A. Messina, quant-ph/0306193, accepted for publication in J.Opt.B: Quantum and Semiclass. Opt.
31. F. Intravaia, S. Maniscalco, J. Piilo, and A. Messina, Phys. Lett. A **308**, 6 (2003).
32. D. Leibfried, R. Blatt, C. Monroe, and D. Wineland, Rev. Mod. Phys. **75**, 281 (2003); D. J. Wineland *et al.*, J. Res. Natl. Inst. Stand. Technol. **103**, 259 (1998).
33. Q. A. Turchette *et al.*, Phys. Rev. A **61**, 063418 (2000); M. Rowe *et al.*, Quantum Inf. Comput. **1**, 57 (2001).
34. D. M. Meekhof *et al.*,Phys. Rev. Lett. **76**, 1796, (1996); D. Leibfried *et al.*, J. Mod. Opt. **44**, 2485 (1997).
35. R. Blatt, Phys. World **9**, 25 (1996); Ch. Roos *et al.*, Phys. Rev. Lett. **83**,

4713 (1999).
36. S. Maniscalco, J. Piilo, F. Intravaia, F. Petruccione, and A. Messina, quant-ph/0307231v2, accepted for pubblication in Phys. Rev. A.

A PURIFICATION SCHEME AND ENTANGLEMENT DISTILLATIONS

HIROMICHI NAKAZATO, MAKOTO UNOKI AND KAZUYA YUASA

Department of Physics, Waseda University, Tokyo 169–8555, Japan
E-mail: hiromici@waseda.jp, yuasa@hep.phys.waseda.ac.jp

A purification scheme which utilizes the action of repeated measurements on a (part of a total) quantum system is briefly reviewed and is applied to a few simple systems to show how it enables us to extract an entangled state as a target pure state. The scheme is rather simple (e.g., we need not prepare a specific initial state) and is shown to have wide applicability and flexibility, and is able to accomplish both the maximal fidelity and non-vanishing yield.

1. Introduction

It is well known that in quantum mechanics, the action of measurement affects the dynamics of the system just measured in an essential way. This has to be contrasted with the situation in classical mechanics, where the effect of measurement can be made as small as one wishes. Typical phenomenon reflecting such a peculiarity in quantum mechanics has been known under the name of "Quantum Zeno Effect (QZE)" [1] and has been extensively studied recently. It states that if a system is frequently measured to confirm that it is in its initial state, the change of the state based on the Hamiltonian of the system is decelerated, or to state differently, the decay of an unstable state is hindered by frequent measurements. It is widely recognized, however, that there is no paradoxical point in such phenomena and the effect is solely understood quantum mechanically [2]. Indeed, the projective measurement is not essential and is just replaced with the generalized spectral decompositions [3], and the effect is due to the peculiar short-time dynamics of quantum systems, known as the "flat derivative" of the survival probability, $\dot{P}(0) = 0$. The effect has also been examined experimentally and the first announcement of its confirmation appeared in an oscillating system [4] and then in a truly unstable system [5]. It is worth while mentioning that Raizen's group has observed the deviation from the exponential law at short times [6] and even the acceleration of decay when the measurements are not frequent enough [5]. The latter effect is called the "Inverse (or Anti) Zeno Effect (IZE)," which has also been studied and discussed [7]. The QZE (and/or IZE) has still been explored extensively in the hope of realizing a protection scheme for system's coherence against possible decoherence by this (or its related) mechanism [8].

Here another interesting consequence of the peculiarity of quantum measurement shall be disclosed. We consider a total system that is composed of two (sub)systems A and B and measurements shall be repeatedly performed only on system A at regular intervals. Our interest lies in the (asymptotic) dynamics of system B, which is indirectly affected by the measurements on A through its interaction with A. It is shown that such indirect measurements can drive system B to a pure state (*purification*), irrespectively of its initial state that is mixed in general [9]. We can say that the effect of measurement is far reaching and profound It can control even the other parts of the system which are not touched directly. In the following sections, the mechanism of this purification is briefly reviewed (Sec. 2) and is applied to a simple qubit system (Sec. 3) to show how it works and how it can be optimized. Since the entangled state, which is one of the key elements in quantum technologies like quantum computation, information, teleportation, etc. [10], is a pure state of a compound system, this method is used to extract entangled states [11,12] in Secs. 4 and 5. A brief summary is presented in Sec. 6.

2. General Framework of Purification Through Repeated Measurements

Let the total system consist of two parts, system A and system B, and the dynamics be described by the total Hamiltonian

$$H = H_A + H_B + H_{int}, \tag{1}$$

where H_{int} stands for the interaction between the two (sub)systems. We initially prepare the system in a product state

$$\rho_0 = |\phi\rangle\langle\phi| \otimes \rho_B(0) \tag{2}$$

at $t = 0$. Such a state can be realized, say, if the system A is found in the state $|\phi\rangle$ after the zeroth measurement. Notice that system B can be in an *arbitrary mixed* state $\rho_B(0)$. We perform measurements on A at regular intervals τ to confirm that it is still in the state $|\phi\rangle$, even though the total system A+B evolves unitarily in terms of the total Hamiltonian H.

Since the measurement is performed only on system A, the action of such a (projective, for simplicity) measurement can be conveniently described by the following projection operator

$$\mathcal{O} \equiv |\phi\rangle\langle\phi| \otimes \hat{1}_B. \tag{3}$$

Thus the state of system A is set back to $|\phi\rangle$ every after τ, while that of B just evolves dynamically on the basis of the total Hamiltonian H. We repeat the same measurement, represented by (3), N times and collect only those events in which system A has been found in state $|\phi\rangle$ consecutively N times; other events are discarded. The state of system B is then described by the density matrix

$$\rho_B^{(\tau)}(N) = \big(V_\phi(\tau)\big)^N \rho_B(0)\big(V_\phi^\dagger(\tau)\big)^N / P^{(\tau)}(N), \tag{4}$$

where

$$V_\phi(\tau) \equiv \langle\phi|e^{-iH\tau}|\phi\rangle \tag{5}$$

is an operator acting on B and

$$\begin{aligned} P^{(\tau)}(N) &= \mathrm{Tr}\Big[(\mathcal{O}e^{-iH\tau}\mathcal{O})^N \rho_0 (\mathcal{O}e^{iH\tau}\mathcal{O})^N\Big] \\ &= \mathrm{Tr}_B\Big[\big(V_\phi(\tau)\big)^N \rho_B(0)\big(V_\phi^\dagger(\tau)\big)^N\Big] \end{aligned} \tag{6}$$

is the success probability for these events to occur (yield). This normalization factor in (4) reflects the fact that only right outcomes are collected in this process.

In order to examine the asymptotic state of system B, consider the spectral decomposition of the operator $V_\phi(\tau)$, which is not hermitian, $V_\phi(\tau) \neq V_\phi^\dagger(\tau)$. We therefore need to set up both the right- and left-eigenvalue problems

$$V_\phi(\tau)|u_n) = \lambda_n|u_n), \qquad (v_n|V_\phi(\tau) = \lambda_n(v_n|. \tag{7}$$

The eigenvalue λ_n is complex valued in general, but its absolute value is bounded [12]

$$0 \leq |\lambda_n| \leq 1, \tag{8}$$

which is a reflection of the unitarity of the time evolution operator $e^{-iH\tau}$. These eigenvectors are assumed to form a complete orthonormal set in the following sense

$$\sum_n |u_n)(v_n| = \hat{1}_B, \qquad (v_n|u_m) = \delta_{nm}. \tag{9}$$

Then the operator $V_\phi(\tau)$ itself is expanded in terms of these eigenvectors

$$V_\phi(\tau) = \sum_n \lambda_n |u_n)(v_n|. \tag{10}$$

It is now easy to see that the Nth power of this operator is expressed as

$$\big(V_\phi(\tau)\big)^N = \sum_n \lambda_n^N |u_n)(v_n| \tag{11}$$

and therefore it is dominated by a single term for large N

$$\big(V_\phi(\tau)\big)^N \xrightarrow{\text{large } N} \lambda_0^N |u_0)(v_0|, \tag{12}$$

when the largest (in magnitude) eigenvalue λ_0 is *discrete, nondegenerate and unique*. If these conditions are satisfied, the density operator of system B is driven to a pure state

$$\rho_B^{(\tau)}(N) \xrightarrow{\text{large } N} |u_0)(u_0|/(u_0|u_0) \tag{13}$$

with the probability

$$P^{(\tau)}(N) \xrightarrow{\text{large } N} |\lambda_0|^{2N}(u_0|u_0)(v_0|\rho_B(0)|v_0). \tag{14}$$

The pure state $|u_0)$, which is nothing but the right-eigenvector of the operator $V_\phi(\tau)$ belonging to the largest (in magnitude) eigenvalue λ_0, is thus distilled in system B. This is the purification scheme proposed in [9].

A few comments are in order. First, the final pure state $|u_0)$ toward which system B is to be driven is dependent on the choice of the state $|\phi\rangle$ on which system A is projected every after measurement, the measurement interval τ and the Hamiltonian H, but does not depend on the initial state of system B at all. In this sense, the purification is accomplished irrespectively of the initial (mixed) state $\rho_B(0)$. Second, as is clear in the above exposition, what is crucial in this purification scheme is the repetition of one and the same measurement (more appropriately, spectral decomposition) and the measurement interval τ need not be very small. It remains to be an adjustable parameter. Third, if we can make other eigenvalues than λ_0 much smaller in magnitude

$$|\lambda_n/\lambda_0| \ll 1 \quad \text{for } n \neq 0, \tag{15}$$

by adjusting parameters, we will need fewer steps (i.e., smaller N) to purify system B.

It is now evident that the purification can be made *optimal*, if the conditions (15) and

$$|\lambda_0| = 1 \tag{16}$$

are satisfied. This condition (16) assures that we can repeat as many measurements as we wish without running the risk of losing the yield (success probability) $P^{(\tau)}(N)$ in order to make the fidelity to the target state $|u_0)$,

$$F^{(\tau)}(N) \equiv \mathrm{Tr}_B \left[\rho_B^{(\tau)}(N)|u_0)(u_0|/(u_0|u_0)\right], \tag{17}$$

higher. Actually, the yield $P^{(\tau)}(N)$ decays like

$$P^{(\tau)}(N) = \sum_{n,m} \lambda_n^N \lambda_m^{*\,N} (v_n|\rho_B(0)|v_m)(u_m|u_n)$$
$$\xrightarrow{\text{large } N} |\lambda_0|^{2N}(u_0|u_0)(v_0|\rho_B(0)|v_0) \tag{18}$$

and the condition (16) can bring us with the non-vanishing yield $(u_0|u_0)(v_0|\rho_B(0)|v_0)$ even in the $N \to \infty$ limit. Therefore the condition (16) makes the two (sometimes not compatible) demands, i.e., *higher fidelity and non-vanishing yield*, achievable, with fewer steps when the condition (15) is met. In this sense, the purification is considered to be optimal.

It would be desirable if an optimal purification can be realized by an appropriate choice of the state $|\phi\rangle$ and/or tuning of the measurement interval τ and parameters in a given system. In the following sections, a few simple systems are examined to show how such optimal purifications are made possible.

3. Purification of a Qubit

As a simplest example, let us consider a total system of interacting two qubits. Systems A and B are represented by two qubits A and B, respectively, and their two levels can be described as the two degrees of freedom of spin-1/2 particle. We measure qubit A at regular intervals τ and examine the state of qubit B, which is in interaction with A. The measurement is conveniently parameterized as that of a spin-1/2 particle along a particular direction $\boldsymbol{n}$. After qubit A has been confirmed that its "spin" is up along $\boldsymbol{n}$, its state is projected to the eigenstate of the spin operator along $\boldsymbol{n}$, i.e., $|\phi\rangle = \boldsymbol{n}\cdot\boldsymbol{\tau}|\phi\rangle$, with $\boldsymbol{\tau} \equiv (\tau_1, \tau_2, \tau_3)$ being the Pauli matrices acting on qubit A. The state $|\phi\rangle$ is parameterized in terms of the two angles θ and φ as

$$|\phi\rangle = \cos(\theta/2)e^{-i\varphi/2}|\uparrow\rangle + \sin(\theta/2)e^{i\varphi/2}|\downarrow\rangle. \tag{19}$$

It is an elementary task to write down the projected operator $V_\phi(\tau) \equiv \langle\phi|e^{-iH\tau}|\phi\rangle$ in the following form

$$V_\phi(\tau) = c_0 + \boldsymbol{c}\cdot\boldsymbol{\sigma}, \tag{20}$$

where $\boldsymbol{\sigma} \equiv (\sigma_1, \sigma_2, \sigma_3)$ are Pauli matrices acting on qubit B and parameters c_0 and $\boldsymbol{c} \equiv (c_1, c_2, c_3)$ are complex valued. Its eigenvalues $\lambda_\pm$ and the corresponding right- and left-eigenvectors are easily found

$$\lambda_\pm = c_0 \pm c, \qquad c \equiv \sqrt{\boldsymbol{c}^2}, \quad c_\pm \equiv c_1 \pm i c_2, \tag{21}$$

$$\begin{aligned} |u_+) &= \frac{1}{\sqrt{2c(c-c_3)}} \Big[c_- |\uparrow) + (c - c_3) |\downarrow) \Big], \\ |u_-) &= \frac{1}{\sqrt{2c(c-c_3)}} \Big[(c_3 - c) |\uparrow) + c_+ |\downarrow) \Big], \end{aligned} \tag{22}$$

$$\begin{aligned} (v_+| &= \frac{1}{\sqrt{2c(c-c_3)}} \Big[c_+ (\uparrow| + (c - c_3)(\downarrow| \Big], \\ (v_-| &= \frac{1}{\sqrt{2c(c-c_3)}} \Big[(c_3 - c)(\uparrow| + c_- (\downarrow| \Big]. \end{aligned} \tag{23}$$

Now let the total Hamiltonian of this system be given by

$$\begin{aligned} H = {} & \frac{\omega_A}{2}(1 + \tau_3) + \frac{\omega_B}{2}(1 + \sigma_3) \\ & + g(\tau_+ \sigma_- + \text{h.c.}) + h(\tau_+ \sigma_+ + \text{h.c.}), \end{aligned} \tag{24}$$

where real parameters g and h are responsible for the interaction between the two qubits, A and B. In this case, the parameters $c_0, \ldots, c_3$ in (20)–(23) are explicitly calculated to be

$$\begin{aligned} c_0 = {} & \frac{1}{2}\left(\cos\frac{\tau\theta_h}{2} + \cos\frac{\tau\theta_g}{2} \right) \\ & - \frac{i}{2}\left(\frac{\omega_+}{\theta_h} \sin\frac{\tau\theta_h}{2} + \frac{\omega_-}{\theta_g} \sin\frac{\tau\theta_g}{2} \right) \cos\theta, \end{aligned} \tag{25}$$

$$c_1 = -i\left(\frac{h}{\theta_h} \sin\frac{\tau\theta_h}{2} + \frac{g}{\theta_g} \sin\frac{\tau\theta_g}{2} \right) \sin\theta\cos\varphi, \tag{26}$$

$$c_2 = i\left(\frac{h}{\theta_h} \sin\frac{\tau\theta_h}{2} - \frac{g}{\theta_g} \sin\frac{\tau\theta_g}{2} \right) \sin\theta\sin\varphi, \tag{27}$$

$$\begin{aligned} c_3 = {} & \frac{1}{2}\left(\cos\frac{\tau\theta_h}{2} - \cos\frac{\tau\theta_g}{2} \right) \cos\theta \\ & - \frac{i}{2}\left(\frac{\omega_+}{\theta_h} \sin\frac{\tau\theta_h}{2} - \frac{\omega_-}{\theta_g} \sin\frac{\tau\theta_g}{2} \right), \end{aligned} \tag{28}$$

where we have introduced

$$\omega_\pm \equiv \omega_A \pm \omega_B, \quad \theta_h \equiv \sqrt{\omega_+^2 + 4h^2}, \quad \theta_g \equiv \sqrt{\omega_-^2 + 4g^2}. \tag{29}$$

In order to illustrate how an optimal purification can be achieved in this system, consider the case where we measure qubit A along the 3-direction, that is, we choose $\theta = 0$ and $|\phi\rangle = |\uparrow\rangle$. Then the eigenvalues $\lambda_\pm$ in (21) and the corresponding eigenvectors are

$$\lambda_+ = \cos\frac{\tau\theta_h}{2} - i\frac{\omega_+}{\theta_h}\sin\frac{\tau\theta_h}{2} \Longleftrightarrow |u_+) = |\uparrow), \tag{30}$$

$$\lambda_- = \cos\frac{\tau\theta_g}{2} - i\frac{\omega_-}{\theta_g}\sin\frac{\tau\theta_g}{2} \Longleftrightarrow |u_-) = |\downarrow). \tag{31}$$

Since we have

$$|\lambda_+|^2 = 1 - \frac{4h^2}{\theta_h^2}\sin^2\frac{\tau\theta_h}{2}, \quad |\lambda_-|^2 = 1 - \frac{4g^2}{\theta_g^2}\sin^2\frac{\tau\theta_g}{2}, \tag{32}$$

the purification can be made optimal, e.g., when the parameters are adjusted so that $h\sin(\tau\theta_h/2) = 0$ is satisfied. In this case, $|\lambda_+| = 1$ and qubit B is driven to a pure state $|\uparrow)$, more quickly for larger g satisfying $\sin^2(\tau\theta_g/2) = 1$. A similar situation can happen; we can extract $|\downarrow)$ in qubit B, more quickly for larger h satisfying $\sin^2(\tau\theta_h/2) = 1$, if we adjust parameters so that $g\sin(\tau\theta_g/2) = 0$ holds.

Needless to say, there are cases where such purifications are not possible. For example, consider a case where we measure qubit A in the 1-2 plane, i.e., $\theta = \pi/2$. In this case, since the parameter c_0 is real, while all the other parameters c_1, c_2 and c_3 become pure imaginary, the eigenvalues $\lambda_\pm = c_0 \pm c$ are degenerated in magnitude

$$|\lambda_+| = \sqrt{c_0^2 + |c|^2} = |\lambda_-| \tag{33}$$

and no purification can occur in this particular case.

4. Entanglement Distillation I

As is mentioned in the Introduction, since entanglement is one of the key elements in quantum technologies [10], it would be useful if the present scheme of purification can be used to extract an entangled state as a target pure state. Notice that since the target system has never been measured directly in the present scheme, it is considered to be suited for extraction of such a fragile pure state as an entangled state. Actually any measurement on its subsystem that consits of entanglement would result in the destruction of the entanglement.

In order to see an entanglement distillation on the basis of the present idea of purification [11,12], we consider a total system composed of a compound system A+B, in which an entangled state is to be extracted, and another system C. Systems A and B interact with system C separately, but do not interact directly with each other. We measure system C repeatedly at regular intervals τ and endeavor to extract an entangled state as a pure state in A+B. In this section, a simple model, in which all systems A, B and C are represented by qubits, is considered with a model Hamiltonian

$$\begin{aligned} H = &\frac{\Omega}{2}(1+\sigma_3^A) + \frac{\Omega}{2}(1+\sigma_3^B) + \frac{\omega}{2}(1+\tau_3) \\ &+ g(\sigma_+^A\tau_- + \sigma_+^B\tau_- + \text{h.c.}) \\ &+ h(\sigma_+^A\tau_+ + \sigma_+^B\tau_+ + \text{h.c.}), \end{aligned} \tag{34}$$

where the Pauli matrices τ_i act on system C. It is assumed here for simplicity that the two systems A and B are the same and the Hamiltonian is symmetric under the exchange A↔B.

In order to find the spectral decomposition (10) of the projected operator $V_\phi(\tau)$ in this case, it turns out to be convenient to introduce the Bell states

$$|\Psi^\pm\rangle = \frac{1}{\sqrt{2}}\Big[|\uparrow\downarrow\rangle \pm |\downarrow\uparrow\rangle\Big], \qquad |\Phi^\pm\rangle = \frac{1}{\sqrt{2}}\Big[|\uparrow\uparrow\rangle \pm |\downarrow\downarrow\rangle\Big] \tag{35}$$

as a complete orthonormal set for A+B, because we are interested in extraction of such entangled states. Indeed, the eigenstates of the total Hamiltonian H can be found after their classification according to the above mentioned A↔B symmetry and a "parity" $\mathcal{P} \equiv \sigma_3^A\sigma_3^B\tau_3$

1) A↔B symmetric and $\mathcal{P} = +$

$$H\begin{pmatrix} |\Phi^+\uparrow\rangle \\ |\Phi^-\uparrow\rangle \\ |\Psi^+\downarrow\rangle \end{pmatrix} = \begin{pmatrix} \Omega+\omega & \Omega & g+h \\ \Omega & \Omega+\omega & -g+h \\ g+h & -g+h & \Omega \end{pmatrix}\begin{pmatrix} |\Phi^+\uparrow\rangle \\ |\Phi^-\uparrow\rangle \\ |\Psi^+\downarrow\rangle \end{pmatrix}, \tag{36}$$

2) A↔B symmetric and $\mathcal{P} = -$

$$H\begin{pmatrix} |\Phi^+\downarrow\rangle \\ |\Phi^-\downarrow\rangle \\ |\Psi^+\uparrow\rangle \end{pmatrix} = \begin{pmatrix} \Omega & \Omega & g+h \\ \Omega & \Omega & g-h \\ g+h & g-h & \Omega+\omega \end{pmatrix}\begin{pmatrix} |\Phi^+\downarrow\rangle \\ |\Phi^-\downarrow\rangle \\ |\Psi^+\uparrow\rangle \end{pmatrix}, \tag{37}$$

3) A↔B anti-symmetric and $\mathcal{P} = -$

$$H|\Psi^-\uparrow\rangle = (\Omega+\omega)|\Psi^-\uparrow\rangle, \tag{38}$$

4) A↔B anti-symmetric and $\mathcal{P} = +$

$$H|\Psi^-\downarrow\rangle = \omega|\Psi^-\downarrow\rangle. \tag{39}$$

Here $|\Phi^+\uparrow\rangle \equiv |\Phi^+) \otimes |\uparrow\rangle$, etc. Thus the time evolution operator $e^{-iH\tau}$ is expressed as

$$e^{-iH\tau} = \sum_s e^{-iE_s\tau}|s\rangle\langle s| + |\Psi^-)(\Psi^-|\Big[e^{-i(\Omega+\omega)\tau}|\uparrow\rangle\langle\uparrow| + e^{-i\Omega\tau}|\downarrow\rangle\langle\downarrow|\Big], \tag{40}$$

where the summation is taken over the six A↔B symmetric eigenstates of H, denoted as $|s\rangle$, that are given as linear combinations of the six states in (36) and (37). Owing to the A↔B symmetry of H, the A↔B anti-symmetric state $|\Psi^-)$ does not mix with the other (A↔B symmetric) eigenstates.

We are now in a position to examine the spectrum of the projected operator $V_\phi(\tau) \equiv \langle\phi|e^{-iH\tau}|\phi\rangle$. If the measurement of C projects its state on

$$|\phi\rangle = \alpha|\uparrow\rangle + \beta|\downarrow\rangle, \tag{41}$$

the operator reads

$$\begin{aligned} V_\phi(\tau) &= \langle\phi|e^{-iH\tau}|\phi\rangle \\ &= \sum_s e^{-iE_s\tau}\Big[|\alpha|^2\langle\uparrow|s\rangle\langle s|\uparrow\rangle + |\beta|^2\langle\downarrow|s\rangle\langle s|\downarrow\rangle \\ &\qquad + \alpha^*\beta\langle\uparrow|s\rangle\langle s|\downarrow\rangle + \text{h.c.}\Big] \\ &\quad + |\Psi^-)(\Psi^-|\Big[|\alpha|^2 e^{-i(\Omega+\omega)\tau} + |\beta|^2 e^{-i\Omega\tau}\Big]. \end{aligned} \tag{42}$$

From this expression, it is evident that the Bell state $|\Psi^-)$ is always one of the eigenstates of this operator and if the measurement interval τ is so adjusted that the condition $\omega\tau = 2\pi$ is met, its eigenvalue λ_{Ψ^-} becomes maximum in magnitude

$$\omega\tau = 2\pi \quad \longrightarrow \quad |\lambda_{\Psi^-}| = 1, \tag{43}$$

irrespectively of the projected state $|\phi\rangle$ of system C because $|\alpha|^2 + |\beta|^2 = 1$. This clearly demonstrates a possibility of entanglement distillation in this simple system, in the sense that one of the conditions for optimal purification (16) can be realized by (43) and the entangled state $|\Psi^-)$ would surely be extracted, only if the other eigenvalues of the operator than λ_{Ψ^-} can be made (much) smaller in magnitude. Further details of the analysis of such conditions for the entanglement distillation and its optimization are found in [11] and [12] in a slightly simplified case.

5. Entanglement Distillation II

The example in the previous section explicitly demonstrates that we can distill an entangled state in the system A+B, through the repeated measurements on the other system C that separately interacts with A and B. The framework is rather simple and the distillation can be made optimal. There is, however, a kind of drawback in this scheme. As is clear in its exposition, it is assumed that system C, on which the measurement is performed, *always and simultaneously interacts* with both A and B and these interactions are crucial for the entanglement distillation. Stating differently, systems A and B (and C) are not (and/or will not be) able to be *separated spatially*, which implies that no entanglement between spatially separated systems is possible by the scheme presented in Sec. 4. It would not be suited to the situations where entanglements among spatially separated systems are required, as in quantum teleportation.

In this section, a resolution to this problem is presented. Since the two systems, A and B, an entanglement between which is to be driven, are considered to be placed at different places, let us consider, instead of system C which can no longer interact simultaneously with A and B, another quantum system, say X, which is assumed to interact with A and B, not simultaneously, but *successively* [13]. System X plays the role of an "entanglement mediator." After such successive interactions with A and then B, system X is measured to confirm that it is in a certain state. If system X is found in this particular state, X is again brought to interaction with A and then with B. This process, i.e., X's interaction with A, that with B and measurement on X, will be repeated many (N) times and we are interested in the asymptotic state of system A+B in the hope of distilling an entangled state.

There are a couple of points to be mentioned here. First, it is clear that in spite of these modifications, the new scheme presented here shares essentially the same idea of purification with the previous ones The dynamics of the system can be affected, in an essential way, by the action of measurement, even if its effect is not direct. Second, such a successive interaction would be conveniently treated in terms of a time-dependent (effective) Hamiltonian $H(t)$. We may thus avoid possible complications caused by the introduction of spatial degrees of freedom, still keeping the essential points.

In order to see how the new scheme works, consider again a three-qubit system, A+B+X, for definiteness and simplicity. We prepare system X, say in up state $|\uparrow\rangle$, while the system A+B can be in an arbitrary mixed state. It is assumed that systems A and B are spatially separated and have no contact with each other and that only system X can interact with them locally for definite time durations. Now consider the following process

(1) System X is first brought to interaction with system A for time duration t_A. The Hamiltonian here is given by $H(t) = H_0 + H_{XA}$. Then the interaction is switched off and the total system evolves freely with the free Hamiltonian H_0 for τ_A.
(2) System X then interacts with system B for time duration t_B, the dynamics of which is now described by another Hamiltonian $H(t) = H_0 + H_{XB}$. After that, the total system again evolves freely with the Hamiltonian H_0 for τ_B.
(3) A (projective) measurement is performed on system X to select only up state $|\uparrow\rangle$. Other states are discarded.

Then this process is repeated N times $1\to2\to3\to1\to\cdots\to1\to2\to3$. It is shown below that the following choice of the Hamiltonians

$$\begin{aligned} H_0 = \frac{\omega}{2}(1+\sigma_3^A) + \frac{\omega}{2}(1+\sigma_3^B) + \frac{\omega}{2}(1+\sigma_3^X), \\ H_{XA} = g_A\sigma_1^X\sigma_1^A, \qquad H_{XB} = g_B\sigma_1^X\sigma_1^B \end{aligned} \tag{44}$$

actually results in an entanglement distillation in system A+B. It is important to notice that the above choice of the interaction Hamiltonians is closely connected to the details of the process $1\to2\to3$ and another choice, e.g., $\sigma_+^X\sigma_-^{A(B)}$ for $H_{XA(B)}$, would result not in an entangled state, but in a product state, in this particular process.

The next task is to find the spectral decomposition of the projected operator $V_\uparrow$ defined, in this case, by

$$\begin{aligned} V_\uparrow = \langle\uparrow| e^{-iH_0\tau_B} e^{-i(H_0+H_{XB})t_B} \\ \times e^{-iH_0\tau_A} e^{-i(H_0+H_{XA})t_A} |\uparrow\rangle. \end{aligned} \tag{45}$$

Since a parity defined by $\mathcal{P} \equiv \sigma_3^A\sigma_3^B$ is conserved in this system, eigenstates of the operator $V_\uparrow$ are easily found. Indeed, we can classify every state of system A+B into two sectors according to the parity $\mathcal{P} = \pm$ and the action of the operator $V_\uparrow$ is closed within each sector. For $\mathcal{P} = +$ states, the action is represented by a matrix $\mathcal{M}$

$$V_\uparrow \begin{bmatrix} |\uparrow\uparrow\rangle \\ |\downarrow\downarrow\rangle \end{bmatrix} = e^{-i\omega(t_A+\tau_A+t_B+\tau_B)} \mathcal{M} \begin{bmatrix} |\uparrow\uparrow\rangle \\ |\downarrow\downarrow\rangle \end{bmatrix}, \tag{46}$$

where its matrix elements read

$$\begin{aligned}
\mathcal{M}_{11} &= e^{-i\omega(t_A+2\tau_A+t_B+2\tau_B)}(\cos\zeta_A - i\sin\zeta_A\cos 2\xi_A) \\
&\qquad \times (\cos\zeta_B - i\sin\zeta_B\cos 2\xi_B), \\
\mathcal{M}_{12} &= -e^{-i\omega t_A}\sin\zeta_A\sin 2\xi_A\sin g_B t_B, \\
\mathcal{M}_{21} &= -e^{-i\omega(t_B+2\tau_B)}\sin g_A t_A\sin\zeta_B\sin 2\xi_B, \\
\mathcal{M}_{22} &= \cos g_A t_A\cos g_B t_B,
\end{aligned} \tag{47}$$

while, for $\mathcal{P} = -$ states, it is represented by another matrix $\mathcal{N}$

$$V_\uparrow \begin{bmatrix} |\uparrow\downarrow) \\ |\downarrow\uparrow) \end{bmatrix} = e^{-i\omega(t_A+t_B+2\tau_B)}\mathcal{N}\begin{bmatrix} |\uparrow\downarrow) \\ |\downarrow\uparrow) \end{bmatrix}, \tag{48}$$

with its matrix elements

$$\begin{aligned}
\mathcal{N}_{11} &= e^{-i\omega(2\tau_A+t_B)}(\cos\zeta_A - i\sin\zeta_A\cos 2\xi_A)\cos g_B t_B, \\
\mathcal{N}_{12} &= -\sin\zeta_A\sin 2\xi_A\sin\zeta_B\sin 2\xi_B, \\
\mathcal{N}_{21} &= -e^{-i\omega(t_A+2\tau_A+t_B)}\sin g_A t_A\sin g_B t_B, \\
\mathcal{N}_{22} &= e^{-i\omega(t_A+2\tau_A)}\cos g_A t_A(\cos\zeta_B - i\sin\zeta_B\cos 2\xi_B).
\end{aligned} \tag{49}$$

Here the angles are defined by

$$\zeta_{A(B)} = t_{A(B)}\sqrt{\omega^2 + g^2_{A(B)}}, \quad \tan 2\xi_{A(B)} = \frac{g_{A(B)}}{\omega}. \tag{50}$$

In order to see the possibility of entanglement distillation in this framework, it is enough to consider a much more simplified case. Let the two systems A and B be treated symmetrically, that is, all parameters are taken to be the same for A and B

$$\begin{aligned}
g_A = g_B \equiv g, \quad t_A = t_B \equiv t, \quad \tau_A = \tau_B \equiv \tau, \\
(\zeta_{A(B)} \to \zeta, \quad \xi_{A(B)} \to \xi).
\end{aligned} \tag{51}$$

It is then easy to see that if the parameters satisfy

$$\cos\zeta - i\sin\zeta\cos 2\xi = -e^{i\omega\tau}\cos gt, \tag{52}$$

an optimal purification of an entangled state $|\Psi)$ of the form

$$|\Psi) \equiv \frac{1}{\sqrt{2}}\Big[|\uparrow\downarrow) + e^{i\chi}|\downarrow\uparrow)\Big] \tag{53}$$

with $\chi = \omega(t+\tau)$ is actually possible, provided

$$\cos gt\sin gt \neq 0, \quad \omega(t+\tau) \neq 2n\pi \quad (n\ \text{integer}). \tag{54}$$

In fact, one can show that $|\Psi)$ is an eigenstate of the operator $V_\uparrow$

$$V_\uparrow|\Psi) = \lambda_\Psi|\Psi). \tag{55}$$

The eigenvalue λ_Ψ is maximum in magnitude

$$\lambda_\Psi = -e^{-3i\omega(t+\tau)}, \quad |\lambda_\Psi| = 1, \tag{56}$$

while all the other eigenvalues remain smaller than unity in magnitude, under the conditions (52) and (54). Therefore, we can repeat the process 1→2→3 as many times as is required to achieve the desired (high) fidelity, without reducing the yield.

This is an example of (optimal) entanglement distillations, where the entanglement between two qubit systems that are (or can be) spatially separated, is extracted through their successive interactions with another qubit, on which one and the same measurement is repeated regularly. Further details of this model and applications to other quantum systems, e.g., extraction of entanglement between two cavity modes at a distance, will be reported elsewhere.

6. Summary

In this paper, a new purification scheme recently proposed [9] is applied to a few simple qubit systems to explicity show its ability of qubit purification (Sec. 3) and entanglement distillations (Secs. 4 and 5) for two-qubit systems. The important and essential idea, on which these particular examples are based, is to utilize the effect caused by the action of measurement on quantum systems.

It should be stressed again that since the basic idea is so simple, that is, one has only to repeat one and the same measurement without being concerned about the preparation of a specific initial (pure) state, this purification scheme is considered to have wide applicability and flexibility. Furthermore, it enables us to make the two demands—*the maximal fidelity and non-vanishing yield*—compatible. The examples presented in this paper just show these characteristics and many variants can be devised according to the actual setups.

Acknowledgments The authors acknowledge useful and helpful discussions with Prof. I. Ohba. Fruitful discussions with P. Facchi and S. Pascazio are also appreciated. One of the authors (H.N.) is grateful for the warm hospitality at Università di Palermo, where he enjoyed the inspiring discussions with A. Messina's group. This work is partly supported by a Grant for The 21st Century COE Program (Physics of Self-Organization Systems) at Waseda University and a Grant-in-Aid for Priority Areas Research (B) from the Ministry of Education, Culture, Sports, Science and Technology, Japan (No. 13135221), by a Grant-in-Aid for Scientific Research (C) (No. 14540280) from the Japan Society for the Promotion of Science, by a Waseda University Grant for Special Research Projects (No. 2002A-567) and by the bilateral Italian-Japanese project 15C1 on "Quantum Information and Computation" of the Italian Ministry for Foreign Affairs.

References

1. B. Misra and E.C.G. Sudarshan, J. Math. Phys. **18**, 756 (1977).
2. T. Petrosky, S. Tasaki and I. Prigogine, Phys. Lett. A **151**, 109 (1990); Physica A **170**, 306 (1991); S. Pascazio and M. Namiki, Phys. Rev. A **50**, 4582 (1994). For reviews, see, for example, H. Nakazato, M. Namiki and S. Pascazio, Int. J. Mod. Phys. B **10**, 247 (1996); D. Home and M.A.B. Whitaker, Ann. Phys. (N.Y.) **258**, 237 (1997); P. Facchi and S. Pascazio, in *Progress in Optics*, edited by E. Wolf (Elsevier, Amsterdam, 2001), Vol. 42, p. 147.
3. E.P. Wigner, Am. J. Phys. **31**, 6 (1963).
4. W.M. Itano, D.J. Heinzen, J.J. Bolinger and D.J. Wineland, Phys. Rev. A **41**, 2295 (1990).
5. M.C. Fischer, B. Gutiérrez-Medina and M.G. Raizen, Phys. Rev. Lett. **87**, 040402 (2001).
6. S.R. Wilkinson, C.F. Bharucha, M.C. Fischer, K.W. Madison, P.R. Morrow, Q. Niu, B. Sundaram and M.G. Raizen, Nature **387**, 575 (1997).
7. A.M. Lane, Phys. Lett. A **99**, 359 (1983); W.C. Schieve, L.P. Horwitz and J. Levitan, Phys. Lett. A **136**, 264 (1989); P. Facchi and S. Pascazio, Phys. Rev. A **62**, 023804 (2000); B. Elattari and S.A. Gurvitz, Phys. Rev. A **62**, 032102 (2000); A.G. Kofman and G. Kurizki, Nature **405**, 546 (2000); P. Facchi, H. Nakazato and S. Pascazio, Phys. Rev. Lett. **86**, 2699 (2001); K. Koshino and A. Shimizu, Phys. Rev. A **67**, 042101 (2003).
8. P. Facchi, D.A. Lidar and S. Pascazio, Phys. Rev. A, in print (2004); P. Facchi, D.A. Lidar, H. Nakazato, S. Pascazio, S. Tasaki and A. Tokuse, in preparation.
9. H. Nakazato, T. Takazawa and K. Yuasa, Phys. Rev. Lett. **90**, 060401 (2003); K. Yuasa, H. Nakazato and T. Takazawa, J. Phys. Soc. Jpn. **72** Suppl. C, 34 (2003).
10. See, for example, M.A. Nielsen and I.L. Chuang, *Quantum Computation and Quantum Information* (Cambridge University Press, Cambridge, 2000); *The Physics of Quantum Information*, edited by D. Bouwmeester, A. Ekert and A. Zeilinger (Springer-Verlag, Heidelberg, 2000).
11. K. Yuasa, H. Nakazato and M. Unoki, "Entanglement purification through Zeno-like measurements," quant-ph/0402184, J. Mod. Opt., in print (2004).
12. H. Nakazato, M. Unoki and K. Yuasa, "Preparation and entanglement purifi-

cation of qubits through Zeno-like measurements," WU-HEP-04-01 (2004), quant-ph/0402182.

13. A. Messina, Eur. Phys. J. D **18**, 379 (2002); D.E. Browne and M.B. Plenio, Phys. Rev. A **67**, 012325 (2003).

GENERALIZED SECTORS AND ADJUNCTIONS TO CONTROL MICRO-MACRO TRANSITIONS

IZUMI OJIMA
RIMS, Kyoto University, Sakyoku, Kyoto 606-8502, Japan
E-mail: ojima@kurims.kyoto-u.ac.jp

1. Unified scheme for micro and macro

A unified understanding of the relations between Micro and Macro is one of the most important issues in quantum physics in general. What I have obtained so far in this context can be summarized in the following scheme:

$$\begin{bmatrix}\text{A) Non-equilibrium}\\ \text{local states in QFT:}\\ \textit{continuous sectors}\end{bmatrix} \longrightarrow \begin{bmatrix}\text{C) Sector structure}\\ \text{of } \textbf{SSB:}\ \textit{discrete}\\ \textit{\& continuous}\end{bmatrix}$$

$$\nearrow \qquad\qquad\qquad \downarrow$$

$$\begin{bmatrix}\text{B) Reformulation of}\\ \text{DHR-DR theory:}\\ \textit{discrete sectors}\end{bmatrix} \qquad \begin{bmatrix}\text{D) Unified scheme for}\\ \text{Micro-Macro based}\\ \text{on } \textit{selection criteria}\end{bmatrix},$$

where

A) General formulation of **non-equilibrium local states** in QFT [1,?],
B) **Reformulation** [3] of DHR-DR sector theory [4] of unbroken internal symmetry,
C) Extension of the above to a **spontaneously broken symmetry** *(SSB)* [3],
D) **Unified scheme for describing Micro-Macro relations** on the basis of *selection criteria* [2,?].

The purpose of this talk is first to explain the fundamental notions underlying the above and then to extend further the proposed scheme so as to incorporate the case of **explicitly broken symmetry** exemplified by the emergence of the temperature as an order parameter of broken scale invariance. Let me start from the following basic points:

(1) General notion of a ***sector*** (applicable not only pure states but

also to *mixed states* such as KMS states) as a natural extension of ***thermodynamics pure phase*** (originally defined as *ergodic* KMS state)
= ***factor state*** $\omega \in E_{\mathfrak{A}}$ with ***trivial centre*** $\mathfrak{Z}_\pi(\mathfrak{A}) := \pi(\mathfrak{A})'' \cap \pi(\mathfrak{A})' = \mathbb{C}\mathbf{1}_{\mathfrak{H}}$ for its GNS representation $(\pi, \mathfrak{H},)$
i.e., Sector = *pure phase* = *factor state* = *trivial centre.*
Most important fact: ***any two factor states are either quasi-equivalent or disjoint!***
$\Downarrow$

(2) ***Mixed phase*** = non-*factor state* = non-*trivial centre* $\mathfrak{Z}_\pi(\mathfrak{A}) \neq \mathbb{C}\mathbf{1}_{\mathfrak{H}}$: allows "simultaneous diagonalization"=***central decomposition*** implying $\exists$ non-trivial sector structure
$\Downarrow$
· $\mathfrak{Z}_\pi(\mathfrak{A})$: set of all ***macroscopic order parameters*** to distinguish among different sectors;
· $Spec(\mathfrak{Z}_\pi(\mathfrak{A}))$: ***classifying space*** to parametrize sectors completely in the sense that ***quasi-equivalent sectors correspond to one and the same point*** and that ***disjoint sectors to the different points***
[$\Longrightarrow$ The most precise definition: ***sector = quasi-equivalence class of factor states***, which is equal to a ***folium of a factor state***]
$\Downarrow$

(3) ***Micro-macro relation***:
Inside a sector: *microscopic situations* prevail (e.g., for a pure state in a sector, as found in the vacuum situations, it represents a "*coherent subspace*" with *superposition principle* being valid);
Intersector level controlled by $\mathfrak{Z}_\pi(\mathfrak{A})$: *macroscopic situations* prevail, which are macroscopically observable and controllable.

(4) ***Selection criterion***:= *physically* and *operationally meaningful* characterization as to ***how and which sectors*** should be gathered for discussing a *specific physical domain.*
E.g., *DHR criterion* for states ω with localizable charges (based upon "Behind-the-Moon" argument) $\pi_\omega \upharpoonright_{\mathfrak{A}(\mathcal{O}')} \cong \pi_0 \upharpoonright_{\mathfrak{A}(\mathcal{O}')}$ in reference to the vacuum representation π_0.
A suitably set up criterion determines the associated ***sector structure*** s.t. natural *physical* ***interpretations*** of a theory in a physical domain specified by it are provided.
In DHR-DR sector theory, we see

(a) Sector structure $\mathfrak{H} = \bigoplus_{\gamma\in\hat{G}} (\mathfrak{H}_\gamma \otimes V_\gamma)$ s.t. $\pi(\mathfrak{A})'' = \bigoplus_{\gamma\in\hat{G}} (\pi_\gamma(\mathfrak{A})'' \otimes \mathbf{1}_{V_\gamma}) = U(G)'$, $U(G)'' = \bigoplus_{\gamma\in\hat{G}} (\mathbf{1}_{\mathfrak{H}_\gamma} \otimes \gamma(G)'') = \pi(\mathfrak{A})'$.

(b) Centre of the "universal" representation & its spectrum: $\mathfrak{Z}_\pi(\mathfrak{A}) = \bigoplus_{\gamma\in\hat{G}} \mathbb{C}(\mathbf{1}_{\mathfrak{H}_\gamma} \otimes \mathbf{1}_{V_\gamma}) = l^\infty(\hat{G})$; $\hat{G} = Spec(\mathfrak{Z}_\pi(\mathfrak{A}))$, which provide the *vocabulary for interpretation* of sectors w.r.t.G-**charges**.

(c) $(\pi_\gamma, \mathfrak{H}_\gamma)$: sector of observable algebra $\mathfrak{A} \overset{1-1}{\longleftrightarrow} (\gamma, V_\gamma) \in \hat{G}$: equiv. class of irred. unitary rep.'s of a *cpt. Lie* group G of *unbroken* internal sym. of *field algebra* $\mathfrak{F} := \mathfrak{A} \underset{\mathcal{O}_d^G}{\otimes} \mathcal{O}_d$.

(d) $(\pi, U, \mathfrak{H})$: covariant irred. vac. rep. of C*-dyn. sys. $\mathfrak{F} \underset{\tau}{\curvearrowleft} G$, $\pi(\tau_g(F)) = U(g)\pi(F)U(g)^*$.

(e) $\mathfrak{A}$, G, $\mathfrak{F}$ constitute a triplet of Galois extension $\mathfrak{F}$ of $\mathfrak{A} = \mathfrak{F}^G$ by Galois group $G = Gal(\mathfrak{F}/\mathfrak{A})$, determining *one* term from *two*. Now, we must ask the most important question as to *how to solve two unknowns* G & $\mathfrak{F}$ from $\mathfrak{A}$? The answer is found in the notion of selection criterion to specify the states of physical relevance:
DHR selection criterion $\Longrightarrow \mathcal{T}$ $(\subset End(\mathfrak{A}))$: DR category $\cong RepG \underset{\text{duality}}{\overset{\text{Tannaka-Krein}}{\Longrightarrow}} G \Longrightarrow \mathfrak{F} \cong \mathfrak{A} \rtimes \hat{G}$:

Invisible micro $\qquad\qquad\qquad\qquad$ Visible macro

$$\begin{bmatrix} Rep\mathcal{T} \cong G \\ \curvearrowright \\ \mathfrak{A} \rtimes \hat{G} \cong \mathfrak{F} \end{bmatrix} \begin{matrix} \nearrow \\ \nwarrow \end{matrix} \begin{matrix} \text{Fourier T-K dual} \\ \rightleftarrows \\ \circlearrowright \\ \rightleftarrows \\ \text{Galois dual} \end{matrix} \begin{bmatrix} \mathcal{T} \cong RepG \\ \curvearrowright \\ \mathfrak{A} = \mathfrak{F}^G \end{bmatrix}.$$

Similar schemes can be found also for B) & C) above:
E.g., C) SSB case:

$$\begin{bmatrix} \text{broken: } G \supset H\text{: unbroken} \\ \curvearrowright \qquad \curvearrowright \\ \hat{\mathfrak{F}} \supset \mathfrak{F} \cong \mathfrak{A}^d \rtimes \hat{H} \\ \| \\ \hat{\mathfrak{F}} \rtimes (\widehat{H\backslash G}) = \mathfrak{A}^d \rtimes \hat{G} \end{bmatrix} \begin{matrix} \rightleftarrows \\ \circlearrowright \\ \rightleftarrows \end{matrix} \begin{bmatrix} \mathcal{T} \cong RepH \sim \hat{H} \hookrightarrow \amalg_{gH\in G/H} g\hat{H}g^{-1} & & \\ \curvearrowright \qquad\qquad\qquad \downarrow & & \text{sector} \\ \mathfrak{A}^d = \mathfrak{F}^H \qquad\qquad G/H & : & \text{bundle} \\ \cup & & \\ \mathfrak{A} = \mathfrak{F}^G & & \end{bmatrix},$$

where the relevant centre is give by $\mathfrak{Z}_{\bar{\pi}}(\mathfrak{A}^d) = L^\infty(H\backslash G; d\dot{g}) \otimes \mathfrak{Z}_\pi(\mathfrak{A}^d) = L^\infty(H\backslash G; d\dot{g}) \otimes l^\infty(\hat{H})$ whose spectrum with bundle structure $Spec(\mathfrak{Z}_{\bar{\pi}}(\mathfrak{A}^d)) = \amalg_{gH\in G/H} g\hat{H}g^{-1} \twoheadrightarrow G/H$(: degenerate vacua) corresponds directly to the "roots" in Galois theory of algebraic equations. Note here that, for *commutative* algebras, factor representations are only 1-dimensional trivial ones, and hence, that $\hat{H}$ is absent!

These observations in A), B), C) naturally lead to

D) "unified scheme for generalized sectors based on selection criteria" [3]:

$$\text{i)}\begin{bmatrix}\text{generic objects}\\ \textit{to be selected}\end{bmatrix} \underset{\uparrow}{\Longrightarrow} \text{ii)}\begin{bmatrix}\textit{standard reference}\text{ sys. with}\\ \textit{classifying space}\text{ of }\textit{sectors}\end{bmatrix}$$

$$\uparrow$$

$$\text{iii) map to } \textit{compare} \text{ i) with ii)}$$

$$\Uparrow \qquad\qquad\qquad \Downarrow$$

$$\text{iv)}\begin{bmatrix}\textit{selection criterion:}\\ \text{ii)} \underset{c\text{-}q}{\Longrightarrow} \text{i)}\end{bmatrix} \underset{\textit{adjunction}}{\overset{\text{categorical}}{\rightleftarrows}} \begin{bmatrix}\textit{interpretation}\text{ of i)}\\ \text{w.r.t. ii): i)} \underset{q\text{-}c}{\Longrightarrow} \text{ii)}\end{bmatrix},$$

as a natural generalization of local charts $\{(U_\lambda, \varphi_\lambda : U_\lambda \to \mathbb{R}^n)\}$ of manifold:

i)= local neighbourhoods U_λ, ii)= $\mathbb{R}^n$, iii)= local charts $\varphi_\lambda : U_\lambda \to \mathbb{R}^n$, iv)= interpretation of the atlas in terms of geometrical invariants.

Non-equilibrium local states in B) are characterized by localizing the following generalized equilibrium states with *fluctuating* thermal parameters:

i) = set E_x of states ω satisfying certain energy bound locally $[\omega((\mathbf{1} + H_{\mathcal{O}})^m) < \infty$ with "local Hamiltonian" $H_{\mathcal{O}}]$,

ii) = the space B_K of thermodynamic parameters (β, μ) and that $M_+(B_K) =: Th$ of probability measures ρ on B_K describes the fluctuations of (β, μ).

iii) = comparison of unknown state ω with members of standard states $\omega_\rho = \mathcal{C}^*(\rho) = \int_{B_K} d\rho(\beta,\mu)\omega_{\beta,\mu}$ with parameters ρ belonging to the reference system, to examine the criterion $\omega \underset{\mathcal{T}_x}{\equiv} \mathcal{C}^*(\rho)$ by means of "quantum fields at x" $\in \mathcal{T}_x$ (justified by the energy bound in i)).

iv) = *adjunction* $E_x/\mathcal{T}_x(\omega, \mathcal{C}^*(\rho)) \overset{q \rightleftarrows c}{\simeq} Th/\mathcal{C}(\mathcal{T}_x)((\mathcal{C}^*)^{-1}(\omega), \rho)$ with *q→c channel* $(\mathcal{C}^*)^{-1}$ as "left adjoint" of *c→q channel* $\mathcal{C}^*$ (from the classical reference system to generic quantum states): as a *localized form of the 0-th law* of thermodynamics, this adjunction achieves simultaneously the two goals of *identify*ing generalized equilibrium local states and of giving the thermal *interpretation* $(\mathcal{C}^*)^{-1}(\omega) \underset{\mathcal{C}^*(\mathcal{T}_x)}{\equiv} \rho$ of a selected generic state ω in the vocabulary of the standard known object $\rho \in Th$.

All the cases above are based upon the mathematical formulations of vacuum and/or KMS states. So, it is important to incorporate the family of all KMS states (including $\beta = \infty$ as a vacuum) in the same scheme. In this direction, we consider the problem of the physical origin of a temperature which is usually regarded as an a priori parameter to be imposed on the physical system from the outside.

2. *Broken* symmetry in *micro-macro composite system*; β=order parameter of *broken scale inv.*

The aim of this section is two-fold:

i) to present a mathematical formalism for consistent and systematic treatments of dynamical systems with **symmetries broken spontaneously or explicitly**,

ii) to show that the (inverse) temperature β appears, in the above context, as an **order parameter of broken scale invariance**. Namely,

[5] 2.1.In algebraic QFT, inverse temperature $\beta := (\beta^\mu \beta_\mu)^{1/2}$ is a macroscopicfor parametrizing thermal sectors arising from the under the renormalization-group transformations, where β^μ is an inverse temperature 4-vector of a relativistic KMS state ω_{β^μ} describing a thermal equilibrium in its rest frame.

In this way, we see a *cross-over* between *thermal* and *geometrical* aspects embodied in QFT. i) will be explained in the course of verifying ii) as formulated above.

2.1. *Is β a priori parameter or physical quantity?*

The relativistic formulation of thermal equilibria requires Lorentz 4-vectors of inverse-temperatures: $\beta^\mu = \beta u^\mu \in V_+$(: forward lightcone), $\beta := (\beta^\mu \beta_\mu)^{1/2} = (k_B T)^{-1}$.

a) Relative velocity u^μ ($u^\mu u_\mu = 1$) specifies the *rest frame* in which a temperature state ω_{β^μ} exhibits its thermal equilibrium and is an *order parameter* of *SSB of Lorentz boosts* [6]

b) What about the inverse temperature β itself?

The most useful hint can be found in the famous Takesaki theorem:

[7] 2.2.For a quantum C*-dynamical system with *type III* representations in its KMS states, any pair of KMS states for $\beta_1 \neq \beta_2$ are mutually *disjoint* $\omega_{\beta_1} \overset{\shortmid}{\circ} \omega_{\beta_2}$.

This mathematical fact implies the following physical picture for quantum systems with infinite degrees of freedom such as QFT: first, the disjointness among different temperatures implie the presence of *continuous sectors* formed by KMS states at different temperatures distinguished mutually by *macroscopic central observables* (in a representation containing all the KMS states) including (inverse) temperature β. Namely, β becomes a *physical macro-variable* running over the space of all possible thermal equilibria, instead of being an a priori given fixed parameter.

Starting from this observation, one can show that β is a physical *order parameter* corresponding to (spontaneous or explicit) *broken scale invariance* under renormalization group, namely, β's not only parametrize continuous sectors of thermal equilibria, but also are interrelated by renormalization-group transformations due to broken scale invariance, through which there emerges a thermodynamic classifying space.

2.2. *Criterion for symmetry breakdown*

Now, for our purpose, we need a precise formulation of *scale transformations* and a clear-cut *criterion for symmetry breakdown.*

Since the usual SSB criterion based on Goldstone commutators is applicable only to spatially homogeneous vacuum states, we give a suitable generalization in the following form of Definition of SSB:

[3] 2.1.A symmetry described by a (strongly continous) automorphic G-action τ: $G \underset{\tau}{\curvearrowright} \mathfrak{F}$(: field algebra), is said to bein a given representation $(\pi, \mathfrak{H})$ of $\mathfrak{F}$ if the spectrum $Spec(\mathfrak{Z}_\pi(\mathfrak{F}))$ of centre $\mathfrak{Z}_\pi(\mathfrak{F}) := \pi(\mathfrak{F})'' \cap \pi(\mathfrak{F})'$ is pointwise invariant (μ-a.e. w.r.t. the central measure μ for the central decomposition of π into factor representations) under the G-action induced on $Spec(\mathfrak{Z}_\pi(\mathfrak{F}))$. If the symmetry is not unbroken in $(\pi, \mathfrak{H})$, it is said to be there.

Remark 2.1. Since $Spec(\mathfrak{Z}_\pi(\mathfrak{F}))$ consists of *macroscopic order parameters* emerging in low-energy infrared regions, SSB means the "**infrared instability**" along the direction of G-action.

Remark 2.2. Since a given π with SSB contains unbroken and broken subrepresentations, $Spec(\mathfrak{Z}_\pi(\mathfrak{F}))$ can be decomposed further into G-invariant domains: each such minimal domain is characterized by G-*ergodicity* which means central ergodicity. So π is decomposed into the direct sum (or integral) of **unbroken factor representations** and **bro-**

ken non-factor representations, each component of which is centrally G-ergodic. Thus we have a phase diagram on $Spec(\mathfrak{Z}_\pi(\mathfrak{F}))$.

Remark 2.3. The essence of SSB lies in *conflict between factoriality and unitary implementability*; the former is respected in the usual approaches at the expense of the latter. With the opposite choice to respect implementability we have a **non-trivial centre** which provides convenient tools for analyzing **sector structure** and flexible treatment of macroscopic **order parameters** to distinguish different sectors.

A *covariant* representation of $(\mathfrak{F} \underset{\tau}{\curvearrowleft} G)$ implementing broken G *minimally* in the sense of *central G-ergodicity* can be constructed as follows:
1) From a covariant rep. $(\pi, U, \mathfrak{H})$ of $\mathfrak{F} \underset{\tau}{\curvearrowleft} H$ of a *maximal unbroken* subgroup H of G in $(\pi, \mathfrak{H})$ s.t. $\pi(\tau_h(F)) = U(h)\pi(F)U(h)^*$, a representation $(\hat{\pi}, \hat{\mathfrak{H}})$ is induced of a crossed product $\hat{\mathfrak{F}} := \mathfrak{F} \rtimes (H\backslash G) = \Gamma(G \times_H \mathfrak{F})$[: algebra of cross sections of $G \times_H \mathfrak{F} \to H\backslash G$] of $\mathfrak{F}$. The action $\hat{\tau}$ of G on $\hat{F} \in \hat{\mathfrak{F}}$ is defined by $[\hat{\tau}_g(\hat{F})](g_1) = \hat{F}(g_1 g)$.
2) On the Hilbert space $\hat{\mathfrak{H}} = \int^{\oplus}_{\xi \in G/H} (d\xi)^{1/2}\mathfrak{H} = \Gamma_{L^2}(G \times_H \mathfrak{H}, d\xi)$ of L^2-sections of $G \times_H \mathfrak{H}$, the above $\hat{\pi}$ and $\hat{U}$ are defined, respectively, by $(\hat{\pi}(\hat{F})\psi)(g) := \pi(\hat{F}(g^{-1}))(\psi(g))$ and $(\hat{U}(g_1)\psi)(g) := \psi(g_1^{-1}g)$ $(\hat{F} \in \hat{\mathfrak{F}}, \psi \in \hat{\mathfrak{H}}, g, g_1 \in G)$. 3) $\mathfrak{F}$ is embedded into $\hat{\mathfrak{F}}$ by $\hat{\imath}_{H\backslash G} : \mathfrak{F} \ni F \longmapsto (G \ni g \longmapsto \tau_g(F)) \in \hat{\mathfrak{F}}$,which intertwines G-actions τ on $\mathfrak{F}$ and $\hat{\tau}$ on $\hat{\mathfrak{F}}$, $\hat{\imath}_{H\backslash G} \circ \tau_g = \hat{\tau}_g \circ \hat{\imath}_{H\backslash G}$ $(\forall g \in G)$, and hence, we have $[\hat{\imath}_{H\backslash G}(\mathfrak{F})]^G = \hat{\imath}_{H\backslash G}(\mathfrak{F}^G) \subset \hat{\imath}_{H\backslash G}(\mathfrak{F}^H) = \hat{\mathfrak{F}}^G$.
4) Combining $\hat{\imath}_{H\backslash G}$ with $\hat{\pi}$, we obtain a covariant rep. $(\bar{\pi}, \hat{U}, \hat{\mathfrak{H}})$, $\bar{\pi} := \hat{\pi} \circ \hat{\imath}_{H\backslash G}$, of $\mathfrak{F} \underset{\tau}{\curvearrowleft} G$ in $\hat{\mathfrak{H}}$ by $(\bar{\pi}(F)\psi)(g) := \pi(\tau_{g^{-1}}(F))\psi(g)$ $(F \in \mathfrak{F}, \psi \in \hat{\mathfrak{H}})$ s.t. $\bar{\pi}(\tau_g(F)) = \hat{U}(g)\bar{\pi}(F)\hat{U}(g)^{-1}$.
5) When $\mathfrak{Z}_\pi(\mathfrak{F}) = \mathbb{C}\mathbf{1}$, we can show [5] $\mathfrak{Z}_{\bar{\pi}}(\mathfrak{F}) = L^\infty(H\backslash G; d\dot{g}) = \mathfrak{Z}_{\hat{\pi}}(\hat{\mathfrak{F}})$.

Applying the above to the GNS rep. $(\pi = \pi_\beta, \mathfrak{H} = \mathfrak{H}_\beta)$ of a KMS state $\omega_{\beta=(\beta^\mu)}$ with $H = \mathbb{R}^4 \rtimes SO(3)$, $G = \mathbb{R}^4 \rtimes L^\uparrow_+$, we can reproduce the statement on SSB of Lorentz boosts, $\mathfrak{Z}_{\bar{\pi}}(\mathfrak{F}) = L^\infty(SO(3)\backslash L^\uparrow_+) = L^\infty(\mathbb{R}^3)$, through the identification $\beta^\mu/\sqrt{\beta^2} = u^\mu = (\frac{1}{\sqrt{1-\mathbf{v}^2/c^2}}, \frac{\mathbf{v}}{\sqrt{1-\mathbf{v}^2/c^2}}) \longleftrightarrow \mathbf{v} \in \mathbb{R}^3$.

2.3. *How to formulate broken scale invariance*

While the above symmetry breaking was a *spontaneous* breakdown of G acting on field algebra $\mathfrak{F}$ by automorphisms, the case of broken scale invariance usually involves *explicit breaking terms* such as *mass*, which seem to *prevent scale transformations from being treated as automorphisms.* However, the results on *scaling algebra* in algebraic QFT due to [8] show that a *scaling net* $\mathcal{O} \to \hat{\mathfrak{A}}(\mathcal{O})$ corresponding to the original local net $\mathcal{O} \to \mathfrak{A}(\mathcal{O})$ of observables is defined as the local net consisting of scale-changed observables under the action of all the possible choice of renormalizaton group transformations. Mathematically $\hat{\mathfrak{A}}(\mathcal{O})$ is defined by the algebra $\Gamma(\mathbb{R}^+ \times \mathfrak{A}(\mathcal{O}))$ of sections $\mathbb{R}^+ \ni \lambda \longmapsto \hat{A}(\lambda) \in \mathfrak{A}_\lambda(\mathcal{O})$ of an algebra bundle $\amalg_{\lambda\in\mathbb{R}^+}\mathfrak{A}_\lambda(\mathcal{O}) \to \mathbb{R}^+$ over the multiplicative group $\mathbb{R}^+$ of scale changes and the *scaling algebra* $\hat{\mathfrak{A}}$ by the C*-inductive limit of all local algebras $\hat{\mathfrak{A}}(\mathcal{O})$.

Algebraic structures making $\hat{\mathfrak{A}}(\mathcal{O})$ a unital C*-algebra are defined in a *pointwise manner* by $(\hat{A} \cdot \hat{B})(\lambda) := \hat{A}(\lambda)\hat{B}(\lambda)$, $(\hat{A}^*)(\lambda) := \hat{A}(\lambda)^*$, etc., and $\| \hat{A} \| := \sup_{\lambda\in\mathbb{R}^+} \| \hat{A}(\lambda) \|$. From scaled actions $\mathfrak{A}_\lambda \underset{\alpha^{(\lambda)}}{\curvearrowleft} \mathcal{P}_+^\uparrow$ of Poincaré group on $\mathfrak{A}_\lambda$ with $\alpha_{x,\Lambda}^{(\lambda)} = \alpha_{\lambda x,\Lambda}$, an action of $\mathcal{P}_+^\uparrow$ is induced on $\hat{\mathfrak{A}}$ by $(\hat{\alpha}_{x,\Lambda}(\hat{A}))(\lambda) := \alpha_{\lambda x,\Lambda}(\hat{A}(\lambda))$.

Then, the scaling net $\mathcal{O} \to \hat{\mathfrak{A}}(\mathcal{O})$ is shown to satisfy all the properties to characterize a relativisitc local net of observables if the original one $\mathcal{O} \to \mathfrak{A}(\mathcal{O})$ does. Scale transformations are defined by an automorphic action $\hat{\sigma}_{\mathbb{R}^+}$ of $\mathbb{R}^+$ on the scaling algebra $\hat{\mathfrak{A}}$, given for $\forall\mu \in \mathbb{R}^+$ by $(\hat{\sigma}_\mu(\hat{A}))(\lambda) := \hat{A}(\mu\lambda), \quad \lambda > 0$, satisfying

$$\hat{\sigma}_\mu(\hat{\mathfrak{A}}(\mathcal{O})) = \hat{\mathfrak{A}}(\mu\mathcal{O}), \quad \mathcal{O} \subset \mathbb{R}^4,$$

$$\hat{\sigma}_\mu \circ \hat{\alpha}_{x,\Lambda} = \hat{\alpha}_{\mu x,\Lambda} \circ \hat{\sigma}_\mu, \quad (x,\Lambda) \in \mathcal{P}_+^\uparrow.$$

Remark 2.4. No miracle in the above "symmetrization method" since we can always restore a broken symmetry by making all explicit breaking parameters (such as m) running variables changed by the broken symmetry transformations!

2.4. *Scale changes on states*

Under the situation of broken scale invariance we have a non-trivial centre $\mathfrak{Z}(\hat{\mathfrak{A}}) = \mathfrak{Z}(\hat{\mathfrak{A}}(\mathcal{O})) = C(\mathbb{R}^+)$. Then, each probability measure μ on $C(\mathbb{R}^+)$ defines canonically a conditional expectation $\hat{\mu} : \hat{\mathfrak{A}} \ni \hat{A} \longmapsto \int_{\mathbb{R}^+} d\mu(\lambda)\hat{A}(\lambda) \in \mathfrak{A}$. By means of this $\hat{\mu}$, each state $\omega \in E_{\mathfrak{A}}$ can be lifted onto $\hat{\mathfrak{A}}$ by $E_{\mathfrak{A}} \ni \omega \longmapsto \hat{\mu}^*(\omega) = \omega \circ \hat{\mu} = \omega \otimes \mu \in E_{\hat{\mathfrak{A}}}$, where we have used $\hat{\mathfrak{A}} \subset C(\mathbb{R}^+, \mathfrak{A}) \cong \mathfrak{A} \otimes C(\mathbb{R}^+)$. The choice $\mu = \delta_{\lambda=1}$(: Dirac measure at $1 \in \mathbb{R}^+$) is called a *canonical lift* $\hat{\omega} := \omega \circ \hat{\delta}_1$ in [8]. Its scale-transform, $\hat{\omega}_\lambda := \hat{\omega} \circ \hat{\sigma}_\lambda = \omega \circ \hat{\delta}_\lambda$, describes the situation at scale λ through the renormalization-group transformation of scale change λ.

Conversely, *central decomposition* of a state $\hat{\omega} \in E_{\hat{\mathfrak{A}}}$ is given as follows: first, define two embedding maps $\iota : \mathfrak{A} \hookrightarrow \hat{\mathfrak{A}}$ $([\iota(A)](\lambda) \equiv A)$ and $\kappa : C(\mathbb{R}^+) \simeq \mathfrak{Z}(\hat{\mathfrak{A}}) \hookrightarrow \hat{\mathfrak{A}}$. Then, the pull-back $\rho_{\hat{\omega}} := \kappa^*(\hat{\omega}) = \hat{\omega} \circ \kappa = \hat{\omega} \restriction_{C(\mathbb{R}^+)}$ of $\hat{\omega}$ by $\kappa^* : E_{\hat{\mathfrak{A}}} \to E_{C(\mathbb{R}^+)}$ is a probability measure on $\mathbb{R}^+$, i.e., $\hat{\omega} \restriction_{C(\mathbb{R}^+)} (f) = \int_{\mathbb{R}^+} d\rho_{\hat{\omega}}(\lambda) f(\lambda)$ for $\forall f \in C(\mathbb{R}^+)$. Because of "absolute continuity" of $\hat{\omega}''$ w.r.t. $\rho''_{\hat{\omega}}$, for the extensions of $\hat{\omega}$ and $\rho_{\hat{\omega}}$, respectively, to $\hat{\pi}_{\hat{\omega}}(\hat{\mathfrak{A}})''$ and $L^\infty(\mathbb{R}^+, d\rho_{\hat{\omega}})$, we can define Radon-Nikodym derivative $\omega_\lambda := \frac{d\hat{\omega}}{d\rho_{\hat{\omega}}}(\lambda)$ of $\hat{\omega}$ w.r.t. $\rho_{\hat{\omega}}$ as a state on $\hat{\pi}_{\hat{\omega}}(\hat{\mathfrak{A}})''$ (similarly to [9]) so that $\hat{\omega}(\hat{A}) = \int d\rho_{\hat{\omega}}(\lambda)\omega_\lambda(\hat{A}(\lambda)) = \int d\rho_{\hat{\omega}}(\lambda)\omega_\lambda(\hat{\delta}_\lambda(\hat{A})) = \int d\rho_{\hat{\omega}}(\lambda) \left[\omega_\lambda \otimes \hat{\delta}_\lambda\right] (\hat{A})$. Then, the pull-back $\iota^*(\hat{\omega}) = \hat{\omega} \circ \iota \in E_{\mathfrak{A}}$ of $\hat{\omega} \in E_{\hat{\mathfrak{A}}}$ by $\iota^* : E_{\hat{\mathfrak{A}}} \to E_{\mathfrak{A}}$ is given by $\iota^*(\hat{\omega}) = \int d\rho_{\hat{\omega}}(\lambda)\omega_\lambda$, owing to the relation $\iota^*(\hat{\omega})(A) = \hat{\omega}(\iota(A)) = \int d\rho_{\hat{\omega}}(\lambda)\omega_\lambda(A) = \left[\int d\rho_{\hat{\omega}}(\lambda)\omega_\lambda\right](A)$. Applying this to the scaled canonical lift, $\hat{\omega}_\lambda := \hat{\omega} \circ \hat{\sigma}_\lambda = (\omega \circ \hat{\delta}_1) \circ \hat{\sigma}_\lambda = \omega \circ \hat{\delta}_\lambda$, of a state $\omega \in E_{\mathfrak{A}}$, we can easily see $\iota^*(\omega \circ \hat{\delta}_\lambda) = \iota^*(\hat{\omega}_\lambda) = \omega_\lambda [= \frac{d\hat{\omega}_\lambda}{d\delta_\lambda}(\lambda)] = \phi_\lambda(\omega)$, where ϕ_λ is the isomorphism between ω and the canonical lift $\hat{\omega}_\lambda \in E_{\hat{\mathfrak{A}}}$ projected onto $\hat{\mathfrak{A}}/\ker(\hat{\pi}_{\hat{\omega}} \circ \hat{\sigma}_\lambda)$ ([8]). Thus, any state $\omega \in E_{\mathfrak{A}}$ can be canonically lifted from $\mathfrak{A}$ to $\hat{\omega} \in E_{\hat{\mathfrak{A}}}$, and then, the scaled state $\hat{\omega} \circ \hat{\sigma}_\lambda$ by $\hat{\sigma}_\lambda$ on $\hat{\mathfrak{A}}$ is sent back onto $\mathfrak{A}$: $\phi_\lambda(\omega) = \omega_\lambda = \iota^*(\omega \circ \hat{\delta}_\lambda)$. So a scale-shifted state $\omega_\lambda \in E_{\mathfrak{A}}$ can be defined for $\omega \in E_{\mathfrak{A}}$ *in spite of the absence of scale invariance on* $\mathfrak{A}$.

Applying this to $\omega = \omega_\beta$ (: any state in the family of relativistic KMS states with the same $(\beta^2)^{1/2}$), we have a genuine KMS state by going to their rest frames. Then $\hat{\omega}_\lambda = (\widehat{\omega_\beta})_\lambda = \omega_\beta \circ \widehat{\delta_\lambda}$ is a KMS state at β/λ:

$$(\omega_\beta \circ \widehat{\delta_\lambda})(\hat{A}\hat{\alpha}_t(\hat{B})) = \omega_\beta(\hat{A}(\lambda)\alpha_{\lambda t}(\hat{B}(\lambda))) = \omega_\beta(\alpha_{\lambda t - i\beta}(\hat{B}(\lambda))\hat{A}(\lambda))$$
$$= \omega_\beta(\alpha_{\lambda(t-i\beta/\lambda)}(\hat{B}(\lambda))\hat{A}(\lambda)) = (\omega_\beta \circ \widehat{\delta_\lambda})(\hat{\alpha}_{t-i\beta/\lambda}(\hat{B})\hat{A}),$$

and hence, $(\widehat{\omega_\beta})_\lambda \in \hat{K}_{\beta/\lambda}$, $\phi_\lambda(\omega_\beta) \in K_{\beta/\lambda}$.

As already remarked, the above discussion applies equally to the spontaneous as well as *explicitly broken* scale invariance with such *explicit breaking* parameters as mass terms. The actions of scale transformations on such variables as x^μ, β^μ and also *conserved charges* are straightforward, because the first and the second ones are of *kinematical* nature and that the second and the third ones appear as *state labels* for specifying the relevant *sectors* in the context of the superselection structures [1,?]. This gives an alternative verification to the so-called *non-renormalization theorem of conserved charges.* In sharp contrast, other such variables as coupling constants (to be read off from correlation functions) are affected by the *scaled dynamics,* and hence, may show non-trivial scaling behaviours with deviations from the canonical (or kinematical) dimensions, in such forms as the running couplings or anomalous dimensions. Thus, the transformations $\hat{\sigma}_\lambda$ (as "exact" symmetry on the augmented algebra $\hat{\mathfrak{A}}$) are understood to play the roles of the renormalization-group transformations (as broken symmetry for the original algebra $\mathfrak{A}$). As a result, we see that *classical macroscopic observable* β naturally emerging from a microscopic quantum system is just an *order parameter* of *broken scale invariance* under the renormalization group.

3. Summary and outlook: method of "variation of natural constants"

To equip such expressions as "*broken scale invariance*" and its "*order parameter*" with their precise formulations, a scheme is formulated above to incorporate *spontaneously as well as explicitly broken* symmetries, in combination with criterion for symmetry breakdown on the basis of an *augmented algebra* with a non-trivial centre defined by $\hat{\mathfrak{F}} = \Gamma(G \times_H \mathfrak{F})$ or $\mathcal{O} \longmapsto \hat{\mathfrak{A}}(\mathcal{O}) = \Gamma(\amalg_{\lambda \in \mathbb{R}^+} \mathfrak{A}(\lambda\mathcal{O}))$. The latter one is just a re-formulation of Buchholz-Verch scaling net of local observables adapted to the former. As an algebra of the *composite system* of a genuine quantum one together with classical macroscopic one (embedded as the centre), the augmented algebra $\hat{\mathfrak{F}}$ or $\hat{\mathfrak{A}}$ can play such important roles that

a) it allows a symmetry *broken explicitly* by breaking terms (like masses$\neq 0$) to be formulated in terms of the *symmetry transformations* acting on $\hat{\mathfrak{A}}$ by automorphisms which is realized by the simultaneous changes of the breaking terms belonging to $\mathfrak{Z}(\hat{\mathfrak{A}})$ to cancel the breaking effects,

a') for a *spontaneously broken* symmetry, this augmented algebra naturally accommodates its *covariant* unitary representation as an induced representation from a subgroup of the unbroken symmetry (at the expense of non-trivial centre characteristic to symmetry breaking),

b) *continuous* behaviours of order parameters under broken symmetry transformations is algebraically expressed at the level of C*-algebraic centre $\mathfrak{Z}(\hat{\mathfrak{A}})$, in sharp contrast to the *discontinuous* ones at the W*-level $\mathfrak{Z}_\pi(\hat{\mathfrak{A}})$ of representations owing to the mutual disjointness among representations corresponding to different values of order parameters (as points on $Spec(\mathfrak{Z}_\pi(\mathfrak{A}))$). To this continuous order parameter some *external fields* can further be coupled, like the coupling between the magnetization and an external magnetic field in the case of a Heisenberg ferromagnet. With this coupling, we can examine, e.g., the mutual relations between the magnetization caused by an external field and the spontaneous one, the latter of which persists in the asymptotic removal of the former in combination with hysteresis effects. Without introducing the augmented algebra $\hat{\mathfrak{A}}$, it seems difficult for this kind of discussions to be adapted to the case of QFT.

Then, the mutual relation between states on $\mathfrak{A}$ and $\hat{\mathfrak{A}}$ is clarified, in use of which the verification of my claim on the behaviour of inverse temperature is just reduced to a simple computation of checking the parameter shift in KMS condition under scale change. What is interesting about the roles of (inverse) temperature β is that a *cross-over occurs between thermal and geometric aspects* expressed in $\beta^\mu = \beta u^\mu$ and in spacetime transformations $\mathcal{P}_+^\uparrow \rtimes \mathbb{R}^+$ including scale one, respectively. At the end, let me mention the possibility inherent to this scheme for a systematic control over different micro-macro domains in nature by means of the *method of "variation of natural constants"* (work in progress; this actually corresponds to what the title of this talk means!!).

References

1. Buchholz, D., Ojima, I. and Roos, H., Ann. Phys. (N.Y.) **297** (2002), 219.
2. Ojima, I., Non-equilibrium local states in relativistic quantum field theory, pp. 48-67 in Proc. of Japan-Italy Joint Workshop on Fundamental Problems in Quantum Physics, 2001, eds. L. Accardi and S. Tasaki, World Scientific (2003); How to formulate non-equilibrium local states in QFT?– General characterization and extension to curved spacetime–, pp.365-384 in "*A Garden of Quanta*", World Scientific (2003).
3. Ojima, I., A unified scheme for generalized sectors based on selection criteria –Order parameters of symmetries and of thermality and physical meanings of adjunctions–, Open Sys. Inf. Dyn. **10** (2003), 235.
4. Doplicher, S., Haag, R. and Roberts, J.E., Comm. Math. Phys. **13** (1969), 1; **15** (1969), 173; **23** (1971), 199; **35** (1974), 49; Doplicher, S. and Roberts, J.E., Comm. Math. Phys. **131** (1990), 51.
5. Ojima, I., Temperature as order parameter of broken scale invariance, Publ. RIMS **40**, 731-756 (2004).
6. Ojima, I., Lett. Math. Phys. **11** (1986), 73.
7. Takesaki, M., Comm. Math. Phys. **17** (1970), 33.
8. Buchholz, D. and Verch, R., Rev. Math. Phys. **7** (1995), 1195.
9. Ozawa, M., Publ. RIMS, Kyoto Univ. **21** (1985), 279.

SATURATION OF AN ENTROPY BOUND AND QUANTUM MARKOV STATES*

DÉNES PETZ

Department for Mathematical Analysis
Budapest University of Technology and Economics
H-1521 Budapest XI., Hungary
E-mail: petz@math.bme.hu

It has been known for a while that the equality in several strong subadditivity inequalities for the von Neumann entropy of the local restriction of states of infinite product chains is equivalent to the Markov property initiated by Accardi. The goal of this paper is to analyse the situation further and to give the structure of states which satisfy strong subadditivity of quantum entropy with equality. This structure has implication for quantum Markov states.

1. Introduction

In this paper two very important topics are connected. The strong subadditivity for the von Neumann entropy was necessary to prove the existence of the entropy density in the van Hove limit and this topic belongs to the fundamentals of quantum statistical mechanics. Markov states of the quantum spin chains form a simple class, this is the class of states which was born together with quantum probability and many ideas of this field appear in connection with Markov states, the mean ergodic theorem or conditional expectation are among them.

It has been known for a while that the equality in several strong subaddtivity inequalities for the von Neumann entropy of the local restriction of states of infinite product chains is equivalent to the Markov property initiated by Accardi (see Proposition 11.5 in [11] or [16]). The goal of this paper is to analyse the situation further, and to give detailed structure for the case of equality in the strong subbadditivity. Our approach is slightly different from the original paper [6]. From this structure, we can deduce the

*written version of the lecture delivered at icqi03, international institute for advanced studies, kyoto, november 6, 2003. the work was supported by the hungarian otka t032662.

form of quantum Markov states which was done in [4,?] by different methods, see these papers concerning the details.

In this paper all Hilbert spaces are finite dimensional. Extension of the results to infinite dimension is under consideration.

2. Strong subadditivity of entropy

The strong subadditivity of entropy is the inequality

$$S(\varphi_{ABC}) + S(\varphi_B) \leq S(\varphi_{AB}) + S(\varphi_{BC})$$

for a system $\mathcal{H}_A \otimes \mathcal{H}_B \otimes \mathcal{H}_C$, where $\varphi_B, \varphi_{AB}, \varphi_{BC}$ are the different restrictions of the state φ_{ABC} of the composite system and S stands for the von Neumann entropy [8]. We have the equivalent form

$$S(D_{ABC}, \tau_{ABC}) + S(D_B, \tau_B) \geq S(D_{AB}, \tau_{AB}) + S(D_{BC}, \tau_{BC})$$

in terms of relative entropy [11], τ denotes the tracial state. This inequality is equivalent to the inequality

$$S(\varphi_{ABC}, \tau_A \otimes \varphi_{BC}) \geq S(\varphi_{AB}, \tau_A \otimes \varphi_B) \tag{1}$$

which, on the other hand, is the consequence of the monotonicity of relative entropy: The states on the RHS are obtained by restricting the corresponding state of the LHS to the subsystem AB. (Our general reference about relative entropy is Chap. 1 of [11].) The characterization of sufficiency tells us that the equality in (1) is equivalent to the existence of a completely positive unital $\beta : B(\mathcal{H}_A \otimes \mathcal{H}_B \otimes \mathcal{H}_C) \to B(\mathcal{H}_A \otimes \mathcal{H}_B)$ which leaves the states φ_{ABC} and $\tau_A \otimes \varphi_{BC}$ invariant ([14,?]). In fact, it is convenient to regard β as a self-mapping of $B(\mathcal{H}_A \otimes \mathcal{H}_B \otimes \mathcal{H}_C)$ since the mean ergodic theorem can be applied in this way. Namely,

$$\frac{1}{n}(a + \beta(a) + \cdots + \beta^{n-1}(a)) \to E(a)$$

as $n \to \infty$, where E is a conditional expectation to the fixed point algebra $\mathcal{B}$ of β.

In fact, the above β can be given explicitly as

$$\begin{aligned}\beta(a \otimes b \otimes c) = {} & (I_A \otimes D_B)^{-1/2} T \\ & \times \Big((I_A \otimes D_{BC})^{1/2} (a \otimes b \otimes c) (I_A \otimes D_{BC})^{1/2} \Big) \\ & \times (I_A \otimes D_B)^{-1/2} \otimes I_C\end{aligned}$$

where

$$T(a \otimes b \otimes c) = (a \otimes b)\tau(c)$$

is the partial trace. The formula gives that $B(\mathcal{H}_A)$ consists of fixed points of β, therefore $\mathcal{B} = B(\mathcal{H}_A) \otimes \mathcal{B}_B$ for a subalgebra $\mathcal{B}_B$ of $B(\mathcal{H}_B)$.

Since β leaves both φ_{ABC} and $\tau_A \otimes \varphi_{BC}$ invariant, it follows that E leaves these states invariant, too. Takesaki's theorem about conditional expectations applies and tells us that

$$D_{ABC}^{\mathrm{it}} \mathcal{B} D_{ABC}^{-\mathrm{it}} \subset \mathcal{B} \,. \tag{2}$$

(See [20] or [2].)

Elements of $\mathcal{B}_B$ have the form

$$b = \oplus_{(m,d)} \left(\oplus_{i=1}^{K(m,d)} \left(\oplus_{t=1}^{m} b\,(m,d,i) \right) \right), \tag{3}$$

where m denotes the multiplicity and d the dimension of the block $b(m,d,i) \in M_d(\mathbb{C})$ and

$$I_B = \sum_{m,d} P(m,d)\,,$$

where the (central) projection $P(m,d)$ corresponds to multiplicity m and dimension d.

The algebra $P(m,d)\mathcal{B}_B P(m,d)$ is isomorphic to $\oplus_{t=1}^{K(m,d)} \mathcal{M}_d(\mathbb{C})$. In this case $\dim P(m,d) = mdK(m,d)$ and elements of $P(m,d)\mathcal{B}_B P(m,d)$ have the form

$$\sum_{i=1}^{K(m,d)} b_i \otimes E_{ii} \otimes I_m \,,$$

where b_i is an element of $\mathcal{M}_d(\mathbb{C})$, $\{E_{ij} \ : \ 1 \le i,j \le K(m,d)\}$ are the standard matrix units of $\mathcal{M}_{K(m,d)}(\mathbb{C})$ and $I_m \in \mathcal{M}_m(\mathbb{C})$ is the identity.

We use these facts to pass to the algebra $\mathcal{B}$. $P'_{m,d} := I_A \otimes P_{m,d} \otimes I_C$ is a central projection and elements of $\mathcal{B}P'_{m,d}$ are of the form

$$\sum_i a_i \otimes E_{ii} \otimes I_m \otimes I_C \,,$$

where a_i is an element of $B(\mathcal{H}_A) \otimes \mathcal{M}_d(\mathbb{C})$.

It can be shown that the unitaries D_{ABC}^{it} commute with $P'_{m,d}$ (see [9]), and we have

$$D_{ABC}^{\mathrm{it}} \left(\mathcal{B}P'_{m,d} \right) D_{ABC}^{-\mathrm{it}} \subset \mathcal{B}P'_{m,d}$$

and this allows us to establish the structure of $D_{ABC}P'(m,d)$:

$$D_{ABC}P'_{m,d} = \sum_i D_{AB_L}(i) \otimes E_{ii} \otimes D_{B_R C}(i)\,, \tag{4}$$

where $D_{AB_L}(i)$ and $D_{B_RC}(i)$ are positive matrices in $B(\mathcal{H}_A)\otimes\mathcal{M}_d(\mathbb{C})$ and $\mathcal{M}_m(\mathbb{C})\otimes B(\mathcal{H}_C)$, respectively. We can conclude the form of D_{ABC} which allows equality in the strong subadditivity for the entropy:

$$D_{ABC}=\sum_{m,d}\sum_{i}^{K(m,d)}\lambda(i,m,d)D_{AB_L}(i,m,d)\otimes E_{ii}(m,d)\otimes D_{B_RC}(i,m,d)\,, \tag{5}$$

where we may assume that D_{AB_L} and D_{B_RC} are density matrices and $\lambda(i,m,d)$ is a probability distribution.

It could be useful to write (5) in a slightly different form, similarly to [6]. One can say that the Hilbert space $\mathcal{H}_B$ decomposes to an orthogonal sum $\oplus_t\mathcal{H}_B^t$ and there is a tensor product decomposition $\mathcal{H}_B^t=\mathcal{H}_{B_L}^t\otimes\mathcal{H}_{B_R}^t$ such that we have density matrices $D_{AB_L}^t$ and $D_{B_RC}^t$ acting on $\mathcal{H}_A\otimes\mathcal{H}_B^{t_L}$ and $\mathcal{H}_B^{t_R}\otimes\mathcal{H}_C$, respectively, and

$$D_{ABC}=\sum_t\lambda(t)D_{AB_L}^t\otimes D_{B_RC}^t \tag{6}$$

with a probability distribution $\lambda(t)$.

It is easy to analyse the situation $\dim\mathcal{H}_B=2$ in details. Then we have the possibilities $D_{AB}\otimes D_C$, $D_A\otimes D_{BC}$ and $pD_A^1\otimes E_{11}\otimes D_C^1+(1-p)D_A^2\otimes E_{22}\otimes D_C^2$ for D_{ABC}.

3. Quantum Markov states

Let $\mathcal{A}_{[1,n]}$ be the tensor product of n copies of a full matrix algebra $\mathcal{M}_k(\mathbb{C})$. When $a_1\otimes a_2\otimes\cdots\otimes a_n\in\mathcal{A}_{[1,n]}$ is identified by $a_1\otimes a_2\otimes\cdots\otimes a_n\otimes I\in\mathcal{A}_{[1,n+1]}$, $\mathcal{A}_{[1,n]}$ is regarded as a subalgebra of $\mathcal{A}_{[1,n+1]}$. Assume that for every n a state φ_n on $\mathcal{A}_{[1,n]}$ is given and they are compatible in the sense that φ_{n+1} restricted to $\mathcal{A}_{[1,n]}$ is φ_n. If this hypothesis is fulfilled then $(\mathcal{A}_{[1,n]},\varphi_n)$ has a direct limit which is the infinite tensorproduct $\mathcal{A}$ endowed with a state φ. Then $(\mathcal{A}_{[1,n]},\varphi_n)\subset(\mathcal{A},\varphi)$. (One can equivalenly start with an arbitray state φ of $\mathcal{A}$ and then φ_n is defined by restriction.) We always assume that φ is *locally faithful*, that is, φ_n is faithful for every n.

A (locally faithful) state φ of $\mathcal{A}$ is called *Markov state* if for every $n\in\mathbb{N}$, there exists a completely positive mapping $F_n:\mathcal{A}_{[1,n+2]}\to\mathcal{A}_{[1,n+1]}$ such that

(a) $F_n(a_n)=a_n$ if $a_n\in\mathcal{A}_n$.
(b) $\varphi_{n+2}\circ F_n=\varphi_{n+1}$.

This definition was given by Accardi and Frigerio [3] after similar trials of Accardi [1].

According to the mean ergodic theorem (due to Kovács and Szűcs)

$$\frac{1}{m}(\mathrm{id} + F_n + F_n^2 + \cdots + F_n^{m-1})$$

converges to a conditional expectation E_n which shares properties (a) and (b).

$$E_n(a_1 \otimes a_2 \otimes \cdots \otimes a_{n+2}) = (a_1 \otimes a_2 \otimes \cdots \otimes a_n) \otimes E^{(n)}(a_{n+1} \otimes a_{n+2}),$$

where $E^{(n)} : M_k(\mathbb{C}) \otimes M_k(\mathbb{C}) \to M_k(\mathbb{C})$ is a conditional expectation.

To make connection to the strong subadditivity of entropy, we set $B(\mathcal{H}_A) = \mathcal{A}_{[1,n]}$, $B(\mathcal{H}_B) = \mathcal{A}_{\{n+1\}}$ and $B(\mathcal{H}_B) = \mathcal{A}_{\{n+2\}}$, and we write D_{ABC} for the density of φ_{n+2}. The strong subbaditivity holds and it is equivalent to the condition

$$S(\varphi_{ABC}, \tau_A \otimes \varphi_{BC}) \geq S(\varphi_{AB}, \tau_A \otimes \varphi_B). \tag{7}$$

For the quantum Markov state we have

$$\begin{aligned}
(\tau_A \otimes \varphi_B)E_n(a_A \otimes a_{n+1} \otimes a_{n+2}) &= (\tau_A \otimes \varphi_B)(a_A \otimes E^{(n)}(a_{n+1} \otimes a_{n+2})) \\
&= \tau_A(a_A)\varphi_B(E^{(n)}(a_{n+1} \otimes a_{n+2})) \\
&= \tau_A(a_A)\varphi_{AB}(I_A \otimes E^{(n)}(a_{n+1} \otimes a_{n+2})) \\
&= \tau_A(a_A)\varphi_{ABC}(I_A \otimes a_{n+1} \otimes a_{n+2}) \\
&= \tau_A \otimes \varphi_{ABC}(a_A \otimes a_{n+1} \otimes a_{n+2}).
\end{aligned}$$

This shows that $(\tau_A \otimes \varphi_B) \circ E_n = \tau_A \otimes \varphi_{ABC}$ and we have equality in the strong subadditivity (7). In this way we have shown that for Markov states the equality case holds. To show the converse, one possibility is to analyse the equality and construct the mapping F_n in the definition of the Markov state. (This was done in [11] long time ago.) However, having the structure (6), we can cunstruct the mapping F_n more directly.

Let Q_t be orthogal projections on $\mathcal{H}_B$ corresponding to the direct sum decomposition

$$\mathcal{H}_B = \oplus_t \mathcal{H}_B^t.$$

Let α be the pinching

$$\alpha(\,\cdot\,) = \sum_t I_A \otimes Q_t \otimes I_C(\,\cdot\,)I_A \otimes Q_t \otimes I_C$$

and define β_t on the operators acting on $\mathcal{H}_A \otimes \mathcal{H}_B^t \otimes \mathcal{H}_C$ as

$$\beta_t(a_{AB_L} \otimes b_{B_RC}) = a_{AB_L} \mathrm{Tr}\,(b_{B_RC} D_{B_RC}^t),$$

where the tensor product decomposition corresponds to that of the Hilbert spaces. Then a convex combination of $\beta_t \circ \alpha$ may play the role of F_n:

$$\sum_t \lambda(t)\beta_t(I_A \otimes Q_t \otimes I_C(\,\cdot\,)I_A \otimes Q_t \otimes I_C)\,. \tag{8}$$

References

1. L. Accardi, A noncommutative Markov Property, (in Russian), Funkcional. Anal. i Prilozen. **9**(1975), 1–8.
2. L. Accardi and C. Cecchini, Conditional expectations in von Neumann algebras and a theorem of Takesaki, J. Funct. Anal. **45**(1982), 245–273.
3. L. Accardi and A. Frigerio, Markovian cocycles, Proc. R. Ir. Acad. **83A**(1983), 251–263
4. L. Accardi and V. Liebscher, Markovian KMS-states for one-dimensional spin chains, Infin. Dimens. Anal. Quantum Probab. Relat. Top. **2**(1999), 645–661.
5. H. Barnum and E. Knill, Reversing quantum dynamics with near optimal quantum and classical fidelity, J. Math. Phys. **43** (2002), 2097–2106.
6. P. Hayden, R. Jozsa, D. Petz and A. Winter, Structure of states which satisfy strong subadditivity of quantum entropy with equality, to be published in Commun. Math. Phys.
7. M. Koashi and N. Imoto, Operations that do not disturb partially known quantum states, Phys. Rev. A, **66**(2002), 022318.
8. E. H. Lieb and M.B. Ruskai, Proof of the strong subadditivity of quantum mechanical entropy, J. Math. Phys. **14**(1973), 1938–1941.
9. M. Mosonyi and D. Petz, Structure of sufficient quantum coarse-grainings, to appear in Lett. Math. Phys.
10. M. A. Nielsen and I. L. Chuang, *Quantum Computation and Quantum Information*, Cambridge University Press, 2000.
11. M. Ohya and D. Petz, *Quantum Entropy and Its Use*, Springer-Verlag, Heidelberg, 1993.
12. H. Ohno, Translation-invariant quantum Markov states, to be published in Interdisc. Inf. Sci.
13. D. Petz, A dual in von Neumann algebras, Quart. J. Math. Oxford **35**(1984), 475–483.
14. D. Petz, Sufficiency of channels over von Neumann algebras, Quart. J. Math. Oxford, **39**(1988), 907–1008.
15. D. Petz, Characterization of sufficient observation channels, in *Mathematical Methods in Statistical Mechanics*, 167–178, Leuven University Press, 1989.
16. D. Petz, Entropy of Markov states, Riv. di Math. Pura ed Appl. **14**(1994), 33–42
17. D. Petz, Monotonicity of quantum relative entropy revisited, Rev. Math. Physics. **15**(2003), 79–91.
18. M. B. Ruskai, Inequalities for quantum entropy: A review with conditions with equality, J. Math. Phys. **43**(2002), 4358–4375.

19. S. Strătilă, *Modular theory in operator algebras*, Abacuss Press, Tunbridge Wells, 1981.
20. M. Takesaki, Conditional expectations in von Neumann algebras, J. Funct. Anal. **9**(1972), 306–321.

AN INFINITE DIMENSIONAL LAPLACIAN ACTING ON SOME CLASS OF LÉVY WHITE NOISE FUNCTIONALS

KIMIAKI SAITÔ

Department of Mathematics
Meijo University
Nagoya 468-8502, Japan
E-mail: ksaito@ccmfs.meijo-u.ac.jp

The Lévy Laplacian is formulated as an operator acting on a class in the Lévy white noise L^2 space. This space includes regular functionals in terms of Gaussian white noise and it is large enough to discuss the stochastic process. This formulation is slightly outside the usual white noise distribution theory, while the Lévy Laplacian has been discussed within the framework of white noise analysis. From Cauchy processes an infinite dimensional stochastic process is constructed, of which the generator is the Lévy Laplacian.

1. Introduction

An infinite dimensional Laplacian was introduced by P. Lévy in his famous book [17]. Since then this exotic Laplacian has been studied by many authors from various aspects see [1–6,18,20,23] and references cited therein. In this paper, generalizing the methods developed in the former works [16,19,24,28,29], we construct a new domain of the Lévy Laplacian acting on some class of Lévy white noise functionals and associated infinite dimensional stochastic processes.

This paper is organized as follows. In Section 1 we summarize basic elements of white noise theory based on a stochastic process given as a difference of two independent Lévy processes. In Section 2, following the recent works Kuo–Obata–Saitô [16], Obata–Saitô [24],Saitô [30] and Saitô–Tsoi [31], we formulate the Lévy Laplacian acting on a Hilbert space consisting of some Lévy white noise functionals and give an equi-continuous semigroup of class (C_0) generated by the Laplacian. This situation is further generalized in Section 3 by means of a direct integral of Hilbert spaces. The space is enough to discuss the stochastic process generated by the Lévy Laplacian. It also includes regular functionals (in the Gaussian sense) as a harmonic functions in terms of the Lévy Laplacian. In Section 4, based on infinitely many Cauchy processes, we give an infinite dimensional stochastic process generated by the Lévy Laplacian.

2. Basic Elememts of White Noise Calculus

Let $E = \mathcal{S}(\mathbf{R})$ be the Schwartz space of rapidly decreasing $\mathbf{R}$-valued functions on $\mathbf{R}$. There exists an orthonormal basis $\{e_\nu\}_{\nu\geq 0}$ of $L^2(\mathbf{R})$ contained in E such that

$$Ae_\nu = 2(\nu+1)e_\nu, \quad \nu = 0,1,2,\ldots, \qquad A = -\frac{d^2}{du^2} + u^2 + 1.$$

For $p \in \mathbf{R}$ define a norm $|\cdot|_p$ by $|f|_p = |A^p f|_{L^2(\mathbf{R})}$ for $f \in E$ and let E_p be the completion of E with respect to the norm $|\cdot|_p$. Then E_p becomes a real separable Hilbert space with the norm $|\cdot|_p$ and the dual space E_p' is identified with E_{-p} by extending the inner product $\langle\cdot,\cdot\rangle$ of $L^2(\mathbf{R})$ to a bilinear form on $E_{-p} \times E_p$. It is known that

$$E = \operatorname*{proj\,lim}_{p\to\infty} E_p, \qquad E^* = \operatorname*{indlim}_{p\to\infty} E_{-p}.$$

The canonical bilinear form on $E^* \times E$ is also denoted by $\langle\cdot,\cdot\rangle$. We denote the complexifications of $L^2(\mathbf{R})$, E and E_p by $L^2_{\mathbf{C}}(\mathbf{R})$, $E_{\mathbf{C}}$ and $E_{\mathbf{C},p}$, respectively.

Let $\{L^1_{\sigma,\lambda}(t)\}_{t\geq 0}$ and $\{L^2_{\sigma,\lambda}(t)\}_{t\geq 0}$ be independent Lévy processes of which the characteristic functions are given by

$$E[e^{izL^j_{\sigma,\lambda}(t)}] = e^{th(z)}, \quad t \geq 0, \quad j = 1,2,$$

$$h(z) = imz - \frac{\sigma^2}{2}z^2 + (e^{i\lambda z} - 1),$$

where $m \in \mathbf{R}, \sigma \geq 0$ and $\lambda \in \mathbf{R}$. Set $\Lambda_{\sigma,\lambda}(t) = L^1_{\sigma,\lambda}(t) - L^2_{\sigma,\lambda}(t)$ for all $t \geq 0$. Then we have

$$\begin{aligned} E[e^{iz\Lambda_{\sigma,\lambda}(t)}] &= e^{t(h(z)+h(-z))} \\ &= \exp\left\{-t\sigma^2 z^2 + t(e^{i\lambda z} + e^{-i\lambda z} - 2)\right\}, \ t \geq 0. \end{aligned}$$

Set $C(\xi) = \exp\left\{\int_{\mathbf{R}}(h(\xi_1(u)) + h(-\xi_2(u)))du\right\}$, $\xi = (\xi_1,\xi_2) \in E\times E$. Then by the Bochner-Minlos Theorem, there exists a probability measure $\mu_{\sigma,\lambda}$ on $E^* \times E^*$ such that

$$\int_{E^*\times E^*} \exp\{i\langle x,\xi\rangle\}\, d\mu_{\sigma,\lambda}(x) = C(\xi), \quad \xi = (\xi_1,\xi_2) \in E \times E,$$

where $\langle x, \xi \rangle = \langle x_1, \xi_1 \rangle + \langle x_2, \xi_2 \rangle$, $x = (x_1, x_2) \in E^* \times E^*, \xi = (\xi_1, \xi_2) \in E \times E$. Let $(L^2)_{\sigma,\lambda} \equiv L^2(E^* \times E^*, \mu_{\sigma,\lambda})$ be the Hilbert space of $\mathbf{C}$-valued square-integrable functions on $E^* \times E^*$ with L^2-norm $\|\cdot\|_{\sigma,\lambda}$ with respect to $\mu_{\sigma,\lambda}$. The Wiener–Itô decomposition theorem says that:

$$(L^2)_{\sigma,\lambda} = \sum_{n=0}^{\infty} \oplus H_n, \tag{1.1}$$

where H_n is the space of multiple Wiener integrals of order $n \in \mathbf{N}$ and $H_0 = \mathbf{C}$. According to (1.1) each $\varphi \in (L^2)_{\sigma,\lambda}$ is represented as

$$\varphi = \sum_{n=0}^{\infty} \mathbf{I}_n(f_n), \qquad f_n \in L^2_{\mathbf{C}}(\mathbf{R})^{\hat{\otimes} n},$$

where $L^2_{\mathbf{C}}(\mathbf{R})^{\hat{\otimes} n}$ denotes the n-fold symmetric tensor power of $L^2_{\mathbf{C}}(\mathbf{R})$ (in the sense of a Hilbert space).

An element of $(L^2)_{\sigma,\lambda}$ is called a *white noise functional.* We denote by $\langle\langle \cdot, \cdot \rangle\rangle$ the canonical bilinear form on $(L^2)_{\sigma,\lambda} \times (L^2)_{\sigma,\lambda}$. Then, for Φ and $\varphi \in (L^2)_{\sigma,\lambda}$ we have

$$\langle\langle \Phi, \varphi \rangle\rangle = \sum_{n=0}^{\infty} n! \langle F_n, f_n \rangle, \qquad \Phi = \sum_{n=0}^{\infty} \mathbf{I}_n(F_n), \quad \varphi = \sum_{n=0}^{\infty} \mathbf{I}_n(f_n),$$

where the canonical bilinear form on $L^2_{\mathbf{C}}(\mathbf{R})^{\otimes n} \times L^2_{\mathbf{C}}(\mathbf{R})^{\otimes n}$ is denoted also by $\langle \cdot, \cdot \rangle$. The $\mathcal{U}$-transform of $\varphi \in (L^2)_{\sigma,\lambda}$ is defined by

$$\mathcal{U}\varphi(\xi) = C(\xi)^{-1} \int_{E^* \times E^*} \varphi(x) \exp\{i \langle x, \xi \rangle\} d\mu(x), \qquad \xi \in E \times E.$$

Theorem 2.1. [26] *(see also* [9,14,22]*) Let F be a complex-valued function defined on $E \times E$. Then F is a $\mathcal{U}$-transform of some white noise functional in $(L^2)_{\sigma,\lambda}$ if and only if there exists a complex-valued function G defined on $E_{\mathbf{C}} \times E_{\mathbf{C}}$ such that*
1) for any ξ and η in $E_{\mathbf{C}} \times E_{\mathbf{C}}$, the function $G(z\xi + \eta)$ is an entire function of $z \in \mathbf{C}$,
2) there exist nonnegative constants K and a such that

$$|G(\xi)| \leq K \exp\left[a|\xi|_0^2\right], \quad \forall \xi \in E_{\mathbf{C}} \times E_{\mathbf{C}},$$

3) $F(\xi) = G(i\sigma^2\xi_1 + i\sigma^2\xi_2 + \lambda(e^{i\lambda\xi_1} - e^{-i\lambda\xi_2}))$ for all $\xi = (\xi_1, \xi_2) \in E \times E$.

3. The Lévy Laplacian Acting on Lévy White Noise Functionals

Consider $F = \mathcal{U}\varphi$ with $\varphi \in (L^2)_{\sigma,\lambda}$. By Theorem 2.1, for any $\xi, \eta \in E \times E$ the function $z \mapsto F(\xi + z\eta)$ admits a Taylor series expansion:

$$F(\xi + z\eta) = \sum_{n=0}^{\infty} \frac{z^n}{n!} F^{(n)}(\xi)(\eta, \ldots, \eta),$$

where $F^{(n)}(\xi) : (E \times E) \times \cdots \times (E \times E) \to \mathbf{C}$ is a continuous n-linear functional. Fixing a finite interval T of $\mathbf{R}$, we take an orthonormal basis $\{\zeta_n\}_{n=0}^{\infty} \subset E \times E$ for $L^2(T) \times L^2(T)$ which is equally dense and uniformly bounded (see e.g. [14,15]). Let $\mathcal{D}_L$ denote the set of all $\varphi \in (L^2)_{\sigma,\lambda}$ such that the limit

$$\widetilde{\Delta}_L(\mathcal{U}\varphi)(\xi) = \lim_{N \to \infty} \frac{1}{N} \sum_{n=0}^{N-1} (\mathcal{U}\varphi)''(\xi)(\zeta_n, \zeta_n)$$

exists for any $\xi \in E \times E$ and $\widetilde{\Delta}_L(\mathcal{U}\varphi)$ is in $\mathcal{U}[(L^2)_{\sigma,\lambda}]$. The Lévy Laplacian Δ_L is defined by

$$\Delta_L \varphi = \mathcal{U}^{-1} \widetilde{\Delta}_L \mathcal{U}\varphi, \qquad \varphi \in \mathcal{D}_L.$$

Given $\sigma \geq 0, \lambda \in \mathbf{R}$, $n \in \mathbf{N}$ and $f \in L^2_{\mathbf{C}}(\mathbf{R})^{\hat{\otimes} n}$, we consider $\varphi \in (L^2)_{\sigma,\lambda}$ of the form:

$$\varphi = \int_{T^n} f(u_1, \ldots, u_n) d\Lambda_{\sigma,\lambda}(u_1) \cdots d\Lambda_{\sigma,\lambda}(u_n). \tag{2.1}$$

The $\mathcal{U}$-transform $\mathcal{U}\varphi$ of φ is given by

$$\mathcal{U}\varphi(\xi) = \int_{T^n} f(u_1, \ldots, u_n) \prod_{j=1}^{n} \Xi_{\sigma,\lambda}(\xi)(u_j) du_1 \ldots du_n, \; \xi \in E \times E.$$

where $\Xi_{\sigma,\lambda}(\xi)(u_j) = \left\{ i\sigma^2 \xi_1(u_j) + i\sigma^2 \xi_2(u_j) + \lambda(e^{i\lambda\xi_1(u_j)} - e^{-i\lambda\xi_2(u_j)}) \right\}$. For any $\sigma \geq 0$, $\lambda \in \mathbf{R}$ and $n \in \mathbf{N}$ let $\mathbf{E}_{\sigma,\lambda,n}$ denote the space of φ which admits an expression as in (2.1), where f belongs to $L^2_{\mathbf{C}}(\mathbf{R})^{\hat{\otimes} n}$ and $\operatorname{supp} f \subset T^n$. Set $\mathbf{E}_{\sigma,\lambda,0} = \mathbf{C}$ for any $\sigma \geq 0, \lambda \in \mathbf{R}$. Then $\mathbf{E}_{\sigma,\lambda,n}$ is a closed linear subspace of $(L^2)_{\sigma,\lambda}$. Using a similar method as in [31], we get the following

Theorem 3.1. [31] *(see also* [16,29]*) For each $\sigma \geq 0$, $n \in \mathbf{N}$ and $\lambda \in \mathbf{R}$ the Lévy Laplacian Δ_L becomes a scalar operator on $\mathbf{E}_{\sigma,0,n} \cup \mathbf{E}_{0,\lambda,n}$ such that*

$$\Delta_L \varphi = 0, \qquad \varphi \in \mathbf{E}_{\sigma,0,n}; \tag{2.2}$$

$$\Delta_L \varphi = -\frac{n\lambda^2}{|T|}\varphi, \qquad \varphi \in \mathbf{E}_{0,\lambda,n}. \tag{2.3}$$

In particular, Δ_L is a self-adjoint operator on $\bar{\boldsymbol{D}}_{n,\lambda}$.

Proposition 3.1. [31] *Let $\lambda \in \mathbf{R}$ be fixed. Consider two white noise functions of the form:*

$$\varphi = \sum_{n=0}^{\infty} \varphi_n, \quad \psi = \sum_{n=0}^{\infty} \psi_n, \qquad \varphi_n, \psi_n \in \mathbf{E}_{0,\lambda,n}.$$

If $\varphi = \psi$ in $(L^2)_{0,\lambda}$, then $\varphi_n = \psi_n$ for all $n \in \mathbf{N} \cup \{0\}$.

Taking (2.2) and (2.3) into account, we put

$$\alpha_N^{\lambda}(n) = \sum_{\ell=0}^{N} \left(\frac{n\lambda^2}{|T|}\right)^{2\ell}.$$

For $N \in \mathbf{N}$ and $\lambda \in \mathbf{R}$ let $\mathbf{D}_N^{0,\lambda}$ be the space of $\varphi \in (L^2)_{0,\lambda}$ which admits an expression

$$\varphi = \sum_{n=1}^{\infty} \varphi_n, \qquad \varphi_n \in \mathbf{E}_{0,\lambda,n},$$

such that

$$\varphi_{N,0,\lambda}^2 = \sum_{n=1}^{\infty} \alpha_N^{\lambda}(n) \|\varphi_n\|_{0,\lambda}^2 < \infty. \tag{2.4}$$

By the Schwartz inequality we see that $\mathbf{D}_N^{0,\lambda}$ is a subspace of $(L^2)_{0,\lambda}$ and becomes a Hilbert space equipped with the new norm $\cdot_{N,0,\lambda}$ defined in (2.4). Moreover, in view of the inclusion relations:

$$(L^2)_{0,\lambda} \supset \mathbf{D}_1^{0,\lambda} \supset \cdots \supset \mathbf{D}_N^{0,\lambda} \supset \mathbf{D}_{N+1}^{0,\lambda} \supset \cdots,$$

we define

$$\mathbf{D}_{\infty}^{0,\lambda} = \operatorname*{proj\,lim}_{N\to\infty} \mathbf{D}_N^{0,\lambda} = \bigcap_{N=1}^{\infty} \mathbf{D}_N^{0,\lambda}.$$

Put

$$\mathbf{D}_{\infty}^{\sigma,0} = \left\{ \sum_{n=0}^{\infty} \varphi_n \,\middle|\, \varphi_n \in \mathbf{E}_{\sigma,0,n}, n = 0, 1, 2, \ldots, \sum_{n=0}^{\infty} \|\varphi_n\|_{\sigma,0}^2 < \infty \right\}.$$

Note that for any $\lambda \in \mathbf{R}$ we have

$$\bigcup_{n=1}^{\infty} \mathbf{E}_{0,\lambda,n} \subset \mathbf{D}_{\infty}^{0,\lambda} \subset (L^2)_{0,\lambda}.$$

Then Δ_L becomes a continuous linear operator defined on $\mathbf{D}_{N+1}^{0,\lambda}$ into $\mathbf{D}_{N}^{0,\lambda}$ satisfying

$$\Delta_L \varphi_{N,0,\lambda} \leq \varphi_{N+1,0,\lambda}, \qquad \varphi \in \mathbf{D}_{\infty}^{0,\lambda}, \quad N \in \mathbf{N}. \tag{2.5}$$

Summing up, we have the following

Theorem 3.2. [16,31] *The operator Δ_L is a self-adjoint operator densely defined in $\mathbf{D}_N^{0,\lambda}$ for each $N \in \mathbf{N}$ and $\lambda \in \mathbf{R}$.*

It follows from (2.5) that Δ_L is a continuous linear operator on $\mathbf{D}_{\infty}^{0,\lambda}$. In view of the action of (2.3), for each $t \geq 0$ and $\lambda \in \mathbf{R}$ we consider an operator G_t^{λ} on $\mathbf{D}_{\infty}^{0,\lambda}$ defined by

$$G_t^{\lambda}\varphi = \sum_{n=1}^{\infty} e^{-tn\lambda^2/|T|}\varphi_n, \qquad \varphi = \sum_{n=1}^{\infty} \varphi_n \in \mathbf{D}_{\infty}^{0,\lambda}.$$

We also define G_t^0 on $\mathbf{D}_{\infty}^{\sigma,0}$ as an identity operator I by $I\varphi = \varphi$, $\varphi \in \mathbf{D}_{\infty}^{\sigma,0}$.

Theorem 3.3. [16,30] *Let $\lambda \in \mathbf{R}$. Then the family of operators $\{G_t^{\lambda}; t \geq 0\}$ on $\mathbf{D}_{\infty}^{0,\lambda}$ is an equi-continuous semigroup of class (C_0) of which the infinitesimal generator is Δ_L.*

4. The Lévy Laplacian Acting on WNF-valued Functions

Let $d\nu(\lambda)$ be a finite Borel measure on $\mathbf{R}$ satisfying

$$\int_{\mathbf{R}} \frac{d\nu(\lambda)}{\lambda^4} < \infty.$$

Fix $N \in \mathbf{N}$. Let $\mathfrak{D}_N^{\sigma}$ be the space of (equivalent classes of) measurable vector functions $\boldsymbol{\varphi} = (\varphi^{\lambda})$ with $\varphi^{\lambda} = \sum_{n=1}^{\infty} \varphi_n^{\lambda} \in \mathbf{D}_N^{0,\lambda}$ for all $\lambda \in \mathbf{R} \setminus \{0\}$, and $\varphi^0 \in \mathbf{D}_N^{\sigma,0}$, such that

$$\boldsymbol{\varphi}_N^2 = \|\varphi^0\|_{\sigma,0}^2 + \sum_{n=1}^{\infty} \int_{\mathbf{R}\setminus\{0\}} \|\varphi_n^{\lambda}\|_{0,\lambda,N}^2 \, d\nu(\lambda) < \infty. \tag{3.1}$$

Then $\mathfrak{D}_N^{\sigma}$ becomes a Hilbert space with the norm given in (3.1).

Let $\mathfrak{D}_0^\sigma$ be the space of (equivalent classes of) measurable vector functions $\boldsymbol{\varphi} = (\varphi^\lambda)$ with $\varphi^\lambda = \sum_{n=1}^\infty \varphi_n^\lambda \in (L^2)_{0,\lambda}$ for all $\lambda \in \mathbf{R} \setminus \{0\}$, and $\varphi^0 \in (L^2)_{\sigma,0}$, such that

$$\boldsymbol{\varphi}_0^2 = \|\varphi^0\|_{\sigma,0}^2 + \sum_{n=1}^\infty \int_{\mathbf{R}\setminus\{0\}} \|\varphi_n^\lambda\|_{0,\lambda}^2 \, d\nu(\lambda) < \infty. \tag{3.2}$$

Then $\mathfrak{D}_0^\sigma$ also becomes a Hilbert space with the norm given in (3.2).

Proposition 4.1. *The map*

$$(\varphi^\lambda) \mapsto \varphi^0 + \sum_{n=1}^\infty \int_{\mathbf{R}\setminus\{0\}} \varphi_n^\lambda \, d\nu(\lambda) \tag{3.3}$$

is a continuous linear map and a bijection from $\mathfrak{D}_N^\sigma$ into $\mathfrak{D}_0^\sigma$.

In view of the natural inclusion: $\mathfrak{D}_{N+1}^\sigma \subset \mathfrak{D}_N^\sigma$ for $N \in \mathbf{N}$, which is obvious from construction, we define

$$\mathfrak{D}_\infty^\sigma = \operatorname*{proj\,lim}_{N\to\infty} \mathfrak{D}_N^\sigma = \bigcap_{N=1}^\infty \mathfrak{D}_N^\sigma.$$

The Lévy Laplacian Δ_L is defined on the space $\mathfrak{D}_\infty^\sigma$ by

$$\Delta_L \boldsymbol{\varphi} = (\Delta_L \varphi^\lambda), \qquad \boldsymbol{\varphi} = (\varphi^\lambda) \in \mathfrak{D}_\infty^\sigma.$$

Then Δ_L is a continuous linear operator from $\mathfrak{D}_\infty^\sigma$ into itself. Similarly, for $t \geq 0$ we define

$$G_t \boldsymbol{\varphi} = (G_t^\lambda \varphi^\lambda), \qquad \boldsymbol{\varphi} = (\varphi^\lambda) \in \mathfrak{D}_\infty^\sigma.$$

Then we have the following:

Theorem 4.1. *The family $\{G_t; t \geq 0\}$ is an equi-continuous semigroup of class (C_0) on $\mathfrak{D}_\infty^\sigma$ whose generator is given by Δ_L.*

5. An Infinite Dimensional Stochastic Process Associated with the Lévy Laplacian

For $p \in \mathbf{R}$ let $E_p^{\mathbf{R}}$ be the linear space of all functions $\lambda \mapsto \xi_\lambda \in E_p \times E_p$, $\lambda \in \mathbf{R}$, which are strongly measurable. An element of $E_p^{\mathbf{R}}$ is denoted by $\boldsymbol{\xi} = (\xi_\lambda)_{\lambda\in\mathbf{R}}$. Equipped with the metric given by

$$d_p(\boldsymbol{\xi}, \boldsymbol{\eta}) = \int_{\mathbf{R}} \frac{|\xi_\lambda - \eta_\lambda|_p}{1 + |\xi_\lambda - \eta_\lambda|_p} \, d\nu(\lambda), \qquad \boldsymbol{\xi} = (\xi_\lambda),\ \boldsymbol{\eta} = (\eta_\lambda),$$

the space $E_p^{\mathbf{R}}$ becomes a complete metric space. Similarly, let $\mathbf{C}^{\mathbf{R}}$ denote the linear space of all measurable function $\lambda \mapsto z_\lambda \in \mathbf{C}$ equipped with the metric defined by

$$\rho(\mathbf{z}, \mathbf{u}) = \int_{\mathbf{R}} \frac{|z_\lambda - u_\lambda|}{1 + |z_\lambda - u_\lambda|} \, d\nu(\lambda), \qquad \mathbf{z} = (z_\lambda), \ \mathbf{u} = (u_\lambda).$$

Then $\mathbf{C}^{\mathbf{R}}$ is also a complete metric space.

In view of $d_p \leq d_q$ for $p \geq q$, we introduce the projective limit space $E^{\mathbf{R}} = \operatorname{proj\,lim}_{p \to \infty} E_p^{\mathbf{R}}$. The $\mathcal{U}$-transform can be extended to a continuous

linear operator on $\mathfrak{D}_\infty^\sigma$ by

$$\mathcal{U}\varphi(\boldsymbol{\xi}) = (\mathcal{U}\varphi^\lambda(\xi_\lambda))_{\lambda \in \mathbf{R}}, \qquad \boldsymbol{\xi} = (\xi_\lambda)_{\lambda \in \mathbf{R}} \in E^{\mathbf{R}},$$

for any $\varphi = (\varphi^\lambda)_{\lambda \in \mathbf{R}} \in \mathfrak{D}_\infty^\sigma$. The space $\mathcal{U}[\mathfrak{D}_\infty^\sigma]$ is endowed with the topology induced from $\mathfrak{D}_\infty^\sigma$ by the $\mathcal{U}$-transform. Then the $\mathcal{U}$-transform becomes a homeomorphism from $\mathfrak{D}_\infty^\sigma$ onto $\mathcal{U}[\mathfrak{D}_\infty^\sigma]$. The transform $\mathcal{U}\varphi$ of $\varphi \in \mathfrak{D}_\infty^\sigma$ is a continuous operator from $E^{\mathbf{R}}$ into $\mathbf{C}^{\mathbf{R}}$. We denote the operator by the same notation $\mathcal{U}\varphi$.

Let $\widetilde{G}_t$ be an operator defined on $\mathcal{U}[\mathfrak{D}_\infty^\sigma]$ by

$$\widetilde{G}_t = \mathcal{U} G_t \mathcal{U}^{-1}, \qquad t \geq 0.$$

Then by Theorem 3.3, $\{\widetilde{G}_t;\, t \geq 0\}$ is an equi-continuous semigroup of class (C_0) generated by the operator $\widetilde{\Delta}_L$.

Let $\{X_t^j\}, j = 1, 2, 3, 4$, be independent Cauchy processes with t running over $[0, \infty)$, of which the characteristic functions are given by

$$\mathbf{E}[e^{izX_t^j}] = e^{-t|z|}, \qquad z \in \mathbf{R}, \quad j = 1, 2, 3, 4.$$

Take a smooth function $\eta_T \in E$ with $\eta_T = 1/|T|$ on T. Set

$$Y_t^\lambda = \begin{cases} (X_{\lambda t}^1 \eta_T, -X_{\lambda t}^2 \eta_T) & \text{if } \lambda \geq 0, \\ (X_{-\lambda t}^3 \eta_T, -X_{-\lambda t}^4 \eta_T), & \text{otherwise.} \end{cases}$$

Define an infinite dimensional stochastic process $\{\mathbf{Y}_t; t \geq 0\}$ starting at $\boldsymbol{\xi} = (\xi_\lambda)_{\lambda \in \mathbf{R}} \in E^{\mathbf{R}}$ by

$$\mathbf{Y}_t = (\xi_\lambda + Y_t^\lambda)_{\lambda \in \mathbf{R}}, \qquad t \geq 0.$$

Then this is an $E^{\mathbf{R}}$-valued stochastic process and we have the following

Theorem 5.1. *If F is the $\mathcal{U}$-transform of an element in $\mathfrak{D}_\infty^\sigma$, we have*

$$\widetilde{G}_t F(\boldsymbol{\xi}) = E[F(\mathbf{Y}_t) | \mathbf{Y}_0 = \boldsymbol{\xi}], \qquad t \geq 0. \tag{4.1}$$

Proof. We first consider the case when $F \in \mathcal{U}[\mathfrak{D}_\infty^\sigma]$ is given by

$$F(\boldsymbol{\xi}) = (F^\lambda(\xi_\lambda))_{\lambda \in \mathbf{R}}, \quad F^0 \in \mathcal{U}[\mathbf{D}_\infty^{\sigma,0}],$$

$$F^\lambda(\xi_\lambda) = \lambda^n \int_{T^n} f(\mathbf{u}) \prod_{j=1}^n \left\{ e^{i\lambda\xi_{1,\lambda}(u_j)} - e^{-i\lambda\xi_{2,\lambda}(u_j)} \right\} d\mathbf{u}, \ (\lambda \neq 0)$$

with $f \in L^2_{\mathbf{C}}(\mathbf{R})^{\hat{\otimes} n}$. Then we have

$$\begin{aligned}
&E[F(\mathbf{Y}_t)|\mathbf{Y}_0 = \boldsymbol{\xi}] \\
&= (E[F^\lambda(\xi_\lambda + Y_t^\lambda)])_{\lambda \in \mathbf{R}} \\
&= \left(F^0(\xi_0)\delta_{\lambda,0} + \lambda^n \int_{T^n} f(\mathbf{u}) E\left[\prod_{j=1}^n \left\{ e^{i\lambda\xi_{1,\lambda}(u_j)} e^{i\frac{\lambda}{|T|}(X^1_{\lambda t} 1_{(0,\infty)}(\lambda) + X^3_{-\lambda t} 1_{(-\infty,0)}(\lambda))} \right.\right.\right. \\
&\quad \left.\left.\left. - e^{-i\lambda\xi_{2,\lambda}(u_j)} e^{i\frac{\lambda}{|T|}(X^2_{\lambda t} 1_{(0,\infty)}(\lambda) + X^4_{-\lambda t} 1_{(-\infty,0)}(\lambda))} \right\} \right] d\mathbf{u} \right)_{\lambda \in \mathbf{R}} \\
&= \left(e^{-tn\lambda/|T|} F^\lambda(\xi_\lambda) \right)_{\lambda \in \mathbf{R}} \\
&= (\widetilde{G_t^\lambda} F^\lambda(\xi_\lambda))_{\lambda \in \mathbf{R}} \\
&= (\widetilde{G_t} F(\boldsymbol{\xi})).
\end{aligned}$$

Next let $F = \left(F^0 \delta_{\lambda,0} + \sum_{n=1}^\infty F_n^\lambda \right)_{\lambda \in \mathbf{R}} \in \mathcal{U}[\mathfrak{D}_\infty^\sigma]$. Then for ν-almost all $\lambda \in \mathbf{R}$ and for any $n \in \mathbf{N}$, F_n^λ is expressed in the following form:

$$F_n^\lambda(\xi_\lambda) = \lim_{N \to \infty} \lambda^n \int_{T^n} f_\lambda^{[N]}(\mathbf{u}) \prod_{j=1}^n \left\{ e^{i\lambda\xi_{1,\lambda}(u_j)} - e^{-i\lambda\xi_{2,\lambda}(u_j)} \right\} d\mathbf{u}.$$

Since $F^0 \in \mathcal{U}[\mathbf{D}_\infty^{\sigma,0}]$ and $F_n^\lambda \in \mathcal{U}[\mathbf{D}_\infty^{0,\lambda}]$, there exist $\varphi^0 \in \mathbf{D}_\infty^{\sigma,0}$ and $\varphi_n^\lambda \in \mathbf{D}_\infty^{0,\lambda}$ such that $F^0 = \mathcal{U}[\varphi^0]$ and $F_n^\lambda = \mathcal{U}[\varphi_n^\lambda]$ for ν-almost all λ and each n. By the Schwarz inequality, we see that

$$\begin{aligned}
&\sum_{n=1}^\infty E[|F_n^\lambda(\xi_\lambda + Y_t^\lambda)|] \\
&\leq \sum_{n=0}^\infty ||\varphi_n^\lambda||_{0,\lambda} E[||\varphi_{\xi_\lambda + Y_t^\lambda}||_{0,\lambda}] \\
&\leq \left\{ \sum_{n=1}^\infty \alpha_N^\lambda(n)^{-1} \right\}^{1/2} \left\{ \sum_{n=1}^\infty \alpha_N^\lambda(n) ||\varphi_n^\lambda||_{0,\lambda}^2 \right\}^{1/2} E[||\varphi_{\xi_\lambda + Y_t^\lambda}||_{0,\lambda}] < \infty,
\end{aligned}$$

where $\varphi_{\xi_\lambda} = C(\xi_\lambda)^{-1}e^{i\langle\cdot,\xi_\lambda\rangle}$ for ν-almost all $\lambda \in \mathbf{R}$ and each $N \in \mathbf{N}$. Therefore by the continuity of $\widetilde{G_t^\lambda}, \lambda \in \mathbf{R}$, we get that

$$\begin{aligned} E[F(\boldsymbol{\xi} + \mathbf{Y}_t)] &= (E[F^\lambda(\xi_\lambda + Y_t^\lambda)])_{\lambda\in\mathbf{R}} \\ &= \left(\sum_{n=1}^{\infty} E[F_n^\lambda(\xi_\lambda + Y_t^\lambda)]\right)_{\lambda\in\mathbf{R}} \\ &= \left(\sum_{n=1}^{\infty} \widetilde{G_t^\lambda} F_n^\lambda(\xi_\lambda)\right)_{\lambda\in\mathbf{R}} \\ &= \left(\widetilde{G_t^\lambda} \sum_{n=1}^{\infty} F_n^\lambda(\xi_\lambda)\right)_{\lambda\in\mathbf{R}} \\ &= \widetilde{G_t} F(\xi). \end{aligned}$$

Thus we obtain the assertion. □

Acknowledgments

This work was written based on a talk at International Conference on Quantum Information 2003. The author would like to express his deep thanks to organizers for their hard work. This work was partially supported by JSPS grant 15540141. The author is grateful for the support.

References

1. L. Accardi and V. Bogachëv: *The Ornstein–Uhlenbeck process associated with the Lévy Laplacian and its Dirichlet form*, Prob. Math. Stat. **17** (1997), 95–114.
2. L. Accardi, P. Gibilisco and I. V. Volovich: *Yang-Mills gauge fields as harmonic functions for the Lévy Laplacian*, Russ. J. Math. Phys. **2** (1994), 235–250.
3. L. Accardi and O. G. Smolyanov: *Trace formulae for Levy-Gaussian measures and their application*, in "Mathematical Approach to Fluctuations Vol. II (T. Hida, Ed.)," pp. 31–47, World Scientific, 1995.
4. D. M. Chung, U. C. Ji and K. Saitô: *Cauchy problems associated with the Lévy Laplacian in white noise analysis*, Infin. Dimen. Anal. Quantum Probab. Rel. Top. **2** (1999), 131–153.
5. M. N. Feller: *Infinite-dimensional elliptic equations and operators of Lévy type*, Russ. Math. Surveys **41** (1986), 119–170.
6. K. Hasegawa: *Lévy's functional analysis in terms of an infinite dimensional Brownian motion I*, Osaka J. Math. **19** (1982), 405–428.
7. T. Hida: "Analysis of Brownian Functionals," Carleton Math. Lect. Notes, No. 13, Carleton University, Ottawa, 1975.

8. T. Hida: *A role of the Lévy Laplacian in the causal calculus of generalized white noise functionals*, in "Stochastic Processes," pp. 131–139, Springer–Verlag, 1993.
9. T. Hida, H.-H. Kuo, J. Potthoff and L. Streit: "White Noise: An Infinite Dimensional Calculus," Kluwer Academic, 1993.
10. T. Hida and K. Saitô: *White noise analysis and the Lévy Laplacian*, in "Stochastic Processes in Physics and Engineering (S. Albeverio et al. Eds.)," pp. 177–184, 1988.
11. E. Hille and R. S. Phillips: "Functional Analysis and Semi-Groups," AMS Colloq. Publ. Vol. 31, Amer. Math. Soc, 1957.
12. I. Kubo and S. Takenaka: *Calculus on Gaussian white noise I–IV*, Proc. Japan Acad. **56A** (1980) 376–380; **56A** (1980) 411–416; **57A** (1981) 433–436; **58A** (1982) 186–189.
13. H.-H. Kuo: *On Laplacian operators of generalized Brownian functionals*, Lect. Notes in Math. Vol. 1203, pp. 119–128, Springer–Verlag, 1986.
14. H.-H. Kuo: "White Noise Distribution Theory," CRC Press, 1996.
15. H.-H. Kuo, N. Obata and K. Saitô: *Lévy Laplacian of generalized functions on a nuclear space*, J. Funct. Anal. **94** (1990), 74–92.
16. H.-H. Kuo, N. Obata and K. Saitô: *Diagonalization of the Lévy Laplacian and related stable processes*, Infin. Dimen. Anal. Quantum Probab. Rel. Top. **5** (2002), 317–331.
17. P. Lévy: "Leçons d'Analyse Fonctionnelle," Gauthier-Villars, Paris, 1922.
18. R. Léandre and I. A. Volovich: *The stochastic Lévy Laplacian and Yang-Mills equation on manifolds*, Infin. Dimen. Anal. Quantum Probab. Rel. Top. **4** (2001) 161–172.
19. K. Nishi and K. Saitô: *An infinite dimensional stochastic process and the Lévy Laplacian acting on WND-valued functions*, to appear in "Quantum Inofrmation and Complexity" World Scientific, 2004.
20. K. Nishi, K. Saitô and A. H. Tsoi: *A stochastic expression of a semi-group generated by the Lévy Laplacian*, in "Quantum Information III (T. Hida and K. Saitô, Eds.)," pp. 105–117, World Scientific, 2000.
21. N. Obata: *A characterization of the Lévy Laplacian in terms of infinite dimensional rotation groups*, Nagoya Math. J. **118** (1990), 111–132.
22. N. Obata: "White Noise Calculus and Fock Space," Lect. Notes in Math. Vol. 1577, Springer-Verlag, 1994.
23. N. Obata: *Quadratic quantum white noises and Lévy Laplacian*, Nonlinear Analysis **47** (2001), 2437–2448.
24. N. Obata and K. Saitô: *Cauchy processes and the Lévy Laplacian*, Quantum Probability and White Noise Analysis **16** (2002), 360–373.
25. E. M. Polishchuk: "Continual Means and Boundary Value Problems in Function Spaces," Birkhäuser, Basel/Boston/Berlin, 1988.
26. J. Potthoff and L. Streit: *A characterization of Hida distributions*, J. Funct. Anal. **101** (1991), 212–229.
27. K. Saitô: *Itô's formula and Lévy's Laplacian I*, Nagoya Math. J. **108** (1987), 67–76; *II*, ibid. **123** (1991), 153–169.
28. K. Saitô: *A (C_0)-group generated by the Lévy Laplacian II*, Infin. Dimen.

Anal. Quantum Probab. Rel. Top. **1** (1998) 425–437.
29. K. Saitô: *A stochastic process generated by the Lévy Laplacian*, Acta Appl. Math. **63** (2000), 363–373.
30. K. Saitô: *The Lévy Laplacian and stable processes*, Chaos, Solitons and Fractals **12** (2001), 2865–2872.
31. K. Saitô and A. H. Tsoi: *The Lévy Laplacian as a self-adjoint operator*, in "Quantum Information (T. Hida and K. Saitô, Eds.)," pp. 159–171, World Scientific, 1999.
32. K. Saitô and A. H. Tsoi: *The Lévy Laplacian acting on Poisson noise functionals*, Infin. Dimen. Anal. Quantum Probab. Rel. Top. **2** (1999), 503–510.
33. K. Saitô and A. H. Tsoi: *Stochastic processes generated by functions of the Lévy Laplacian*, in "Quantum Information II (T. Hida and K. Saitô, Eds.)," pp. 183–194, World Scientific, 2000.
34. K. Yosida: "Functional Analysis (3rd Edition)," Springer-Verlag, 1971.

STRUCTURE OF LINEAR PROCESSES

SI SI
Graduate school of Information Science and Technology
Aichi Prefectural University
Aichi-ken 480-1198, Japan
E-mail: sisi@ist.aichi-pu.ac.jp

WIN WIN HTAY
Department of Computational mathematics
University of Computer Studies
Yangon, Myanamr
E-mail: win@totalmail.com

Following P. Lévy, we consider a particular linear process given by

$$X(t) = X_B(t) + X_P(t),$$

where $X_B(t) = \int_0^t F(t,u)\dot{B}(u)du$, $X_P(t) = \int_0^t G(t,u)\dot{Y}(u)du$, and they are independent each other, in which $\dot{B}(t)$ and $\dot{Y}(t)$ are white noise and a compound Poisson noise, respectively. By the effective use of characteristic functional of $X(t)$ and observing the values of $X(t)$, we discuss how to determine the probability structure of $X(t)$, namely, how to identify the kernels $F(t,u)$ and $G(t,u)$. .

1. Introduction

The characteristic functional of a stochastic process (including generalized stochastic process) determines the probability ditribution of the process, and hence the analysis of the characteristic functional tells us the probabilistic structure of the process in question.

Following P. Lévy [7] we consider a linear process expressed in the form

$$\Phi(t) = \int_0^t F(t,u)X(u)du + \int_0^t G(t,u)Y(u)du + \int_0^t H(t,u)Z(u)du, \quad (1)$$

where the three terms are Gaussian, Compound Poisson and a fixed discontinuity part, and where $X(u)du$, $Y(u)du$ and $Z(u)du$ are random measures.

The rigorous meaning of the three integrals will be given when they are necessary.

Interesting question is that observing the values of $\Phi(t)$, how to determine the probabilitistic structure of $\Phi(t)$, namely how to identify the kernels $F(t,u), G(t,u)$ and $H(t,u)$. For this purpose we need detail properties of sample functions. Further, we use not only L^2-theory, but also analytic properties of characteristic functionals, which we are going to use effectively.

2. An illustrative example, Linear functionals of Poisson noise

Let a stochastic process $X_P(t)$ be given by an integral

$$X_P(t) = \int_0^t G(t,u)\dot{P}(u)du. \tag{1}$$

It may be viewed simply as a linear functional of $\dot{P}(t)$, however there are two ways of understanding the meaning of the integral in such a way that

i) the integral is defined in the Hilbert space by taking $\dot{P}(t)dt$ to be a *random measure*, and the stochastic integral is defined. Like the homogeneous Chaos, multiple integral can also be defined.
ii) on the other hand the integral is understood as a continuous bilinear form of a test function $G(t,\cdot)$ and a sample function of $\dot{P}(t)$ (the *path-wise* integral). This can be done if the kernel is a smooth function of u over the interval $[0,t]$ since a sample function of $\dot{P}(\cdot)$ is a generalized function. Namely, the integral is defined as a bilinear form, and $X_P(t)$ is called a linear process.

Assume that $G(t,t)$ never vanishes and that it is not a canonical kernel, that is, it is not a kernel function of an invertible integral operator. Then, we can claim that for the integral in the first sense $X_P(t)$ has less information compared to $P(t)$. Because there is a linear function of $P(s), s \leq t$ which is orthogonal to $X_P(s), s \leq t$. On the other hand, if $X_P(t)$ is defined in the second sense, we can prove the following proposition.

Note that we always assume that the variation $\delta X_P(t)$ exists.

Proposition 2.1. *Under the assumptions stated above, if the $X_P(t)$ above is defined sample function-wise, we have the following equality for sigma-fields:*

$$\mathbf{B}_t(X_P) = \mathbf{B}_t(P), \quad t \geq 0. \tag{2}$$

Proof. By assumption it is easy to see that $X_P(t)$ and $P(t)$ share the same jump points, which means that the information is fully transfered from $P(t)$ to $X_P(t)$.This proves the equality. □

The above argument tells us that we are led to introduce a space $(\mathbf{P})$ of random variables that come from separable stochastic processes for which existence of variance may not be expected. This sounds to be a vague statement, however we can rigorously defined. There the topology is defined by either the almost sure convergent or the convergence in probability, and there is no need to think of mean square topology. On the space $(\mathbf{P})$ filtering and prediction for strictly stationary process can naturally be discussed.

For the linear process given by (2.1), we can define N−ple Markov property in the same manner as in the Gaussian case.

Proposition 2.2. *If $X_P(t)$ given by (2.1), we can define N−ple Markov then the kernel $G(t,u)$ is a Goursat kernel of degree N.*

Also we prove that the conditional expectation is given by the formula

$$E[X_P(t)|\mathbf{B}_s(X_P)] = \int_0^s G(t,u)\dot{P}(u)du, \, s \le t. \tag{3}$$

The rest of the proof is exactly the same as in the Gaussian case.

Remark 2.1. In the above proposition the kernel G is not necessarily a canonical (an invertible) kernel, unlike the Gaussian case.

3. Innovation of a linear process

We now come to define a linear process in the folowing way.

Definition 3.1. Given a system of random variables $\{X, \{X_\alpha\}\}$. The system X is said to be *linearly correlated* with $\{X_\alpha\}$, if X is expressed as a sum

$$X = U + V,$$

where V is a linear function of the X_α's and where U and V are independent.

Definition 3.2. A stochastic process $\{X(t)\}$ is called a *linear process* if for any t and $h > 0$, the variable $X(t+h)$ is linealy correlated with the system $\{X(s), s \le t\}$.

We are interested in the following linear process which is an important class of linear process ,

$$X(t) = \int_0^t F(t,u)\dot{B}(u)du + \int_0^t G(t,u)\dot{Y}(u)du \tag{1}$$

where $F(t,u)$ and $G(t,u)$ be in C^2 as functions of u for any fixed t and where $B(t)$ and $Y(u)$ are Brownian motion and a compound Poisson process with stationary increments, respectively and they are independent.

The sample functions $\dot{B}(u,\omega)$ and $\dot{Y}(u,\omega)$, for fixed ω, are generalized functions in the sense of Gel'fand. However, as we observed in Section 2, the integrals in (3.1) are considered as the path-wise integrals.

Now, assume that $G(t,t) \neq 0$ for every t. Observing the sample functions of $X(t,\omega)$ we can obtain all the jumps of $Y(t,\omega)$ by noting the equation:

$$\begin{aligned}\delta X(t) = F(t,t)\dot{B}(t)dt + dt \int_0^t \frac{\partial}{\partial t}F(t,u)\dot{B}(u)du \\ + G(t,t)\dot{Y}(t)dt + dt \int_0^t \frac{\partial}{\partial t}G(t,u)\dot{Y}(u)du.\end{aligned} \tag{2}$$

Then, $G(t,s)$, for $t > s$, is obtained by taking the covariance of $X(t)$ and $\dot{Y}(t)$. Thus we obtain

$$\int_0^t F(t,u)\dot{B}(u)du$$

which is equal to

$$X(t) - \int_0^t G(t,u)\dot{Y}(u)du.$$

(1) If F is canonical then $\dot{B}(t)$ is obtained in a usual way.
(2) For the case of non-canonical F, it is known how to obtain the innovation $B_1(t)$ for

$$\int_0^t F(t,u)\dot{B}(u)du.$$

(See Accardi-Hida-Si Si [1].)

Thus, we can obtain the innovation for both cases . The innovation is obtained as $\dot{B}_1(t) + \dot{Y}(t)$ of $X(t)$ where $\dot{B}(t)$ is the original $\dot{B}(t)$ for the case 1, above.
Summing up

Proposition 3.1. *For a linear process given by (3.1) we have*

(1) **Proposition 3.2.**

1) The innovation of the process $X(t)$ is given by

$$\dot{B}(t) + \dot{Y}(t),$$

or equivalently by the system $\{\dot{B}(t), \dot{Y}(t)\}$, where $\dot{B}(t)$ is the original $\dot{B}(t)$ if the kernel F is canonical.

2) The Gaussian part $\int_0^t F(t,u)\dot{B}(u)du$ and the Poisson part $\int_0^t G(t,u)\dot{Y}(u)du$ can be descriminated.

4. Determination of kernels

Let $X(t)$ be a linear process given by (3.1). Assume that both Gaussian part and Poisson part have canonical representation. Namely, $X(t)$ can be written as

$$X(t) = X_B(t) + X_P(t), \tag{1}$$

where

$$X_B(t) = \int_0^t F(t,u)\dot{B}(u)du$$

and

$$X_P(t) = \int_0^t G(t,u)\dot{Y}(u)du$$

by using the same notation.

Since Gaussian part $X_B(t)$ and $X_P(t)$ are independent, the characteristic functional of $X(t)$ is expressed in the form

$$C_X(\xi) = E[e^{i\langle X,\xi\rangle}] = C_B(\xi)C_P(\xi), \tag{2}$$

where

$$C_B(\xi) = \exp\left[-\frac{1}{2}\int_0^\infty (F*\xi)(s)^2 ds\right] \tag{3}$$

in which

$$(F*\xi)(s) = \int_0^\infty F(t,s)\xi(t)dt, \; F(t,s) = 0 \; for \; t < s, \tag{4}$$

and where

$$C_P(\xi) = \exp\left[\int_{-\infty}^\infty \int_0^\infty \left(e^{iu(G*\xi)(s)} - 1 - iu(G*\xi)(s)\right) dsdn(u)\right] \tag{5}$$

in which ξ is a test function, $dn(u)$ is Lévy's measure satisfying

$$\int_{-\infty}^{\infty} \frac{u^2}{1+u^2} dn(u) < \infty, \; dn(0) = 0.$$

We are now going to find out a characterization of a linear process $X(t)$ expressed in (4.1) by specifying it's characteristic function (4.2).

To fix the idea, we take the compound Poisson noise to be just a single noise with $dn(u) = \delta_1(u)$. Then, we have

$$C_P(\xi) = \exp\left[\int_0^{\infty} \left(e^{i(G*\xi)(s)} - 1 - i(G*\xi)(s)\right) ds\right]. \tag{6}$$

Here we note that the second term is non-random.

It can be seen that $C_X(\xi)$ never vanishes, so that we can define

$$c_X(\xi) = \log C_X(\xi) \tag{7}$$

by taking a branch that is 0 at $\xi = 0$. Then, we have

$$c_X(\xi) = c_B(\xi) + c_P(\xi), \tag{8}$$

where

$$c_B(\xi)) = \log C_B(\xi) = -\frac{1}{2}\int_0^t (F*\xi)(s)^2 ds \tag{9}$$

and

$$c_P(\xi) = \log C_P(\xi) = \int_0^t (e^{iG*\xi(s)} - 1 - i(G*\xi)(s))ds. \tag{10}$$

We have the functional derivative

$$(c_B)'_\xi(t) = -\int_0^t (F*\xi)(s)F(t,s)ds.$$

Similar computation leads us to prove

$$(c_B)''_{\xi\xi'}(t,t') = -\int_0^{t\wedge t'} F(t,s)F(t',s)ds.$$

This is equal to the covariance function, $\Gamma_B(t,t')$, of the process $X_G(t)$ up to minus sign.

Note 1. If $\Gamma_B(t,t')$ is known, we can find the canonical kernel by factorization assumming that the multiplicity of X_B is equal to one.

As for the functional $c_P(\xi)$, we have the functional derivative

$$(c_P)'_\xi(t) = i\int_0^t G(t,s)e^{i(G*\xi)(s)}ds - G(t,s).$$

Hence, we have

$$(c_P)''_{\xi\xi'}(t,t') = -\int_0^{t\wedge t'} G(t,s)G(t',s)e^{i(G*\xi)(s)}ds.$$

Set $\xi = 0$, then we have the covariance function $\Gamma_P(t,t')$ of the component process $X_P(t)$, up to minus sign.

Note 2. Here we can see that Poisson part is like the case of Gaussian part as is expressed in Note 1.

We are now given the second functional derivative of $c_X(\xi)$, which can be theoretically written as

$$\begin{aligned}(c_X)''_{\xi\xi'} &= (c_B)''_{\xi\xi'} + (c_P)''_{\xi\xi'} \\ &= -\int_0^{t\wedge t'} F(t,s)F(t',s)ds - \int_0^{t\wedge t'} G(t,s)G(t',s)\exp[i(G*\xi)](s)ds.\end{aligned}$$

Although the characteristic functional $C_X(\xi)$ is given, $(c_X)''_{\xi\xi'}$ is known however we do not yet know both $(c_B)''_{\xi\xi'}$ and $(c_P)''_{\xi\xi'}$ separately.

We, however, claim that the two terms in the above equation can be separated. The trick is as follows.

Let

$$\begin{aligned}\varphi(\xi,t) &= c_B(\xi) + c_P(\xi) \\ &= -\frac{1}{2}\int (F*\xi)^2(s)ds + \int \left(e^{i\int(G*\xi)(s)} - 1\right)ds \qquad (11)\end{aligned}$$

which is known. Its functional derivative is given by

$$\varphi'_\xi(\xi,t) = -\int F(t,s)(F*\xi)(s)ds + i\int G(t,s)e^{i(G*\xi)(s)}ds. \qquad (12)$$

Take a convolution with ξ,

$$\int \varphi'_\xi(\xi,t)\xi(t)dt = -\int (F*\xi)^2(s)ds + i\int (G*\xi)(s)e^{i(G*\xi)(s)}ds. \qquad (13)$$

From (4.11) and (4.13)

$$\varphi(\xi,t) - \frac{1}{2}\varphi'_\xi(\xi,t) = \int \left(e^{i(G*\xi)(s)} - 1 - \frac{i}{2}(G*\xi)(s)\right)ds$$

is obtained.

Let us denote it by $h(\xi, ut)$, and replace ξ with $u\xi, u \in R$. Then

$$\frac{d^2}{du^2}h(\xi, ut)\,|_{u=0} = -\int (G * \xi)^2(s)ds,$$

from which

$$\int (G * \xi_1)(s)(G * \xi_2)(s)ds$$

is obtained. Consequently

$$\int (G(t,s)G(t',s)ds$$

which is the covariance function $\Gamma_P(t,t')$ of Poisson part is also obtained.

According to the assumption that $G(t,s)$ is canonical kernel, it can be calculated by factorizing the covariance function $\Gamma_P(t,t')$. Thus we obtain the Poisson part and then naturally the remaining part which is Gaussian.

Summing up we have the following assertion.

Theorem 4.1. *If a linear process $X(t)$ is given, then Gaussian and Poisson parts, $X_B(t)$ and $X_P(t)$, are discriminated and the canonical kernels F and G are obtained.*

If the original kernels F and G, expressed in (4.1) are non-canonical kernels, the obtained kernels may not be the same as the original kernels. In this case, there is no problem for Gaussian part since the probability distribution is the same for both canonical and non-canonical kernels. But, for the Poisson part, the probability distributions may be different. Thus we need to obtain the original non-canonical kernel for Poisson part. For this purpose, we ignore the Gauusian part and use the sample function, as is discussed in Section 3, to obtain the original non-canonical kernel of the Poisson part.

References

1. L. Accardi, T. Hida and Si Si, Innovations for some stochastic processes. Volterra Center Notes, 2002.
2. T. Hida, Canonical representation of Gaussian processes and their applications, Memoirs Coll. Sc., Kyoto A33 (1960) 109-155
3. T. Hida, and Si Si, Innovations for random fields, Infinite Dimensional Analysis, Quantum Probability and Related Topics Vol 1 (1998), World Scientific, 499-509.

4. T. Hida and Si Si, Elemental random variables in White Noise Theory: beyond Reductionism, Quantum Information III, ed. T. Hida et. Al, World Scientific 2001, pp59-66
5. T. Hida and Si Si, Innovation approach to some random fields, Application of white noise theory, World Scientific (to appear)
6. T. Hida, Si Si and Win Win Htay, Variational calculus for random fields parametrized by a curve or surface, to appear in " Infinite Dimensional Analysis, Quantum Probability and Related Topics",World Scientific
7. P. Lévy, Fonctions aléatoires à corrélation linéaire, Illinois J. of Math. 1 (1957), 217-258.
8. T. Hida, Canonical representations of Gaussian processes and their applications Memoirs Coll. Sci., Univ. Kyoto A 33 (1960) 109-155.
9. T. Hida, Stationary Stochastic Processes, Math. Notes, Princeton Univ. Press. 1970.
10. T. Hida and Si Si, Innovation for random fields. Infinite Dimensional Analysis, Quantum Probability and Related Topics. 1 (1998), 499-509. Soc.Translations of Mathematical Monographs vol.12. 1993.
11. P. Lévy Théorie de L'addition des variables aléatoires, Gauthier-Villars, Paris.1937.
12. P. Lévy, Processus stochastiques et mouvement brownien. 2ème ed. (1965) Gauthier-Villars.
13. P. Lévy, A special problem of brownian motion, and a general theory of Gaussian random functions. Proc. Third Berkeley Symposium on Mahtematical Statistics and Probability, vol. II, (1956) 133 - 175.
14. Si Si, Topics on random fields, Quantum Information I, ed. T. Hida and K. Saito, World Scientific,1999, pp. 179-194.
15. Si Si, Gaussian processes and Gaussian random fields, Quantum Information III, ed. T. Hida and K. Saito. World Scientific 2001, pp. 195-204.
16. Si Si and Win Win Htay, Entropy in subordination and Filtering, Acta Applicandae Mathemathecae Vol.63, Kluwer Academic Publishers, pp 433-439.
17. Si Si, Innovation of the Lévy Brownian motion. Volterra Center Notes, N.475, May, 2001.
18. Selected Papers of Takeyuki Hida. (2001) World Scientific Pub. Co.

GROUP THEORY OF DYNAMICAL MAPS

E. C. G. SUDARSHAN
Department of Physics, University of Texas,
Austin, Texas 78712-1081USA
E-mail: sudarshan@physics.utexas.edu

Quantum stochastic processes are characterized in terms of completely nonnegative quantum dynamical maps which form a convex set. The canonical form of such a map is in terms of $R \leq N$, $N \times N$ matrices with their phase completely arbitrary. The maps constitute a convex set of which the extremal elements can be identified. Every such map may be viewed as the product of a unitary transformation in the adjoint representation, a classical stochastic semigroup in N variables followed by another unitary transformation.

The states of a quantum system with finitely many states is given by a density matrix ρ which is hermitian, nonnegative and of trace unity [1]. Among them pure states are distinguished by the density matrix reducing to a projection. Since every density matrix has a canonical decomposition into a probabilistic combination of pure density matrices, we see that the density matrices constitute a bounded convex set:

$$\rho = \sum_{\alpha} \lambda_\alpha \Pi^{(\alpha)} \quad ; \quad \Pi^{(\alpha)}\Pi^{(\beta)} = \delta_{\alpha,\beta}\Pi^{(\alpha)} \quad ; \quad \sum_{\alpha} \lambda_\alpha = 1 \ , \ \lambda_\alpha > 0. \tag{1}$$

The extremal elements are the pure states, which generate the convex set. Since every pure state corresponds to a ray generated by N complex numbers, $z_1, z_2, \ldots, z_N$ with

$$|z_1|^2 + |z_2|^2 + \ldots |z_N|^2 = 1, \tag{2}$$

$|z_N|$ is not independent of $z_1, z_2, \ldots, z_{N-1}$. Since the phase of z_N is arbitrary, it follows that we have a projective space of $N - 1$ complex variables: so the extremal states constitute the CP^{N-1} manifold. The volume of the CP^{N-1} manifold of states can easily be computed to be [2],

$$\text{Vol } CP^N = \frac{\pi^n}{n!}. \tag{3}$$

An isolated system will have a unitary evolution

$$\rho(t) = U(t)\rho U^\dagger(t) \quad ; \quad U(t) = \mathcal{T}\left\{\exp\left(-i\int_0^t H(t')\,dt'\right)\right\}. \tag{4}$$

But the most general evolution is by a stochastic dynamic map [3,4].

$$\rho(t) = \sum_{r,s} B_{rr',s's}(t)\rho_{r's'}. \tag{5}$$

This is a linear evolution and appropriate for an open quantum system. The requirements on the density matrix impose the following restriction on the dynamical matrix [6,7,8,5]:

$$B^*_{rr',s's} = B_{ss',r'r} \quad \text{(Hermiticity);} \tag{6}$$

$$\sum_r B_{rr',s'r} = \delta_{r's'} \quad \text{(Normalization);} \tag{7}$$

$$\sum_{rr',ss'} x^*_r y_{r'} B_{rr',s's} x_s y^*_{s'} \geq 0 \qquad \text{(Non-negativity).} \tag{8}$$

It follows that, considered as a matrix in the composite indices rr', $s's$, the dynamical matrix is hermitian and of trace N. It must therefore have a decomposition [3,4]

$$B_{rr',s's} = \sum_\alpha \mu_\alpha \xi_{rr'}(\alpha)\xi^*_{ss'}(\alpha) \tag{9}$$

with

$$\sum_\alpha \mu_\alpha = 1 \quad ; \quad \mathrm{tr}[\xi^\dagger(\alpha)\xi(\beta)] = 1. \tag{10}$$

The non-negativity constraint is not sufficient to assure that B itself is nonnegative. If this is to be satisfied, we must have

$$\sum_{rr',ss'} z^*_{rr'} B_{rr',s's} z_{ss'} \geq 0 \qquad \text{(Complete positivity).} \tag{11}$$

In this case $\mu_\alpha \geq 0$ and we could absorb them into the eigenvectors of B by redefining [9]

$$C_{rr'}(\alpha) = \sqrt{\mu_\alpha} \cdot \xi_{rr'}(\alpha). \tag{12}$$

The normalization condition for these completely positive maps takes the simple form

$$\sum_\alpha C^\dagger(\alpha)C(\alpha) = 1. \tag{13}$$

These completely positive maps may then be displayed in the Choi form [3,6,7,10]

$$\rho \longrightarrow \rho' = \sum_{\alpha} C(\alpha)\rho C^{\dagger}(\alpha). \tag{14}$$

Since the density matrices from a convex set and the completely nonnegative maps form a bounded convex set, it is then interesting to find the extremal maps in terms of which all maps could be expressed as convex combinations. Clearly the unitary maps

$$\rho \longrightarrow U\rho U^{\dagger} \tag{15}$$

belong to this distinguished extremal set. (Anti-unitary maps are not completely positive.) But there are many other non-unitary maps. The simplest of them is the pin map: it maps all density matrices to a fixed pure density matrix. It can be shown that all extremal maps may be characterized by $R \leq N$ matrices $C(\alpha)$ with the R^2 matrices $C^{\dagger}(\alpha)C(\beta)$ being linearly independent.

$$\sum_{\alpha,\beta} f_{\alpha\beta} C^{\dagger}(\alpha)C(\beta) = 0 \quad \Rightarrow \quad f_{\alpha\beta} = 0. \tag{16}$$

So $R \leq N$ determines the class of dynamical maps. A systematic construction of all such classes is given elsewhere [11,12]. We are now interested in studying the group properties of dynamical maps.

The generic density matrix may be written as

$$\rho = \frac{1}{N}\left(1 + \sum_{\alpha=1}^{N^2-1} \lambda(\alpha)x(\alpha)\right) \tag{17}$$

in terms of any set of N^2-1 linearly independent $N \times N$ hermitian traceless matrices that together with the unit matrix $\mathbf{1}$ spans all $N \times N$ hermitian matrices. These may be chosen as the generalized GellMann-Tilma [13] matrices. In this characterization all the $\lambda(\alpha)$ except for $\alpha = n^2-1$, $2 \leq n \leq N$ are defined as follows: For every $i, j = 1, 2, 3, \ldots N$; $i < j$ we define two $N \times N$ matrices

$$\begin{aligned} [\lambda^{\{1\}}(i,j)]_{\mu\nu} &= \delta_{j\mu}\delta_{i\nu} + \delta_{j\nu}\delta_{i\mu}, \\ [\lambda^{\{2\}}(i,j)]_{\mu\nu} &= -i(\delta_{i\mu}\delta_{j\nu} - \delta_{i\nu}\delta_{j\mu}). \end{aligned} \tag{18}$$

This furnishes $N(N-1)$ of the N^2-1 matrices that are needed. The remaining $N-1$ matrices are labelled $\lambda(n^2-1)$ and they have n diagonal elements, The first $n-1$ diagonal elements are all

$$[\lambda(n^2-1)]_{k,k} = \sqrt{\frac{2}{n^2-n}} \quad k < n \tag{19}$$

the nth element being

$$[\lambda(n^2-1)]_{n,n} = \sqrt{\frac{2}{n^2-n}} \times -(n-1); \tag{20}$$

the other elements being 0. All the $\lambda(\alpha)$ are normalized by $\text{tr}[(\lambda(\alpha))^2] = 2$.

The dynamical map is of the form

$$\rho \longrightarrow \rho' = \left(1 + \sum_{\alpha=1}^{N^2-1} \lambda(\alpha) y(\alpha)\right) \tag{21}$$

with

$$y(\alpha) = g_{\alpha\beta} x(\beta) + h_\alpha \tag{22}$$

where $g_{\alpha\beta}$ and h_α are real. Any density matrix ρ can be diagonalized by a unitary transformation to have at most N nonnegative eigenvalues which sum up to one. It is more useful to include $\lambda(N^2) = \left(\frac{2}{N}\right)^{1/2} \cdot \mathbf{1}$ among the set $\{\lambda(\alpha)\}$ so that we may write the transformation as a homogeneous linear transformation. At this stage it is also convenient to change the normalization to $\text{tr}[\mu(\alpha)\mu(\beta)] = \delta_{\alpha\beta}$; so $\mu(\alpha) = \frac{1}{\sqrt{2}}\lambda(\alpha)$, $\mu(N^2) = N^{-1/2}\mathbf{1}$. We rewrite the map as

$$\sum_1^{N^2} q(\alpha)\mu(\alpha) \rightarrow \sum_1^{N^2} r(\alpha)\mu(\alpha) \quad ; \quad r(N^2) = q(N^2) = \left(\frac{N}{2}\right)^{\frac{1}{2}} \tag{23}$$

We could diagonalize ρ and ρ' by suitable unitary transformations from $SU(N)$, which act as the adjoint representations on $\lambda(\alpha)$. So we may write

$$\mu(\alpha) \rightarrow M_{\alpha,\beta}\mu_\beta \quad ; \quad M_{N^2,N^2} = 1. \tag{24}$$

which can be rewritten as

$$M_{\alpha,\beta} = (uDv^\dagger)_{\alpha,\beta} \tag{25}$$

where u, v are elements of $SU(N)$ and

$$D_{\alpha,\beta} = 0 \quad ; \quad \alpha \neq \{3, 8, \ldots n^2-1\}. \tag{26}$$

The restricted transformations $D_{\alpha,\beta}$ connects only the N orthogonal projection $\Pi_1, \Pi_2, \ldots, \Pi_N$ to the diagonal matrices.

The matrices $D_{\alpha,\beta}$ are thus effectively a stochastic matrix [14] in N variables S_{jk}:

$$\sum_k S_{jk} = 1 \quad ; \quad S_{jk} \geq 0 \quad ; \quad 1 \leq j,k \leq N \tag{27}$$

These transformations yield a semigroup: the product of two classical stochastic matrices is a stochastic matrix.

The classical stochastic maps are completely defined by their action on the N projections Π_j. In fact, the Π_j with $\Pi_j \geq 0$ and $\sum \Pi_j = 1$ form homogeneous area (or hyper-volume) coordinates on a point in $N-1$ regular polytope with vertices at $(1,0,0,\ldots)$, $(0,1,0,\ldots)$, ..., $(0,0,\ldots,0)$. It may also be thought of as the intersection of the hyperplane $x_1+x_2+\ldots x_N = 1$ with the positive axes.

We are thus in a position to identify the group-structure of the quantum stochastic maps:

$$\mathcal{M} = u\Delta v^{\dagger} \quad ; \quad u, v \in SU(N) \tag{28}$$

and Δ is the classical stochastic semigroup:

$$\Pi_j \longrightarrow \sum_k S_{jk}\Pi_k \quad ; \quad \sum_k S_{jk} = 1\ ,\ S_{jk} \geq 0 \tag{29}$$

where Δ can be realized by a dynamical map.

The generic quantum dynamical maps have no inverse, but the product of two maps is a map. Hence the quantum dynamical maps form a semigroup. What we have shown is that it consists of two separate unitary maps and a linear classical stochastic map in N variables.We have thus filtered the semigroup of quantum process on the $N \times N$ density matrices into two unitary transformations and a classical stochastic semigroup on N dimensional probability vectors.

Stochastic Maps of Density Matrices

As we have said, the most general linear evolution of the density matrix of a system (elementary or composite) is given by a stochastic map:

$$\rho_{rs} \to \rho'_{rs} = B_{rr',ss'}\rho_{r's'} \tag{30}$$

with the properties

$$\begin{aligned} B^*_{rr',ss'} &= B_{ss',rr'}, \\ \sum_n B_{nr',ns'} &= \delta_{r',s'}, \\ x_r y^*_{r'} B_{rr',ss'} x^*_s y_{s'} &\geq 0 \end{aligned} \tag{31}$$

which are consequences of the hermicity, trace normalization, and non-negativity of the density matrices. Since B is hermitian it has an eigenvector decomposition

$$B_{rr',ss'} = \sum_{\alpha} \mu(\alpha)\xi_{rr'}(\alpha)\xi^*_{ss'}(\alpha). \tag{32}$$

If all the $\mu(\alpha)$ are positive the map is said to be completely positive. In that case we can define [9]

$$C_{rr'}(\alpha) = \sqrt{\mu(\alpha)} \cdot \xi_{rr'}(\alpha) \tag{33}$$

so that such maps may be displayed in the Choi form [3,6,7,10]:

$$\rho \to \rho' = \sum_{\alpha} C(\alpha)\rho C(\alpha)^\dagger. \tag{34}$$

Any completely positive map maybe viewed as the contraction of a unitary map

$$\rho \otimes q \to V\rho \otimes qV^\dagger, \tag{35}$$

with respect to the auxiliary system τ being traced over.

In these considerations, ρ may be an elementary system or a composite system consisting of one or more subsystems. For the case that it is composed of two qubits it has dimension 4 and the matrices are 4×4 matrices for which a basis can be chosen as the matrices in the adjoint representation of $U(4)$, the unitary group in 4 dimensions.

From Maps to Kossakowski Semigroups

The maps $\rho \to B\rho = \rho'$ are maps labeled by a continuous time parameter with the relation

$$B(t_1) \cdot B(t_2) = B. \tag{36}$$

This is a map, but it is not a one-parameter group:

$$B(t_1) \cdot B(t_2) \neq B(t_1 + t_2). \tag{37}$$

All these maps are contractive. One could ask whether we can derive a contractive parameterized semigroup. For this purpose we wish to construct the related semigroup. This construction *ab initio* was carried out by Kossakowski and independently by Lindblad [17,18,15,6,16]. We shall give a simple method of generating the semigroup from maps in the neighborhood of the identity.

Consider the map

$$\rho \to \sum_{\alpha} C(\alpha)\rho C(\alpha)^{\dagger} \tag{38}$$

in the vicinity of the identity. Then we could have

$$C(1) = 1 + D(1) \quad ; \quad C(n) = D(n) \quad n \geq 2. \tag{39}$$

Collecting terms up to the second degree in the quantity $D(\alpha)$ we get

$$\rho \to (1 + D(1))\rho(1 + D(1))^{\dagger} + \sum_{n\geq 2} D(n)\rho D(n)^{\dagger}. \tag{40}$$

with the 'trace' condition

$$(1 + D(1)^{\dagger})(1 + D(1)) + \sum_{n\geq 2} D(n)^{\dagger} D(n) = 1 \tag{41}$$

which may be rewritten as

$$D(1)\rho D(1)^{\dagger} + \sum_{n\geq 2} D(n)\rho D(n)^{\dagger} = \rho' - \rho = \Delta\rho. \tag{42}$$

Rewrite

$$\begin{aligned} D(1) &\to -iH + K(1) + \frac{(iH)^2}{2}, \\ D(1)^{\dagger} &\to iH + K(1) + \frac{(-iH)^2}{2}. \end{aligned} \tag{43}$$

The change in ρ is then

$$\begin{aligned} \Delta\rho = &-i[H,\rho] + H\rho H - \frac{1}{2}(H^2\rho + \rho H^2) + K(\alpha)\rho K(\alpha)^{\dagger} \\ &-\frac{1}{2}K(\alpha)^{\dagger}K(\alpha)\rho - \frac{1}{2}\rho K(\alpha)^{\dagger}K(\alpha) \\ = &\ i[H,\rho] + \frac{1}{2}[K(\alpha)\rho, K(\alpha)^{\dagger}] + \frac{1}{2}[K(\alpha), \rho K(\alpha)^{\dagger}] \end{aligned} \tag{44}$$

which is the Kossakowski formula [17,18,15,6,16,19].

We now have a simple characterization of the extremal maps. The maps constitute a convex set and the extremal maps have rank $2 \leq R \leq N$. The semigroups have the structure of a 'convex cone', the extremal rays being associated, one to one, with each of the extremal convex maps. The connection is straightforward - similarly all the Kossakowski semigroups are completely positive.

Having found the generic form for the semigroup we could specialize it to the case of two coupled qubits in interaction with an external bath:

$$\rho \equiv \rho_{ab},$$
$$\frac{\Delta\rho}{\Delta t} = -i[H,\rho] + D(\alpha)\rho D(\alpha)^\dagger - \frac{1}{2}(D(\alpha)^\dagger D(\alpha)\rho + \rho D(\alpha)^\dagger D(\alpha)) \quad (45)$$

where H, $D(\alpha)$ are $SU(4)$ matrices acting on the state space of AB. If we denote the matrices of A and B by σ and τ, these all have the form:

$$\vec{a}\cdot\vec{\sigma}\otimes 1 + 1\otimes\tau\cdot\vec{b} + \vec{\sigma}\cdot c\cdot\vec{\tau} + d1\otimes 1. \quad (46)$$

What if we restrict our attention to one qubit and consider the other qubit as a part of an extended reservoir? The density matrix of one qubit is obtained by taking the partial trace with respect to the second (reservoir) qubit:

$$\rho \to \tilde{\rho} = tr_B\rho. \quad (47)$$

The Kossakowski semigroup acting on the reduced density matrix

$$\frac{\Delta\tilde{\rho}}{\Delta t} = -i[h,\rho] + d(\beta)\tilde{\rho}d(\beta)^\dagger - \frac{1}{2}d(\beta)^\dagger d(\beta)\tilde{\rho} - \frac{1}{2}\tilde{\rho}d(\beta)d(\beta)^\dagger \quad (48)$$

where all the operators are 2×2 matrices of $U(2)$ acting only on the first qubit. How may we obtain $\{h, d(\beta)\}$ from $\{H, D(\alpha)\}$?

If H and $K(\alpha)$ are simply separable, that is the $U(4)$ matrices are direct products of $U(2)\otimes U(2)$ matrices, all we need to do is to take the partial traces:

$$h = tr_B H,$$
$$d(\beta) = tr_B D(\beta). \quad (49)$$

If the operators are separable but not simply separable we have the following structures

$$H = \sum_n h^{(n)} \otimes j^{(n)},$$
$$D(\alpha) = \sum_n d(\alpha)^{(n)} \otimes e(\alpha)^{(n)} \quad (50)$$

it is not sufficient to take partial traces, but we may replace

$$h = \sum_n h^{(n)} tr_B j^{(n)},$$
$$d(\alpha, n) = \sum_n d(\alpha)^{(n)} tr_B e(\alpha)^{(n)}. \quad (51)$$

So the number of Kossakowski generators is labeled by (n, α) in place of α.

But the more interesting case is to consider the case of non-separable generators H and $D(\alpha)$. In this case, the computation is more elaborate for the Hamiltonian H can be written as

$$H_{ra,sb} \rightarrow H_{rs,ab} = F(H_{ra,sb}) \tag{52}$$

where F is a Fierz recoupling matrix [20]. This maybe obtained by using a complete set of matrices $\{1, \sigma\}$ which satisify:

$$1_{ra} \times 1_{sb} + \vec{\sigma}_{ra} \otimes \vec{\sigma}_{sb} = 2 * 1_{rs} \otimes 1_{ab}. \tag{53}$$

This enables us to write ρ, H, and $D(\alpha)$ in the exchanged indices by Fierz recoupling

$$\rho_{ra,sb} = R_{rs,ab} \ , \ H_{ra,sb} = H_{rs,ab},$$

thus

$$D_{ra,sb}(\alpha) \rightarrow D_{rs,ab}. \tag{54}$$

Now we may rewrite the semigroup generator and density matrix in the index-exchanged form. This new form allows us to take the partial trace with respect to $a = b$ in terms of the *dynamical map*

$$\rho_{rs} \rightarrow B_{rr',ss'} \rho_{r's'} \tag{55}$$

in the Ritz form [21]

$$\rho_{rs} \rightarrow B_{rs,r's'} \rho_{r's'}. \tag{56}$$

Explicit computation with special Hamiltonians can now be carried out, and will be presented elsewhere [22].

Recoupling Identity and Protocol

Consider $R \times R$ matrices. Let $\{e_{ab}(n)\}$ be a set of R^2 matrices which are linearly independent. They would then span the $R \times R$ matrix space. There must exist R^2 matrices $\{f_{rs}(n)\}$ which are dual to them:

$$tr\{f(n')e(n)\} = \delta_{nn'} \equiv f_{ab}(n')R_{ba}(n). \tag{57}$$

Then any $R \times R$ matrix M can be expanded as

$$M_{pr,sq} = \sum a_{p,q}(n)e_{r,s}(n). \tag{58}$$

Then the expansion coefficients, $a(n)$, may be calculated by computing

$$M_{rs}f_{sr}(n') = a(n)e_{rs}(n)f_{sr}(n') = a(n)\delta_{nn'} = a_{n'}. \tag{59}$$

Hence

$$M_{rs} = \sum e_{rs}(n) M_{ab} f_{ba}(n). \tag{60}$$

This is true for every M. It follows that

$$\sum e_{rs}(n) f_{ba}(n) = \delta_{ar}\delta_{sb}. \tag{61}$$

This is the recoupling identity. Using it we can replace

$$\rho_{rs} \to B_{ra,sb}\rho_{ab} \tag{62}$$

with

$$\rho_{rs} \to A_{ab,rs}\rho_{ab}. \tag{63}$$

In this form we have (ab) or (rs) treated as matrix indices. Further, we can use this form in bipartite systems to recouple indices:

$$M_{ar,bs} \to N_{ab,rs}. \tag{64}$$

In this form, the partial traces are easily done; one simply takes the trace with respect to one set of indices.

References

1. J. von Neumann, *Mathematical Foundations of Quantum Mechanics.* Princeton University Press, Princeton NJ (1955)
2. L.J. Boya, E. C. G. Sudarshan, and T. Tilma, *Volumes of compact manifolds*, Rev. of Math. Phys. **52**, 401 (2003)
3. E. C. G. Sudarshan, P. M. Mathews, and J. Rau, *Stochastic dynamics of quantum mechanical systems*, Phys. Rev. **121**, 920 (1961)
4. K. Kraus, *General state changes in quantum theory*, Ann. Phys. **64**, 311 (1971)
5. E. Stromer, Acta. Math. **110**, 232 (1963)
6. M. D. Choi, *Positive linear maps of C^* algebras*, Can. J. Math. **24**, 520 (1972)
7. M. D. Choi, *A Schwarz inequality for positive linear maps on C^* algebras*, Ill. J. Math. **18**, 565 (1974)
8. E. B. Davis, *Quantum Theory of Open Systems.* Academic Press, London (1976)
9. E. C. G. Sudarshan, and A. Shaji, *Structure and parameterization of stochastic maps of density matrices*, J. Phys. A: Math. Gen. **36**, 5073 (2003)
10. K. Kraus, *States Effects and Operators: Fundamental Notions of Quantum Theory.* Springer Verlag, Berlin (1983)
11. E. C. G. Sudarshan *Quantum Measurements and Dynamical Maps* in "From SU(3) to Gravity" ed. E. Gotsman and G. Tauber, Cambride Univerisy Press, Cambridge (1986)

12. E. C. G. Sudarshan, *Evolution and decoherence in finite level systems*, Chaos, Solitons, and Fractals **16**, 369 (2003)
13. T. Tilma, Ph.D. Thesis, Univeristy of Texas (2002) as well as T. Tilma and E. C. G. Sudarshan *Generalized Euler angle parameterization for SU(N)*, J. Phys. A: Math. Gen. **35**, 10467 (2002)
14. A. Ramakrishnan, and N. R. Ranganathan, J. Math. Anal. Appl. (1958)
15. A. Kossakowski, *On quantum statistical mechanics of non-Hamiltonian systems*, Rep. Math. Phys. **3**, 247 (1972)
16. G. Lindblad, *Completely positive maps and entropy inequalities*, Commun. Math. Phys. **40**, 147 (1975)
17. V. Gorini, A. Kossakowski, and E. C. G. Sudarshan, *Completely positive dynamical semigroups of N-level systems*, J. Math. Phys. **17**, 821 (1976)
18. V. Gorini, F. Verri, and A. Kossakowski, *Properties of quantum Markovian master equations*, Rep. Math. Phys. **13**, 149 (1978)
19. E. B. Davis, *Quantum stochastic processes I*, Commun. Math. Phys. **15**, 277 (1969); *Quantum stochastic processes II*, Commun. Math. Phys. **19**, 83 (1970); *Quantum stochastic processes III*, Commun. Math. Phys. **22**, 51 (1971)
20. M. Fierz, *Attraction of conducting planes in a vacuum*, Helv. Phys. Acta. **33**, 855 (1960)
21. E. C. G. Sudarshan, *Quantum dynamics, metastable states, and contractive semigroups*, Phys. Rev. A **46**, 37 (1992)
22. T. Jordan, A. Shaji, and E. C. G. Sudarshan, *Dynamics of initially entangled open quantum systems*, arXiv:quant-ph/0407083

ON QUANTUM ANALYSIS, QUANTUM TRANSFER-MATRIX METHOD, AND EFFECTIVE INFORMATION ENTROPY

MASUO SUZUKI

Department of Applied Physics, Tokyo University of Science,
1-3 Kagurazaka, Shinjuku-ku, Tokyo 162-8601
E-mail: msuzuki@rs.kagu.tus.ac.jp

The purpose of the present paper is to review the quantum analysis and exponential product formulas, together with the quantum transfer-matrix method and to propose the concept of effective information "entropy".

1. Introduction

The quantum analysis and the quantum transfer-matrix method (QTM) are briefly reviewed in Sections 2 and 3, respectively. In Section 4, we introduce effective information "entropy".

2. Quantum Analysis

We have introduced the following quantum derivative[1-3] of an operator function $f(A)$ with respect to the operator A itself:

$$\frac{df(A)}{dA} = \frac{f(A) - f(A - \delta_A)}{\delta_A} = \frac{\delta_{f(A)}}{\delta_A} = \int_0^1 f^{(1)}(A - t\delta_A)dt, \tag{2.1}$$

where $f^{(n)}(x)$ denotes the nth derivative of $f(x)$ with respect to an ordinary number x. Here δ_A is the inner derivation defined by

$$\delta_A B = [A, B] = AB - BA. \tag{2.2}$$

Then, the Gâteaux differential $df(A)$ of $f(A)$ defined by

$$df(A) = \lim_{h\to 0} \frac{f(A + hdA) - f(A)}{h} \tag{2.3}$$

is expressed as

$$df(A) = \frac{df(A)}{dA} \cdot dA\ . \tag{2.4}$$

Namely the quantum derivative $df(A)$ is a linear hyper-operator to map the operator dA into $df(A)$.

Similarly the nth quantum derivative $d^n f(A)/dA^n$ is expressed by

$$\frac{d^n f(A)}{dA^n} = n! \int_0^1 dt_1 \int_0^{t_1} dt_2 \cdots$$

$$\int_0^{t_{n-1}} dt_n f^{(n)}(A - t_1\delta_1 - t_2\delta_2 \cdots - t_n\delta_n), \quad (2.5)$$

where δ_j is a hyper-operator defined as

$$\delta_j : B^n = \delta_j : B \cdot \cdots \cdot B = B^{j-1}(\delta_A B)B^{n-j}. \quad (2.6)$$

The following general operator Taylor expansion formula has been derived

$$f(A + xB) = \sum_{n=0}^{\infty} \frac{x^n}{n!} \frac{d^n f(A)}{dA^n} \ : \ B^n$$

$$= f(A) + \sum_{n=1}^{\infty} x^n \int_0^1 dt_1 \int_0^{t_1} dt_2 \cdots$$

$$\int_0^{t_{n-1}} dt_n f^{(n)}(A - t_1\delta_1 - t_2\delta_2 \cdots - t_n\delta_n) \ : \ B^n. \quad (2.7)$$

From this general formula, we can easily obtain the wellknown Feynman expansion formula on $e^{t(A+xB)}$ as

$$e^{tA+xB)} = \sum_{n=0}^{\infty} \frac{x^n}{n!} \frac{d^n e^{tA}}{dA^n} \ : \ B^n$$

$$= e^{tA} \sum_{n=0}^{\infty} x^n \int_0^t dt_1 \int_0^{t_1} dt_2 \cdots \int_0^{t_{n-1}} dt_n B(t_1)B(t_2)\cdots B(t_n), \quad (2.8)$$

where

$$B(t) = e^{-t\delta_A} B = e^{-tA} B e^{tA}. \quad (2.9)$$

Many other explicit operator expansion formulas have been derived [1–3] and been applied to physical systems[4,5].

The above quantum analysis is useful in evaluating commutation relations such as $[f(A),\, h(B)]$:

$$[f(A),\, h(B)] = \delta_{a(A)} h(B) = \frac{df(A)}{dA} \delta_A h(B)$$

$$= -\frac{df(A)}{dA}[h(B),\, A] = -\frac{df(A)}{dA}\delta_{h(B)}A$$

$$= -\frac{df(A)}{dA}\cdot\frac{dh(B)}{dB}\delta_B\cdot A = \frac{df(A)}{dA}\frac{dh(B)}{dB}[A,\, B]$$

$$= \int_0^1 dt \int_0^1 ds f^{(1)}(A - t\delta_B)h^{(1)}(B - s\delta_B)[A,\, B]. \tag{2.10}$$

where we have used repeatedly the formula (2.1), namely

$$\delta_{f(A)} = \frac{df(A)}{dA}\delta_A. \tag{2.11}$$

In particular, we have

$$[e^{\lambda A},\, e^{\mu B}] = \int_0^\lambda dt \int_0^\mu ds\; e^{(\lambda-t)A}e^{(\mu-s)B}[A,\, B]e^{sB}e^{tA}. \tag{2.12}$$

The foumula (2.12) was derived previously[6)] in an adhoc way. The above derivation is quite systematic.

3. Quantum Transfer-Matrix Method based on Exponential Product Formulas and its Connection with Quantum Information

The transfer-matrix method has been used very often in studying rigorously classical systems such as the Ising model. Here we review the quantum transfer-matrix method (QTM) from a viewpoint of quantum information propagation.

In order to study theoretically how quantum information propagates, we have to evaluate the wave function at time t:

$$\Psi(t) = \exp(-it\mathcal{H}/\hbar)\Psi(0) \tag{3.1}$$

when the initial wave function $\Psi(0)$ is given. Here, $\mathcal{H}$ denotes the Hamiltonian of the relevant system to describe quantum information propagation. If the Hamiltonian $\mathcal{H}$ is composed of the non-commutative subhamiltonian $\mathcal{H}_{12}, \mathcal{H}_{23}, \cdots \mathcal{H}_{N-1,N}$, then we can make use of the exponential product formulas[7−10)]:

$$\mathrm{e}^{-\beta\mathcal{H}} = \lim_{m\to\infty}\left(\mathrm{e}^{-\frac{\beta}{m}\mathcal{H}_{12}}\mathrm{e}^{-\frac{\beta}{m}\mathcal{H}_{23}}\cdots\mathrm{e}^{-\frac{\beta}{m}\mathcal{H}_{N-1,N}}\right)^m \tag{3.2}$$

with $\beta = it\mathcal{H}/\hbar$. As was first discussed analytically by the present author[11,12], the following quantum transfer matrix $\mathcal{T}_m$ can be defined as

$$Z(\beta) = \text{Tr } e^{-\beta\mathcal{H}} = \lim_{m\to\infty} \text{Tr } \mathcal{T}_m^N \tag{3.3}$$

by rearranging the ST-transformed[8-12] $(d+1)$-dimensional calssical system from the real space to the virtual space[11] in which the quantum transfer matrix $\mathcal{T}_m$ is constructed. Namely, this QTM transfers quantum information in the virtual space to the direction of the real space. The most dominant quantum information in this system is expressed by the maximum eigenvalue $\lambda_{\max}^{(m)}$ of the QTM $\mathcal{T}_m$:

$$\lim_{N\to\infty} \frac{1}{N} \log Z(\beta) = \lim_{m\to\infty} \log \lambda_{\max}^{(m)} , \tag{3.4}$$

using the exchangeability theorem[11,12] on the two limits $N \to \infty$ and $m \to \infty$.

By the way, we remark here the recent development of our study on generalization of exponential product formulas[13-15,3,5]. For non-commutative operators A and B, we have

$$\begin{aligned} e^{x(A+B)} &= e^{xA}e^{xB} + O(x^2) \\ &= e^{\frac{x}{2}A}e^{xB}e^{\frac{x}{2}A} + O(x^3) \\ &= S_m(x) + O(x^{m+1}) \end{aligned} \tag{3.5}$$

A systematic method to construct the m-th order decomposition formula was first discovered by the present author[13-15] using the recursion formula:

$$S_m(x) = S_{m-1}(p_1x)\, S_{m-1}(p_2x) \cdots S_{m-1}(p_rx) \tag{3.6}$$

with the conditions on the decomposition parameters $\{p_j\}$:

$$p_1 + p_2 \cdots + p_r = 1 \quad \text{and} \quad p_1^m + p_2^m + \cdots + p_r^m = 0 \ . \tag{3.7}$$

In particular, we have the following higher-order formulas based on the recursion formulas:

$$S_{2m}(x) = S_{2m-2}(k_mx)\, S_{2m-2}\Big((1-2k_m)x\Big)\, S_{m-2}(k_mx) \tag{3.8}$$

with[14]

$$k_m = \frac{1}{2 - 2^{1/(2m-1)}} \tag{3.9}$$

and the standard scheme

$$S_{2m}^*(x) = \Big(S_{2m-2}^*(p_m x)\Big)^2 S_{2m-2}^*\Big((1-4p_m)x\Big)\Big(S_{2m-2}^*(p_m x)\Big)^2 \qquad (3.10)$$

with[13,15]

$$p_m = \frac{1}{4 - 4^{1/(2m-1)}} \quad ,$$

starting with the second-order symmetric decomposition

$$S_2^*(x) = S_2(x) = \mathrm{e}^{\frac{x}{2}A}\mathrm{e}^{xB}\mathrm{e}^{\frac{x}{2}A} \quad . \qquad (3.11)$$

The standard scheme (3.10) is stable and consequently more effective in practical applications, because $0 < p_m < 1$ and $|1 - 4p_m| < 1$.

The above general schemes proposed by the present author have been used not only in physics but also in the field of differential equations and their numerical solutions[16].

4. Effective Information Entropy

Shannon's information "entropy"[17] S_{I} takes the form

$$S_{\mathrm{I}} = -\sum_i p_i \log p_i \quad , \qquad (4.1)$$

where p_i denotes the probability for the phenomenon i to occur. This form is the same as Boltzmann's entropy, but they are quite different in their contents. The latter is related to the energy of the relevant system, or a physical force, while Shannon's "entropy" is not necessarily related to such quantities, but it may be related to information power to controle communities.

As is clear for physicists, Boltzmann's entropy is defined and used *only for large system-size* ($N \to \infty$). On the other hand, Shannon's information entropy S_{I} is more effective *for its smaller values*, because S_{I} denotes uncertainty of a stochastic system desceribed by the probability set $\{p_j\}$. When the system is determined to be in a specific state i, namely $p_i = 1,\ p_j = 0$ for $j \neq i$, Shannon's entropy $S_{\mathrm{I}} = 0$. This indicates the decrease of uncertainty, namely gain of information, as in Shannon's paper[17].

The folowing remarks will be instructive for people in the field of information theory.

The increase of Boltzmann's entropy S_{B}, namely dS is accompanied by the heat increase dQ through the relation $dS = dQ/dT$ in a quasi-static process, as is well known. Here it should be emphasized that the increase of it has a positive meaning in an infinite system or in the thermodynamic limit $N \to \infty$. Each specific state in thermal equilibrium has no meaning from a viewpoint of information, but it contributes only to a thermal average by an infinitesimally small amount. In this sense, Boltzmann's entropy S_{B} is often called to give a measure of complexity of the relevant system.

Thus, S_{B} describes the whole situation of *particles* as a background, while S_{I} describes the behaviour of holes in the sea of *particles*. If the number of *holes* is small enough to be *accessible in appropriate time scale*, then these holes can play a role of information. Here we may define information as knowledge available and accessible in appropriate time scale.

Next we introduce the concept of effective information "entropy" S_{EI}, which is defined by the product of the information "entropy" S_{I} and some searching factor $S(N)$, namely

$$S_{\mathrm{EI}} = S_{\mathrm{I}}\, S(N) \tag{4.2}$$

Here, $S(N)$ has a qualitative factor which depends on the procedure to search information and which satisfies the conditions $S(1) = 1$ and $S(\infty) = 0$. The second condition $S(\infty) = 0$ means that an infinite number of information corresponds to zero information practically. (We often experience such situtations in which too much information may spoil creativity of scientists.) Then, the effective information will be optimized for a certain range of N.

>From the above arguments, we may say that the above two concepts (of "entropy") are *complementary* to each other, even when S_{I} is embedded in materials such as mesoscopic elements and consequently the two concepts of "entropy" have the common region of applicability. This remark may be related to L. Brillouin's arguments (1956) on constrained information and negentropy.

5. Summary and Discussion

In Section 2, we have reviewed briefly quantum analysis, namely quantum derivative on operator-valued functions. This is useful in deriving many practical formulas such as (i) the time-evolution of the entropy function[2–4], (ii) Kubo's indentity[5] which is useful in deriving the KMS condition and the linear response theory, (iii) the operator Taylor expansion formula, and (iv) generalized higher-order exponential product formulas.

In particular, the ST-transformation gives[8] the equivalence theorem that a d-dimensional quantum system *with finite-range interactions* is transformed into the corresponding $(d+1)$-dimensional classical system *with finite-range interactions.* This equivalence theorem has been used frequently in computational physics and also even conceptually inspired Parisi and Wu to formulate stochastic quantization (1981).

We have proposed the concept of effective information "entropy". In practical applications of continuous information, relative information "entropy" is effectively used[18].

Acknowledgements The author would like to thank Professor H-G. Matuttis for informing Ref.15, and also thank Professor H. DeRaedt for useful comments on the draft.

References

1. M. Suzuki: *Quantum analysis - Noncommutative differential and intergral calculi*, Commun. Math. Phys. **183** (1997), 339-363.
2. M. Suzuki: *General Formulation of Quantum Analysis*, Rev. Math. Phys. **11** (1999) 243-265.
3. M. Suzuki: *Quantum analysis, exponential product formulas and stochastic processes*, in"Trends in Contemporary Infinite Dimensional Analysis and Quantum Probability (L. Accardi, Hui-H Kuo, N. Obata, K. Saito, S. Si and L. Streit, eds.)", Institute Italiano di Culture, Kyoto, 2000.
4. M. Suzuki: *Nonlinear Responses in Magnetic Systems*, in Fromtiers in Magnetism, J. Phys. Soc. Jpn **69** (2000) Suppl. A. 156-159.
5. M. Suzuki: *Separation of non-commutative procedures-exponential product formulas and quantum analysis*, ed. N. Obata, T. Matsui, and A. Hora, in The Crossroad of Non-Commutativity, Infinite-Dimensionality and Probability (World Scientific, Singapore, 2002).
6. M. Suzuki: *Decomposition formulas of exponential operators and Lie exponentials with some applications to quantum mechanics and statistical physics*, J. Math. Phys. **26** (1985) 601-612.
7. M. Suzuki: *Generalized Trotter's Formula and Systematic Approximants of Exponential Operators and Inner Derivations with Applications to Many-Body Problems*, Commun. Math. Phys. **51** (1976) 183.
8. M. Suzuki: *Relationship between d-Dimensional Quantal Spin systems and (d+1)- Dementsional Ising Systems — Equivalence, Critical Exponents and Systematic Approximants of the Partition Functions and Spin Correlations* —, Prog. Theor. Phys. **56** (1976) 1454.
9. M. Suzuki: *Quantum Monte Corlo Methods*, ed., Solid State Sciences, Vol.

74, Springer, Berlin, 1986.

10. M. Suzuki: *Quantum Monte Carlo Methods in Condensed Matter Physics*, ed., World Scientific, Singapore, 1993.
11. M. Suzuki: *Transfer-Matrix Method and Monte Carlo Simulation in Quanum Spin systmes*, Phys. Rev. **B31** (1985) 2957.
12. M. Suzuki and M. Inoue: *The ST-Transformation Approach to Analytic Solutions of Quantum Systems. I — General Formulations and Basic Limit Theorems —*, Prog. Theor. Phys. **78** (1987) 787.
13. M. Suzuki: *Fractal Decomposition of Exponential Operators with Many-Body Theories and Monte Carlo Simulations*, Phys. Lett. **A146** (1990) 319.
14. M. Suzuki: *General theory of fractal path integrals with applications to many-body theories and statistical physics*, J. Math. Phys. **32** (1991) 400.
15. M. Suzuki: *General Theory of Higher-Order Decomposition of Exponential Operators and Symplectic Integrators*, Phys. Lett. **A165** (1992) 387.
16. E. Hairer, C. Lubich and G. Wanner: *Geometric Numerical Integration — Structure-Preserving Algorithms for Ordinaly Differential Equations*, Springer (2002). In this book, the present recursive schemes are explained in detail. However, there is an incorrect statement in the fifth line from below in page 45. (The recursive sequence (4.4) in page 40 was frist found in Refs. 12 and 13 of the present paper. Namely the fractal structure of the decomposition was found only by the present author.
17. C. E. Shannon: *A Mathematical Theory of Communication*, Bell system Tech. J. **27** (1948) 379 and 623.
18. M. Phya and D. Pez: *Quantum Entropy and its Use*, Springer (1993) and their articles in the present book.

NONEQUILIBRIUM STEADY STATES FOR A HARMONIC OSCILLATOR INTERACTING WITH TWO BOSE FIELDS - STOCHASTIC LIMIT APPROACH AND C* ALGEBRAIC APPROACH -

S. TASAKI

Advanced Institute for Complex Systems and Department of Applied Physics, Waseda University, 3-4-1 Okubo, Shinjuku-ku, Tokyo 169-8555, Japan
E-mail: stasaki@waseda.jp

L. ACCARDI

Centro Vito Volterra, Universita' di Roma Tor Vergata, 00133 Rome, Italy
E-mail: accardi@volterra.uniroma2.it

Recently, nonequilibrium steady states (NESS) are intensively studied by several methods. In this article, NESS of a harmonic oscillator interacting with two free-boson reservoirs at different temperatures and/or chemical potentials are studied by the stochastic limit approach and C* algebraic approach. And their interrelation is investigated.

1. Introduction

The understanding of irreversible phenomena including nonequilibrium steady states (NESS) is a longstanding problem of statistical mechanics. Various theories have been developed so far[1]. One of promising approaches deals with infinitely extended dynamical systems[2,3,4]. Not only equilibrium properties, but also nonequilibrium properties has been rigorously investigated. The latter include analytical studies of nonequilibrium steady states, e.g., of harmonic crystals[5,6], a one-dimensional gas[7], unharmonic chains[8], an isotropic XY-chain[9], systems with asymptotic abelianness[10,11], a one-dimensional quantum conductor[12], an interacting fermion-spin system[13] and junction systems[14]. Entropy production has been rigorously studied as well (see also Refs.15-19 and the references therein).

The stochastic limit approach[20,21,22] provides convenient tools to investigate return-to-equilibrium, where the density matrices reduced to the system obey Pauli's master equation, and was generalized[21,22] so that it can deal with external degrees of freedom, which turned out to be quantum white noises. Recently, this approach was extended to systems arbitrarily far from equilibrium[23,24] and is expected to provide a new practically useful tool for dealing with nonequilibrium behaviors. However, the relation between the two appraches is not clear and, in this paper, we compare them for a harmonic oscillator linearly coupled with two harmonic reservoirs at different temperatures and/or chemical potentials.

The paper is arranged as follows In the next section, the model is described. In Sec. 3, the results of the stochastic limit approach is reviewed. In Sec. 4, nonequilibrium steady states (NESS) are derived with the aid of the C*-algebraic approach. In Sec. 5, their weak coupling limits are investigated. The last section is devoted to the summary.

2. Model

To illustrate the relation between the C*-algebraic and stochastic limit approaches, we consider a harmonic oscillator linearly coupled with two harmonic reservoirs[24], which is a typical junction system studied in Ref. 14.

The model is defined on a tensor product of a Hilbert space $L^2(\mathbf{R})$ and two Fock spaces $\mathcal{H}_n$ $(n = 1, 2)$, both of which are constructed from $L^2(\mathbf{R}^3)$ $\mathcal{H} \equiv L^2(\mathbf{R}) \otimes \mathcal{H}_1 \otimes \mathcal{H}_2$. In terms of standard annihilation operators a and $a_{n,k}$ satisfying canonical commutation relations

$$[a, a^\dagger] = 1, \quad [a_{n,k}, a^\dagger_{n',k'}] = \delta_{n,n'}\delta(k - k') \ , \tag{1}$$

the Hamiltonian is given by

$$H = H_0 + \lambda \sum_{n=1,2} V_n \tag{2}$$

where

$$H_0 = \Omega a^\dagger a + \sum_{n=1,2} \int dk \omega_k a^\dagger_{n,k} a_{n,k} \ , \tag{3}$$

$$V_n = \int dk \left(g^*_n(k) a^\dagger a_{n,k} + g_n(k) a a^\dagger_{n,k} \right) \ . \tag{4}$$

In the above, $\omega_k = |k|^2$, k-integrations are taken over $\mathbf{R}^3$ and the functions $g_n(k)$ $(n = 1, 2)$ are square integrable. Strictly speaking, the free field parts $\int dk \omega_k a^\dagger_{n,k} a_{n,k}$ $(n = 1, 2)$ should be understood as the second quantization of the multiplicative self-adjoint operator $\varphi(k) \to \omega_k \varphi(k)$ defined on the Fourier transform of $L^2(\mathbf{R}^3)$ to the Fock spaces $\mathcal{H}_n$ $(n = 1, 2)$, respectively.

At initial states, the two fields and the harmonic oscillator are assumed to be independent of each other. Moreover, the two fields are assumed to obey Gibbsian distributions with different temperatures $\beta_1^{-1}, \beta_2^{-1}$ and chemical potentials μ_1, μ_2. Namely, one has

$$\omega_0(a_{n,k}^\dagger a_{n',k'}) = \delta_{n,n'} \mathcal{N}_n(\omega_k) \delta(k - k') , \tag{5}$$

where

$$\mathcal{N}_1(\omega_k) = \frac{1}{e^{\beta_1(\omega_k - \mu_1)} - 1}, \quad \mathcal{N}_2(\omega_k) = \frac{1}{e^{\beta_2(\omega_k - \mu_2)} - 1}. \tag{6}$$

In this paper, we only consider the case where $\mathcal{N}_n(\omega_k)$ $(n = 1, 2)$ are bounded.

Provided $g_n(-k) = g_n(k)^*$, the system is invariant under the time reversal operation ι, which is an antilinear involution satisfying

$$\iota a \iota = a , \quad \iota a^\dagger \iota = a^\dagger \tag{7}$$

$$\iota a_{n,k} \iota = a_{n,-k} , \quad \iota a_{n,k}^\dagger \iota = a_{n,-k}^\dagger . \tag{8}$$

Under the time evolution generated by H, the mass and energy are conserved. Indeed, in the Heisenberg representation, one has

$$\frac{d}{dt} a^\dagger a = \sum_{n=1,2} J_n , \tag{9}$$

where J_n $(n = 1, 2)$ are given by

$$J_n = i\lambda \int dk \left(g_n(k) a_{n,k}^\dagger a - g_n(k)^* a^\dagger a_{n,k} \right) , \tag{10}$$

and they correspond to the mass flows from the reservoirs since formally $J_n = -\frac{d}{dt} \int dk a_{n,k}^\dagger a_{n,k}$ holds. Similarly, the energy conservation holds

$$\frac{d}{dt} \Omega a^\dagger a = \sum_{n=1,2} \Omega J_n , \tag{11}$$

$$\frac{d}{dt} V_n = J_n^\epsilon - \Omega J_n , \tag{12}$$

where J_n^ϵ $(n = 1, 2)$ are given by

$$J_n^\epsilon = i\lambda \int dk \omega_k \left(g_n(k) a_{n,k}^\dagger a - g_n(k)^* a^\dagger a_{n,k} \right) , \tag{13}$$

and they correspond to the energy flows from the reservoirs since formally $J_n^\epsilon = -\frac{d}{dt} \int dk \omega_k a_{n,k}^\dagger a_{n,k}$ holds. From Eqs.(11) and (12), one finds that J_1^ϵ, ΩJ_1, $-\Omega J_2$ and $-J_2^\epsilon$ correspond, respectively, the energy flows from the first reservoir to the first joint, from the first joint to the oscillator, from the oscillator to the second joint and from the second joint to the second reservoir.

3. Stochastic Limit Approach

Here we summarize the results of the stochastic limit approach obtained by Accardi, Imafuku and Lu[24].

In the stochastic limit approach[21,22], one studies the rescaled evolution operator in the interaction picture

$$U^{(\lambda)}_{t/\lambda^2} \equiv \exp\left(itH_0/\lambda^2\right)\exp\left(-itH/\lambda^2\right) , \tag{14}$$

which is shown to converge, in the sense of correlations under appropriate assumptions on the model[22], to the solution of

$$\frac{d}{dt}U_t = -ih_tU_t, \quad U(0) = 1 \tag{15}$$

where

$$h_t = \lim_{\lambda\to 0}\frac{1}{\lambda}\exp\left(itH_0/\lambda^2\right)\sum_{n=1,2}V_n\exp\left(-itH_0/\lambda^2\right) . \tag{16}$$

In the present case, two steps are necessary to write down (15). Firstly, in order to deal with the finite temperature situation, field operators $a_{n,k}$ are represented in terms of thermal fields $\{\xi_k^{(n)}, \tilde{\xi}_k^{(n)}\}$[22] as

$$a_{n,k} = \sqrt{1+\mathcal{N}_n(\omega_k)}\xi_k^{(n)} + \sqrt{\mathcal{N}_n(\omega_k)}\tilde{\xi}_k^{(n)\,\dagger} \tag{17}$$

$${a_{n,k}}^{\dagger} = \sqrt{1+\mathcal{N}_n(\omega_k)}\xi_k^{(n)\,\dagger} + \sqrt{\mathcal{N}_n(\omega_k)}\tilde{\xi}_k^{(n)} \tag{18}$$

where $\xi_k^{(n)}$ and $\tilde{\xi}_k^{(n)}$ satisfy the commutation relations

$$[\xi_k^{(n)}, \xi_{k'}^{(n')\,\dagger}] = \delta_{n,n'}\delta(k-k') , \quad [\tilde{\xi}_k^{(n)}, \tilde{\xi}_{k'}^{(n')\,\dagger}] = \delta_{n,n'}\delta(k-k') ,$$

and the initial state is represented as the vacuum state with respect to $\xi_k^{(n)}$ and $\tilde{\xi}_k^{(n)}$ $(n = 1, 2)$. Secondly, the stochastic limit is taken and one obtains the evolution equation

$$\frac{d}{dt}U_t = -i\sum_{n=1,2}\left\{a\left(c_t^{(n)\,\dagger} + d_t^{(n)}\right) + a^{\dagger}\left(c_t^{(n)} + d_t^{(n)\,\dagger}\right)\right\}U_t , \tag{19}$$

where $c_t^{(n)}$ and $d_t^{(n)}$ are quantum white noises defined by

$$c_t^{(n)} = \lim_{\lambda\to 0}\frac{1}{\lambda}\int dk\ \sqrt{1+\mathcal{N}_n(\omega_k)}g_n(k)\xi_k^{(n)}e^{-i\frac{t}{\lambda^2}(\omega_k-\Omega)} \tag{20}$$

$$d_t^{(n)} = \lim_{\lambda\to 0}\frac{1}{\lambda}\int dk\ \sqrt{\mathcal{N}_n(\omega_k)}g_n^*(k)\tilde{\xi}_k^{(n)}e^{+i\frac{t}{\lambda^2}(\omega_k-\Omega)}. \tag{21}$$

Based on the evolution equation (19), Accardi, Imafuku and Lu[24] studied nonequilibrium steady states and found

(i) The reduced density matrix of NESS is given by a function of the system Hamiltonian $\Omega a^\dagger a$

$$\rho_{\rm sys} = \frac{1}{Z} e^{-\beta' \Omega a^\dagger a} \tag{22}$$

where the parameter β' is given by

$$\beta' = \frac{1}{\Omega} \log \left(\frac{\gamma^{(1)} \frac{e^{\beta_1(\Omega-\mu_1)}}{e^{\beta_1(\Omega-\mu_1)}-1} + \gamma^{(2)} \frac{e^{\beta_2(\Omega-\mu_2)}}{e^{\beta_2(\Omega-\mu_2)}-1}}{\gamma^{(1)} \frac{1}{e^{\beta_1(\Omega-\mu_1)}-1} + \gamma^{(2)} \frac{1}{e^{\beta_2(\Omega-\mu_2)}-1}} \right) , \tag{23}$$

and

$$\gamma^{(n)} = \pi \int dk |g_n(k)|^2 \delta\left(\omega_k - \Omega\right) . \tag{24}$$

(ii) NESS carries nonvanishing mass and energy flows, which are consistent with the second law of thermodynamics. Particularly, the mass flow is

$$\langle \hat{J}(+\infty) \rangle = 2 \frac{\gamma^{(1)}\gamma^{(2)}}{\gamma^{(1)} + \gamma^{(2)}} \left(\frac{1}{e^{\beta_1(\Omega-\mu_1)} - 1} - \frac{1}{e^{\beta_2(\Omega-\mu_2)} - 1} \right) , \tag{25}$$

where the mass flow operator $\hat{J}$ is defined as[a]

$$\hat{J}(t) = \frac{1}{2} \frac{d}{dt} \left\{ U_{-t}(N_2 - N_1) U_{-t}^\dagger \right\} , \tag{26}$$

with N_n $(n = 1, 2)$ the number operator corresponding to $\int dk a_{n,k}^\dagger a_{n,k}$.

As the system Hamiltonian is time reversal symmetric and the flows are not, the finding (i) implies that the reduced NESS is time-reversal symmetric, while the second finding (ii) asserts that NESS is not. However, as we will see, the two apparently inconsistent observations are both valid and imply the necessity of a careful treatment of observables in the stochastic limit.

4. Nonequilibrium Steady States in C*-algebraic Approach

In the C*-algebraic approach, the Heisenberg time evolution is considered and steady states are obtained as the weak limits of the initial states[10]. Although the transports of the similar systems, namely junction systems,

[a]In Ref. 24, $2\hat{J}(t)$ was identified with the mass flow. However, because of the mass conservation (9), $\hat{J}(t)$ should be identified with the mass flow. See the arguments in the next section.

were studied in detail by Fröhlich, Merkli and Ueltschi[14], it is instructive to show the explicit derivations. To proceed, we additionally assume the followings

(A) The initial state satisfies

$$|\omega_0(a^{\natural_1}a^{\natural_2}\cdots a^{\natural_n})| \le n!K_n \tag{27}$$

where $a^{\natural_j} = a$ or $a^\dagger$ and $K_n(>0)$ satisfies $\lim_{n\to\infty} K_{n+1}/K_n = 0$.

(B) The form factor $g_n(k)$ is a function of $|k|$ and the function

$$\frac{\gamma^{(n)}(\omega)}{\pi} = \int dk|g_n(k)|^2\delta(\omega_k-\omega) = \begin{cases} 2\pi\sqrt{\omega}|g_n\left(\sqrt{\omega}\right)|^2 & (\omega\ge 0) \\ 0 & (\omega<0)\ , \end{cases} \tag{28}$$

is in $L^2(\mathbf{R})$ and uniformly Hölder continuous with index $\alpha\in(0,1)$, i.e.,

$$\left|\gamma^{(n)}(x)-\gamma^{(n)}(y)\right| \le K|x-y|^\alpha \quad ({}^\exists K>0)$$

(C) There exists no real solution for $\eta(z)=0$, where

$$\eta(z) \equiv z-\Omega-\lambda^2\sum_{n'=1,2}\int dk'\frac{|g_{n'}(k')|^2}{z-\omega_{k'}}\ , \tag{29}$$

and $1/\eta_-(\omega)\equiv 1/\eta(\omega-i0)$ $(\omega\in\mathbf{R})$ is bounded.

The assumptions (A) and (B) are posed in order to simplify the investigation, while the first half of the assumption (C) plays an essential role as it guarantees the existence of the steady states.

First we note that the Hilbert space $\mathcal{H}$, where our model is defined, is a boson Fock space over a Hilbert space

$$\ell^2 \equiv \left\{ f=\begin{pmatrix} c \\ \psi_1(k) \\ \psi_2(k)\end{pmatrix} |c|^2+\sum_{n=1,2}\int dk|\psi_n(k)|^2<+\infty \right\}\ ,$$

equipped with the inner product

$$(f,\widetilde{f}) = c^*\widetilde{c}+\sum_{n=1,2}\int dk\psi_n(k)^*\widetilde{\psi}_n(k)\ .$$

The CCR algebra $\mathcal{A}$ is, then, generated by Weyl operators (see Theorem 5.2.8 of Ref. 2)

$$W(f)=\exp\left(i\Phi(f)\right) \tag{30}$$

where Φ is a map from the Hilbert space $\ell^2(\ni f = (c, \psi_1, \psi_2))$ to the space $(\ni \Phi(f))$ of (unbounded) operators on $\mathcal{H}$ and is formally defined as $\Phi(f) = \Phi_S(f) + \sum_{n=1,2} \Phi_{b,n}(f)$ with

$$\Phi_S(f) = \overline{\frac{1}{\sqrt{2}}\{c^* a + c a^\dagger\}} \tag{31}$$

$$\Phi_{b,n}(f) = \overline{\int \frac{dk}{\sqrt{2}} \left(\psi_n(k)^* a_{n,k} + \psi_n(k) a_{n,k}^\dagger\right)} \,. \tag{32}$$

In the above, the overline stands for the closure. In other words, any element of $\mathcal{A}$ can be approximated with arbitrary precision by a finite linear combination of finite products of Weyl operators. On the other hands, by repeatedly applying the identity (Theorem 5.2.4 of Ref. 2) $W(f)W(g) = W(f + g)\exp(-i\mathrm{Im}(f,g))$, a product of Weyl operators is found to reduce to a single Weyl operator. Therefore, any element of $\mathcal{A}$ can be approximated with arbitrary precision by a finite linear combination of Weyl operators and, in order to investigate the time evolution of the states, it is enough to investigate the average values of the Weyl operators.

One, then, obtains the following proposition and corollary.

Proposition For the Weyl operator, we have

$$\lim_{t\to+\infty} \omega_0 \{\tau_t (W(f))\} = \exp\left(-\frac{1}{2} \sum_{n=1,2} \int dk \, |\varphi_n(k,f)|^2 \left\{\mathcal{N}_n(\omega_k) + \frac{1}{2}\right\}\right)$$
$$\equiv \omega_{+\infty} (W(f)) \tag{33}$$

where $\varphi_n(k,f)$ is given by

$$\varphi_n(k,f) = \psi_n(k) + \frac{\lambda g_n(k)}{\eta_-(\omega_k)} \left\{c + \lambda \sum_{n'=1,2} \int dk' \frac{g_{n'}^*(k')\psi_{n'}(k')}{\omega_k - \omega_{k'} - i0}\right\} \,. \tag{34}$$

This implies that NESS $\omega_{+\infty}$ exists and that it is quasi-free with a two-point function

$$\omega_{+\infty} \left(\Phi(f)^2\right) = \sum_{n=1,2} \int dk \, |\varphi_n(k,f)|^2 \left\{\mathcal{N}_n(\omega_k) + \frac{1}{2}\right\} \,. \tag{35}$$

Corollary

(i) The average values of the harmonic oscillator variables are

$$\omega_{+\infty}(a^\dagger a) = \sum_{n=1,2} \int dk \frac{\lambda^2 |g_n(k)|^2}{|\eta_+(\omega_k)|^2} \mathcal{N}_n(\omega_k) \,, \qquad \omega_{+\infty}(aa) = 0 \,.$$

This implies that the reduced state is described by the density matrix

$$\rho_{\rm sys}^{\lambda} \equiv \frac{1}{Z_\lambda} e^{-\beta_\lambda \Omega a^\dagger a}$$

$$\beta_\lambda = \frac{1}{\Omega} \log \frac{\sum_{n=1,2} \int dk \frac{\lambda^2 |g_n(k)|^2}{|\eta_+(\omega_k)|^2} \{\mathcal{N}_n(\omega_k) + 1\}}{\sum_{n=1,2} \int dk \frac{\lambda^2 |g_n(k)|^2}{|\eta_+(\omega_k)|^2} \mathcal{N}_n(\omega_k)} \tag{36}$$

where Z_λ is the normalization constant.

(ii) The average mass flow is

$$\omega_{+\infty}(J_1) = -\omega_{+\infty}(J_2)$$
$$= 2\pi\lambda^4 \int dk \int dk' \frac{|g_1(k)|^2 |g_2(k')|^2}{|\eta_+(\omega_k)|^2} \delta(\omega_k - \omega_{k'}) \{\mathcal{N}_1(\omega_k) - \mathcal{N}_2(\omega_k)\} .$$

The rest of this section is devoted to the proof of the proposition.

4.1. *Time Evolution of Weyl Operators*

Note that the Hamiltonian H is the second quantization of the Hamiltonian h densely defined on ℓ^2

$$hf \equiv h \begin{pmatrix} c \\ \psi_1(k) \\ \psi_2(k) \end{pmatrix} = \begin{pmatrix} \Omega c + \lambda \sum_{n=1,2} \int dk g_n(k)^* \psi_n(k) \\ \omega_k \psi_1(k) + \lambda g_1(k) c \\ \omega_k \psi_2(k) + \lambda g_2(k) c \end{pmatrix} .$$

The group τ_t of time-evolution automorphisms generated by H satisfies (cf. the argument before Proposition 5.2.27 of Ref. 2)

$$\tau_t(W(f)) = W\left(e^{iht} f\right) ,$$

and, under the condition (C), one has

$$e^{iht} f = \begin{pmatrix} c(t) \\ \psi_1(k,t) \\ \psi_2(k,t) \end{pmatrix} , \tag{37}$$

where

$$c(t) = \sum_{n=1,2} \int dk \frac{\lambda g_n^*(k)}{\eta_+(\omega_k)} e^{i\omega_k t} \varphi_n(k,f)$$

$$\psi_n(k,t) = e^{i\omega_k t} \varphi_n(k,f) + \sum_{n'=1,2} \int dk' \frac{\lambda^2 g_n(k) g_{n'}^*(k') e^{i\omega_{k'} t} \varphi_{n'}(k',f)}{\eta_+(\omega_{k'})(\omega_k - \omega_{k'} - i0)}$$

Then, since the two fields and the harmonic oscillator are independent at the initial state, the average value of the Weyl operator at time t is evaluated as

$$\omega_0\left(\tau_t\left(W(f)\right)\right)=\omega_0\left(\exp\left(i\Phi_S(e^{iht}f)\right)\right)\prod_{n=1,2}\omega_0\left(\exp\left(i\Phi_{b,n}(e^{iht}f)\right)\right) \quad (38)$$

4.2. *On the limit* $\lim_{t\to\pm\infty}\omega_0\left(\exp\left(i\Phi_S(e^{iht}f)\right)\right)$

As shown in Appendix A, $\eta_+(\omega)$ is continuous and $\varphi_n(k,f)\in L^2(\mathbf{R}^3)$ under the assumption (B). Because of $\omega_k=k^2$ and $g_n(k)=g_n(\sqrt{\omega})$, $c(t)$ can be rewritten as

$$c(t)=\frac{\lambda}{2}\sum_{n=1,2}\int_0^\infty d\omega e^{i\omega t}v_n(\omega)\ , \quad (39)$$

$$v_n(\omega)=\frac{\sqrt{\omega}g_n^*(\sqrt{\omega})}{\eta_+(\omega)}\left[\int d\hat{k}\ \varphi_n(k,f)\right]_{|k|=\sqrt{\omega}}\ , \quad (40)$$

where $d\hat{k}$ stands for the angular integral. Since $g_n(k)$ and $\varphi_n(k,f)$ are $L^2(\mathbf{R}^3)$, one has

$$\begin{aligned}\bar{c}_n&\equiv\int_0^\infty d\omega\,|v_n(\omega)|\\&\le\sup_{0<\omega}\frac{2}{|\eta_+(\omega)|}\sqrt{\int dk|g_n(k)|^2}\sqrt{\int dk|\varphi_n(k,f)|^2}<+\infty\ ,\end{aligned}$$

or $v_n\in L^1(\mathbf{R}^+)$. Hence the Riemann-Lebesgue theorem[25] leads to $\lim_{t\to+\infty}c(t)=0$.

The assumption (A) implies

$$\left|\omega_0\left(\Phi_S(e^{iht}f)^m\right)\right|\le m!(2|c(t)|)^mK_m\ , \quad (41)$$

and, because $|c(t)|\le\lambda(\bar{c}_1+\bar{c}_2)/2\equiv\bar{c}$,

$$\begin{aligned}&\left|\omega_0\left(\exp\left(i\Phi_S(e^{iht}f)\right)\right)-1\right|\le\sum_{m=1}^\infty\frac{1}{m!}\left|\omega_0\left(\Phi_S(e^{iht}f)^m\right)\right|\\&\le\sum_{m=1}^\infty(2|c(t)|)^mK_m\le\frac{|c(t)|}{\bar{c}}\sum_{m=1}^\infty(2\bar{c})^mK_m\to 0\quad(\text{as } t\to+\infty)\ .\end{aligned}$$

The sum in the right-hand-side is finite as a result of $\lim_{j\to+\infty}K_{j+1}/K_j=0$.

4.3. *On the limit* $\lim_{t\to\pm\infty}\omega_0\left(\exp\left(i\Phi_{b,n}(e^{iht}f)\right)\right)$

Since the initial state ω_0 is quasi-free with respect to reservoir degrees of freedom, we have

$$\omega_0\left(\exp\left(i\Phi_{b,n}(e^{iht}f)\right)\right)=\exp\left\{-\omega_0\left(\Phi_{b,n}(e^{iht}f)^2\right)/2\right\}\ , \tag{42}$$

and

$$\omega_0\left(\Phi_{b,n}(e^{iht}f)^2\right)=\int dk\,|\varphi_n(k,f)|^2\left\{\mathcal{N}_n(\omega_k)+\frac{1}{2}\right\}$$
$$+2\lambda^2\mathrm{Re}I_n^{(1)}(t)+\lambda^4 I_n^{(2)}(t)\ ,$$

where

$$I_n^{(1)}(t)\equiv\int dk\varphi_n(k,f)^*g_n(k)I(\omega_k;t)\left\{\mathcal{N}_n(\omega_k)+\frac{1}{2}\right\}$$
$$I_n^{(2)}(t)\equiv\int dk|g_n(k)|^2|I(\omega_k;t)|^2\left\{\mathcal{N}_n(\omega_k)+\frac{1}{2}\right\}\ ,$$
$$I(\omega_k;t)=\sum_{n'=1,2}\int dk'\frac{g_{n'}^*(k')e^{i(\omega_{k'}-\omega_k)t}\varphi_{n'}(k',f)}{\eta_+(\omega_{k'})(\omega_k-\omega_{k'}-i0)}\ .$$

The time-dependent terms $I_n^{(j)}(t)$ $(j=1,2)$ are shown to vanish in the limit of $t\to+\infty$ with the aid of the following Lemma. Its proof is given in Appendix B.

Lemma 1 If $w_n\in L^1(\mathbf{R}^+)\cap L^2(\mathbf{R}^+)$ $(n=1,2)$, then

$$\lim_{t\to+\infty}\int_0^\infty d\omega w_1^*(\omega)\int_0^\infty d\omega'\frac{w_2(\omega')e^{i(\omega'-\omega)t}}{\omega-\omega'-i0}=0 \tag{43}$$

As easily seen, one has

$$I_n^{(1)}(t)=\frac{1}{4}\sum_{m=1,2}\int_0^\infty d\omega u_n(\omega)^*\int_0^\infty d\omega'\frac{v_m(\omega')e^{i(\omega'-\omega)t}}{\omega-\omega'-i0}$$

where $v_m(\omega)$ is given by (40) and

$$u_n(\omega)=\sqrt{\omega}\ g_n^*(\sqrt{\omega})\left\{\mathcal{N}_n(\omega)+\frac{1}{2}\right\}\left[\int d\hat{k}\ \varphi_n(k,f)\right]_{|k|=\sqrt{\omega}}$$

In the previous subsection, we have seen $v_m\in L^1(\mathbf{R}^+)$. On the other hand,

$$\int_0^\infty d\omega|v_m(\omega)|^2\le\sup_{\omega>0}\frac{4}{|\eta_+(\omega)|}\left(\sup_{\omega>0}\frac{\gamma^{(m)}(\omega)}{\pi}\right)\int dk|\varphi_m(k,f)|^2<+\infty$$

or $v_m \in L^2(\mathbf{R}^+)$. Exactly in the same way, one has $u_n \in L^1(\mathbf{R}^+) \cap L^2(\mathbf{R}^+)$. Hence, Lemma 1 implies $I_n^{(1)}(t) \to 0$ (as $t \to +\infty$).

Moreover, one has

$$I_n^{(2)}(t) = 2\pi \int_0^\infty d\omega \sqrt{\omega} \left|g_n(\sqrt{\omega})\right|^2 \left\{\mathcal{N}_n(\omega) + \frac{1}{2}\right\} |I(\omega;t)|^2$$
$$\le \sup_{0<\omega} \frac{\gamma^{(n)}(\omega)}{\pi} \sup_{0<\omega} \left\{\mathcal{N}_n(\omega) + \frac{1}{2}\right\} \int_{-\infty}^{+\infty} d\omega \, |I(\omega;t)|^2$$

and

$$\int_{-\infty}^{+\infty} d\omega \, |I(\omega;t)|^2 = \frac{\pi}{2i} \sum_{n,m=1,2} \int_0^\infty d\omega_1 \int_0^\infty d\omega_2 \frac{v_m(\omega_1) v_n^*(\omega_2)}{\omega_2 - \omega_1 - i0} e^{i(\omega_1 - \omega_2)t} .$$

Then, because of v_m, $v_n \in L^1(\mathbf{R}^+) \cap L^2(\mathbf{R}^+)$ and Lemma 1, we have

$$\lim_{t \to +\infty} \int_{-\infty}^{+\infty} d\omega \, |I(\omega;t)|^2 = 0$$

and, thus, $I_n^{(2)}(t) \to 0$ (for $t \to +\infty$).

In short, we derive

$$\lim_{t \to +\infty} \omega_0 \left(\Phi_{b,n}(e^{iht} f)^2\right) = \int dk \, |\varphi_n(k, f)|^2 \left\{\mathcal{N}_n(\omega_k) + \frac{1}{2}\right\} , \tag{44}$$

which implies the desired results.

5. Weak coupling limits

In order to compare the results of the C*-algebraic approach with those of the stochastic limit approach, we consider weak limits of the former with the aid of the following lemma

Lemma 2 Suppose $\sum_{n=1,2} \gamma^{(n)}(\omega)$ is uniformly Hölder continuous on $\mathbf{R}$, square integrable on $\mathbf{R}^+$ and $\sum_{n=1,2} \gamma^{(n)}(\Omega) > 0$, then, for any continuous bounded function $F(\omega)$,

$$\lim_{\lambda \to 0} \sum_{n=1,2} \int dk \frac{\lambda^2 |g_n(k)|^2}{|\eta_-(\omega_k)|^2} F(\omega_k) = F(\Omega) \tag{45}$$

Lemma 2 gives

$$\lim_{\lambda \to 0} \int \frac{\lambda^2 |g_n(k)|^2}{|\eta_+(\omega_k)|^2} \mathcal{N}_n(\omega_k) = \lim_{\lambda \to 0} \sum_{m=1,2} \int \frac{\lambda^2 |g_m(k)|^2}{|\eta_+(\omega_k)|^2} \frac{|g_n(k)|^2 \mathcal{N}_n(\omega_k)}{\sum_{m'=1,2} |g_{m'}(k)|^2}$$
$$= \frac{\gamma^{(n)}}{\gamma^{(1)} + \gamma^{(2)}} \mathcal{N}_n(\Omega) ,$$

where $|g_n(k)|^2_{k^2=\Omega} = |g_n(\sqrt{\Omega})|^2 = \gamma^{(n)}/(2\pi^2\sqrt{\Omega})$ is used. Then, Corollary (i) leads to

$$\lim_{\lambda\to 0}\rho^{\lambda}_{\rm sys} = \frac{1}{Z}e^{-\beta'\Omega a^\dagger a}$$
$$\beta' = \lim_{\lambda\to 0}\beta_\lambda = \frac{1}{\Omega}\log\frac{\sum_{n=1,2}\gamma^{(n)}\left\{\mathcal{N}_n(\Omega)+1\right\}}{\sum_{n=1,2}\gamma^{(n)}\mathcal{N}_n(\Omega)}$$
$$Z = \lim_{\lambda\to 0} Z_\lambda\ ,$$

which agrees with the the reduced distribution (22)-(23) derived in the stochastic limit approach.

Next we consider the flows. From Corollary (ii), one finds that the limit of $\lambda \to 0$ leads to a contradictory result, i.e., the absence of the flow

$$\lim_{\lambda\to 0}\omega_{+\infty}(J_1) = 0\ . \tag{46}$$

This is, however, consistent with the physical situation When the system-reservoir interactions are vanishingly weak, the flows induced by them are vanishingly small and the accumulation over a longer time interval is necessary to have observable values. Indeed, well-defined weak-coupling limit of the flow can be obtained after dividing them by a factor of λ^2

$$\lim_{\lambda\to 0}\omega_{+\infty}(J_1)/\lambda^2 = \frac{2\gamma^{(1)}\gamma^{(2)}}{\gamma^{(1)}+\gamma^{(2)}}\left\{\mathcal{N}_1(\Omega)-\mathcal{N}_2(\Omega)\right\}\ . \tag{47}$$

This limit agrees with NESS flow (25) derived by the stochastic limit approach. Note that the division by λ^2 precisely corresponds to the scaling of the time variable $t \to \lambda^2 t$.

In short, the stochastic limit approach successfully provides the weak coupling limits of NESS averages of approriately rescaled observables.

6. Summary and Discussion

We have compared the nonequilibrium steady states obtained by the stochastic limit approach and C*-algebraic approach and shown that the stochastic limit approach does provide the weak coupling limits of the nonequilibrium-steady-state averages of appropriately rescaled observables.

We note that the apparent inconsistency of the results addressed at the end of Sec.3 is not a problem. Indeed, as shown in Corollary (i), the exact reduced density matrix can be time-reversal symmetric even though the original state is not.

Here a remark is in order. Generally speaking, the reduced density matrix *is not* time-reversal symmetric when the original state is not, but its weak coupling limit *is* time-reversal symmetric[24]. For example, when one reduces the state onto the subalgebra generated by a and

$$b \equiv \frac{1}{\alpha}\int dk g_1^*(k) a_{1,k} \quad \left(\alpha^2 \equiv \int dk |g_1(k)|^2 \text{ the normalization constant}\right) ,$$

the reduced density matrix is given by

$$\rho_{\rm red} = \frac{1}{Z_{\rm red}} \exp\left(-\beta_1 a^\dagger a - \beta_2 b^\dagger b - \beta_{12} a^\dagger b - \beta_{12}^* b^\dagger a\right)$$

where $Z_{\rm red}$ is the normalization constant and the real parameters β_1, β_2 and a complex parameter β_{12} are related to the correlation matrix C as

$$\begin{pmatrix} \beta_1 , & \beta_{12} \\ \beta_{12}^* , & \beta_2 \end{pmatrix} = \log(E + C^{-1}) ,$$

$$C \equiv \begin{pmatrix} \omega_{+\infty}(a^\dagger a) , & \omega_{+\infty}(a^\dagger b) \\ \omega_{+\infty}(b^\dagger a) , & \omega_{+\infty}(b^\dagger b) \end{pmatrix} ,$$

with E the 2×2 unit matrix. The matrix elements of C are given by

$$\omega_{+\infty}(a^\dagger a) = \sum_{m=1,2} \int dk \frac{\lambda^2 |g_m(k)|^2}{|\eta_+(\omega_k)|^2} \mathcal{N}_m(\omega_k)$$

$$\omega_{+\infty}(b^\dagger a) = \sum_{m=1,2} \int dk \frac{\lambda |g_m(k)|^2}{\alpha \eta_+(\omega_k)} \mathcal{N}_m(\omega_k) \left\{ \delta_{1,m} + \frac{\lambda^2}{\eta_-(\omega_k)} \int dk' \frac{|g_1(k')|^2}{\omega_k - \omega_{k'} -}\right.$$

$$\omega_{+\infty}(b^\dagger b) = \sum_{m=1,2} \int dk \frac{|g_m(k)|^2}{\alpha^2} \mathcal{N}_m(\omega_k) \left| \delta_{1,m} + \frac{\lambda^2}{\eta_-(\omega_k)} \int dk' \frac{|g_1(k')|^2}{\omega_k - \omega_{k'} - i(}\right.$$

Since the symmetry of $g_1(k)$ leads to $\iota b \iota = b$, the reduced density matrix is time-reversal symmetric if and only if β_{12} is real. As easily seen, this condition is equivalent to

$$0 \neq \omega_{+\infty}(b^\dagger a) - \omega_{+\infty}(a^\dagger b) = -i \frac{\omega_{+\infty}(J_1)}{\lambda \alpha}$$

where J_1 stands for the mass flow operator (10). Therefore, when the steady state admits nonvanishing flow, the reduced state $\rho_{\rm red}$ is not timre-reversal symmetric. However, in the weak coupling limit, the correlation matrix reduces to

$$\lim_{\lambda \to 0} C = \begin{pmatrix} 1/\{e^{\beta' \Omega} - 1\} , & 0 \\ 0 , & \int dk |g_1(k)|^2 \mathcal{N}_1(\omega_k)/\alpha^2 \end{pmatrix}$$

which corresponds to the time-reversal symmetric reduced state

$$\widetilde{\rho}_{\text{red}} = \frac{1}{\widetilde{Z}_{\text{red}}} \exp\left(-\beta'\Omega a^\dagger a - \tilde{\beta}_2 b^\dagger b\right)$$

where $\widetilde{Z}_{\text{red}}$ is the normalization constant, β' is given by (23) and

$$\tilde{\beta}_2 = \log \frac{\int dk |g_1(k)|^2 \{\mathcal{N}_1(\omega_k)+1\}}{\int dk |g_1(k)|^2 \mathcal{N}_1(\omega_k)} .$$

The reason of the time-reversal symmetry of the exact reduced density matrix $\rho_{\text{sys}}^\lambda$ can be understood as follows. From Corollary, the steady state $\omega_{+\infty}$ and, hence, the reduced state are invariant under the gauge transformation. As the reduced state is quasi-free, it can be expressed as an exponential function of a bilinear form of the creation and annihilation operators of the oscillator. On the other hand, $a^\dagger a$ is the only quadratic operator which is invariant under the gauge transformation. Therefore, the reduced density matrix $\rho_{\text{sys}}^\lambda$ should be an exponential function of $a^\dagger a$ and, thus, is time-reversal symmetric.

The present results suggest that, when one considers the weak coupling limit, it is necessary to classify observables into the ones such as the number operator $a^\dagger a$ possessing nontrivial $\lambda \to 0$ limits, the ones such as the mass flow J_1 which should be divided by λ^2 before taking the $\lambda \to 0$ limit, and so on. The average values of the former observables can be characterized by the reduced density matrix in the weak coupling limit. However, we do not know reduced states which provide the average values of the latter observables. This aspect will be investigated elsewhere.

Acknowledgments

The authors thank Professor T. Hida for the invitation and hospitality at the Meijo Winter School (Meijo University, 6-10 January, 2003), where this collaboration starts. They thank Dr. K. Imafuku, who participated to the early stage of this work, for discussions and comments. Also they are grateful to Professors T. Hida, M. Ohya, N. Watanabe, and T. Matsui for fruitful discussions and valuable comments. This work is partially supported by Grant-in-Aid for Scientific Research (C) from the Japan Society of the Promotion of Science, by a Grant-in-Aid for Scientific Research of Priority Areas "Control of Molecules in Intense Laser Fields" and the 21st Century COE Program at Waseda University "Holistic Research and Education Center for Physics of Self-organization Systems" both from the Ministry of Education, Culture, Sports, Science and Technology of Japan.

Appendix A On functions $\eta_+(\omega)$, $\varphi_n(k,f)$ and $\psi_n(k,t)$

Here we show that $\varphi_n(k,f)$, $\psi_n(k,t) \in L^2(\mathbf{R}^3)$ and that $\eta_+(\omega) - \omega + \Omega$ is uniformly Hölder continuous and square integrable on the positive real line.

Since $g_n(k)$ is spherical symmetric, in terms of the function $\gamma^{(n)}(\omega)$ defined by (28), one has

$$\eta(\omega + i\epsilon) - \omega - i\epsilon + \Omega = -\frac{\lambda^2}{\pi} \sum_{n=1,2} \int_0^\infty d\omega' \frac{\gamma^{(n)}(\omega')}{\omega - \omega' + i\epsilon} .$$

Then, as $\gamma^{(n)}(\omega)$ is square integrable and uniformly Hölder continuous (cf. assumption B), Theorem 106 of Section 5.15 of Ref. 25 implies that the limit of the right-hand side for $\epsilon \to 0+$ exists and is again square integrable and uniformly Hölder continuous with the same index α as $\gamma^{(n)}$. Because of the same theorem, one finds

$$\lim_{\omega\to+\infty} \gamma^{(n)}(\omega) = 0 , \tag{48}$$

$$\lim_{\omega\to+\infty} \{\omega - \Omega - \eta_+(\omega)\} = \lim_{\omega\to+\infty} \frac{\lambda^2}{2\pi} \sum_{n=1,2} \int_0^\infty d\omega' \frac{\gamma^{(n)}(\omega')}{\omega - \omega' + i0} = 0 \tag{49}$$

Now we show that $\varphi_n(k,f)$ given by (34) is square integrable. Since $\psi_n(k)$ and $g_n(k)$ are in $L^2(\mathbf{R}^3)$ and $1/\eta_-(\omega_k)$ is bounded, $\varphi_n(k,f) \in L^2(\mathbf{R}^3)$ follows if

$$\frac{g_n(k)}{\eta_-(\omega_k)} \int dk' \frac{g^*_{n'}(k')\psi_{n'}(k')}{\omega_k - \omega_{k'} - i0} \tag{50}$$

is square integrable. On the other hand, if $w(\omega) \in L^2(\mathbf{R}^+)$, then,

$$\frac{g_n(k) w(\omega_k)}{\eta_-(\omega_k)} \in L^2(\mathbf{R}^3) .$$

Indeed, (48) implies $\sup_{0<\omega} \gamma^{(n)}(\omega) < +\infty$ and one has

$$\int dk \left| \frac{g_n(k) w(\omega_k)}{\eta_-(\omega_k)} \right|^2 = \frac{1}{\pi} \int_0^\infty d\omega \frac{\gamma^{(n)}(\omega)}{|\eta_+(\omega)|^2} |w(\omega)|^2$$
$$\leq \frac{1}{\pi} \sup_{\omega>0} \gamma^{(n)}(\omega) \ \sup_{\omega>0} \frac{1}{|\eta_+(\omega)|^2} \int_0^\infty d\omega |w(\omega)|^2 < +\infty$$

Therefore, it is enough to show that the integral in (50)

$$\int dk' \frac{g^*_{n'}(k')\psi_{n'}(k')}{\omega - \omega_{k'} - i0} = \frac{1}{2} \int_0^\infty d\omega' \frac{\sqrt{\omega'} g^*_{n'}(\sqrt{\omega'})}{\omega - \omega' - i0} \left[\int d\hat{k}' \psi_{n'}(k') \right]_{|k'|=\sqrt{\omega'}}$$

defines a square integrable function of $\omega \in \mathbf{R}^+$. It is the case because of Theorem 101 of Section 5.10 of Ref. 25 and

$$\int_0^\infty d\omega' \left| \sqrt{\omega'} g^*_{n'}(\sqrt{\omega'}) \int d\hat{k}' \psi_{n'}(k') \right|^2 \le \frac{4}{\pi} \sup_{0<\omega} \gamma^{(n')}(\omega) \int dk |\psi_{n'}(k)|^2 < +\infty \,.$$

Exactly in the same way, one can show that $\psi_n(k,t)$ defined after (37) is in $L^2(\mathbf{R}^3)$.

Appendix B Proof of Lemma 1

We have

$$\begin{aligned}
&\int_0^\infty d\omega w_1^*(\omega) \int_0^\infty d\omega' \frac{w_2(\omega')e^{i(\omega'-\omega)t}}{\omega-\omega'-i0} \\
&= \lim_{\epsilon\to 0+} \int_0^\infty d\omega w_1^*(\omega) \int_0^\infty d\omega' \frac{w_2(\omega')e^{i(\omega'-\omega)t}}{\omega-\omega'-i\epsilon} \\
&= \lim_{\epsilon\to 0+} \frac{1}{i} e^{\epsilon t} \int_0^\infty d\omega w_1^*(\omega) \int_0^\infty d\omega' w_2(\omega') \int_0^\infty ds e^{i(\omega'-\omega+i\epsilon)(s+t)} \\
&= \lim_{\epsilon\to 0+} \frac{1}{i} e^{\epsilon t} \int_t^\infty ds e^{-\epsilon s} \widetilde{w}_1^*(s) \widetilde{w}_2(s) = \frac{1}{i} \int_t^\infty ds \widetilde{w}_1^*(s)\widetilde{w}_2(s) \qquad (51)
\end{aligned}$$

where $\widetilde{w}_n(s) \equiv \int_0^\infty d\omega w_n(\omega) \exp(i\omega s)$ stands for the Fourier transform.

In the above, the first equality holds because the ω'-integral exists for $w_2 \in L^2(\mathbf{R}^+)$ as an element of $L^2(\mathbf{R}^+)$ (cf. Theorem 101 of Section 5.10 of Ref. 25).

The second and third equalities follow from Fubini's theorem and the absolute integrability of the integrand (remind that $w_1, w_2 \in L^1(\mathbf{R}^+)$).

Since $w_1, w_2 \in L^2(\mathbf{R}^+)$, their Fourier transformations are square integrable as well. Thus, $\widetilde{w}_1^*(s)\widetilde{w}_2(s)$ is integrable and Lebesgue's dominated convergence theorem leads to the last equality.

The desired result also follows from the integrability of $\widetilde{w}_1^*(s)\widetilde{w}_2(s)$

$$\begin{aligned}
&\lim_{t\to+\infty} \int_0^\infty d\omega w_1^*(\omega) \int_0^\infty d\omega' \frac{w_2(\omega')e^{i(\omega'-\omega)t}}{\omega-\omega'-i0} \\
&= \lim_{t\to+\infty} \frac{1}{i} \int_t^\infty ds \widetilde{w}_1^*(s)\widetilde{w}_2(s) = 0 \qquad (52)
\end{aligned}$$

Appendix C Proof of Lemma 2

Let $\gamma(\omega) \equiv \sum_{n=1,2} \gamma^{(n)}(\omega)$, then, since $\gamma(\Omega) > 0$ and $\gamma(\omega)$ is continuous, there exists $\kappa > 0$ such that $\gamma(\omega) \ge \gamma(\Omega)/2$ for $|\omega - \Omega| \le \kappa$. As shown in

Appendix A, the integral involved in $\eta_+(\omega)$

$$I(\omega) \equiv \frac{1}{\pi}\int_0^\infty d\omega' \frac{\gamma(\omega')}{\omega - \omega' + i0}$$

is bounded and, thus, for sufficient small λ,

$$|\mathrm{Re}\eta_+(\omega) - \omega + \Omega| = \lambda^2 |I(\omega)| \leq \frac{\kappa}{2} .$$

Then, we divide the given integral into four terms

$$\begin{aligned}
\int_0^\infty d\omega F(\omega)\frac{\lambda^2\gamma(\omega)}{|\eta_+(\omega)|^2} &= \int_0^{\Omega-\kappa} d\omega F(\omega)\frac{\lambda^2\gamma(\omega)}{|\eta_+(\omega)|^2} + \int_{\Omega+\kappa}^\infty d\omega F(\omega)\frac{\lambda^2\gamma(\omega)}{|\eta_+(\omega)|^2} \\
&+ \int_{\Omega-\kappa}^{\Omega+\kappa} d\omega F(\omega)\left\{\frac{\lambda^2\gamma(\omega)}{|\eta_+(\omega)|^2} - \frac{\lambda^2\gamma(\Omega)}{(\omega-\Omega-\Delta_\lambda)^2 + \lambda^4\gamma(\Omega)^2}\right\} \\
&+ \int_{\Omega-\kappa}^{\Omega+\kappa} d\omega F(\omega)\frac{\lambda^2\gamma(\Omega)}{(\omega-\Omega-\Delta_\lambda)^2 + \lambda^4\gamma(\Omega)^2} .
\end{aligned} \tag{53}$$

where $\Delta_\lambda = \lambda^2 \mathrm{Re}I(\Omega)$.

In a standard way, one can easily show

$$\lim_{\lambda\to 0}\int_{\Omega-\kappa}^{\Omega+\kappa} d\omega F(\omega)\frac{\lambda^2\gamma(\Omega)}{(\omega-\Omega-\Delta_\lambda)^2 + \lambda^4\gamma(\Omega)^2} = \pi F(\Omega) . \tag{54}$$

When $\omega \leq \Omega - \kappa$, one has

$$|\eta_+(\omega)| \geq |\mathrm{Re}\eta_+(\omega)| \geq \Omega - \omega - |\mathrm{Re}\eta_+(\omega) - \omega + \Omega| \geq \Omega - \omega - \frac{\kappa}{2}$$

and the first integral of (53) is evaluated as

$$\begin{aligned}
\left|\int_0^{\Omega-\kappa} d\omega F(\omega)\frac{\lambda^2\gamma(\omega)}{|\eta_+(\omega)|^2}\right| &\leq \lambda^2 \sup_{0<\omega}|F(\omega)| \sup_{0<\omega}|\gamma(\omega)| \int_{-\infty}^{\Omega-\kappa}\frac{d\omega}{(\Omega-\omega-\kappa/2)^2} \\
&= \frac{2\lambda^2}{\kappa}\sup_{0<\omega}|F(\omega)| \sup_{0<\omega}|\gamma(\omega)| \to 0 \quad (\text{for } \lambda \to 0)
\end{aligned}$$

The second integral of (53) can be evaluated in the same way

$$\lim_{\lambda\to 0}\int_{\Omega+\kappa}^\infty d\omega F(\omega)\frac{\lambda^2\gamma(\omega)}{|\eta_+(\omega)|^2} = 0 .$$

The second integral is rewritten as

$$\int_{\Omega-\kappa}^{\Omega+\kappa} d\omega F(\omega)\left\{\frac{\lambda^2\gamma(\omega)}{|\eta_+(\omega)|^2}-\frac{\lambda^2\gamma(\Omega)}{(\omega-\Omega-\Delta_\lambda)^2+\lambda^4\gamma(\Omega)^2}\right\}$$
$$=\lambda^4\int_{\Omega-\kappa}^{\Omega+\kappa} d\omega F(\omega)\gamma(\omega)\ \frac{2\mathrm{Re}\eta_+(\omega)\mathrm{Re}\{I(\omega)-I(\Omega)\}}{|\eta_+(\omega)|^2\left[(\omega-\Omega-\Delta_\lambda)^2+\lambda^4\gamma(\Omega)^2\right]}$$
$$+\lambda^6\int_{\Omega-\kappa}^{\Omega+\kappa} d\omega F(\omega)\gamma(\omega)\ \frac{2\mathrm{Re}I(\omega)\mathrm{Re}\{I(\omega)-I(\Omega)\}+|I(\omega)|^2-|I(\Omega)|^2}{|\eta_+(\omega)|^2\left[(\omega-\Omega-\Delta_\lambda)^2+\lambda^4\gamma(\Omega)^2\right]}$$
$$+\int_{\Omega-\kappa}^{\Omega+\kappa} d\omega F(\omega)\left\{\frac{\gamma(\omega)}{\gamma(\Omega)}-1\right\}\frac{\lambda^2\gamma(\Omega)}{(\omega-\Omega-\Delta_\lambda)^2+\lambda^4\gamma(\Omega)^2}\ .$$

Then, because of $|\eta_+(\omega)|\geq\lambda^2\gamma(\omega)\geq\lambda^2\gamma(\Omega)/2$ and $|\mathrm{Re}\eta_+(\omega)|/|\eta_+(\omega)|\leq 1$, one has

$$\left|\int_{\Omega-\kappa}^{\Omega+\kappa} d\omega F(\omega)\left\{\frac{\lambda^2\gamma(\omega)}{|\eta_+(\omega)|^2}-\frac{\lambda^2\gamma(\Omega)}{(\omega-\Omega-\Delta_\lambda)^2+\lambda^4\gamma(\Omega)^2}\right\}\right|$$
$$\leq\int_{\Omega-\kappa}^{\Omega+\kappa} d\omega H(\omega)\frac{\lambda^2\gamma(\Omega)}{(\omega-\Omega-\Delta_\lambda)^2+\lambda^4\gamma(\Omega)^2}\ , \tag{55}$$

where

$$H(\omega)=\frac{4\gamma(\omega)}{\gamma(\Omega)^3}|F(\omega)|\left|2\mathrm{Re}I(\omega)\mathrm{Re}\{I(\omega)-I(\Omega)\}+|I(\omega)|^2-|I(\Omega)|^2\right|$$
$$+\frac{4\gamma(\omega)}{\gamma(\Omega)^2}|F(\omega)|\,|\mathrm{Re}\{I(\omega)-I(\Omega)\}|+|F(\omega)|\left|\frac{\gamma(\omega)}{\gamma(\Omega)}-1\right|\ .$$

Since $H(\omega)$ is continuous and $H(\Omega)=0$, (54) implies that the second integral of (53) vanishes in the $\lambda\to 0$ limit.

$$\lim_{\lambda\to 0}\left|\int_{\Omega-\kappa}^{\Omega+\kappa} d\omega F(\omega)\left\{\frac{\lambda^2\gamma(\omega)}{|\eta_+(\omega)|^2}-\frac{\lambda^2\gamma(\Omega)}{(\omega-\Omega-\Delta_\lambda)^2+\lambda^4\gamma(\Omega)^2}\right\}\right|$$
$$\leq\pi H(\Omega)=0\ .$$

In short, one obtains

$$\lim_{\lambda\to 0}\int_0^\infty d\omega F(\omega)\frac{\lambda^2\gamma(\omega)}{|\eta_+(\omega)|^2}=\pi F(\Omega)\ ,$$

and, thus,

$$\lim_{\lambda\to 0}\sum_{n=1,2}\int dk F(\omega_k)\frac{\lambda^2|g_n(k)|^2}{|\eta_-(\omega_k)|^2}=\lim_{\lambda\to 0}\int_0^\infty d\omega F(\omega)\frac{\lambda^2\gamma(\omega)}{\pi|\eta_+(\omega)|^2}=F(\Omega)\ .$$

References

1. For example, see, M. Toda, R. Kubo and N. Saito, *Statistical Physics I* (Springer, New York, 1992); R. Kubo, M. Toda and N. Hashitsume, *Statistical Physics II* (Springer, New York, 1991).
2. O. Bratteli and D.W. Robinson, *Operator Algebras and Quantum Statistical Mechanics* vol.1 (Springer, New York, 1987); vol.2, (Springer, New York, 1997).
3. I.P. Cornfeld, S.V. Fomin and Ya.G. Sinai, *Ergodic Theory*, (Springer, New York, 1982); L.A. Bunimovich *et al.*, *Dynamical Systems, Ergodic Theory and Applications*, Encyclopedia of Mathematical Sciences **100**, (Springer, Berlin, 2000).
4. D. Ruelle, *Statistical Mechanics Rigorous Results*, (Benjamin, Reading, 1969); Ya. G. Sinai, *The Theory of Phase Transitions Rigorous Results*, (Pergamon, Oxford, 1982).
5. H. Spohn and J.L. Lebowitz, *Commun. Math. Phys.* **54**, 97 (1977) and references therein.
6. J. Bafaluy and J.M. Rubi, *Physica* A**153**, 129 (1988); *ibid.* **153**, 147 (1988).
7. J. Farmer, S. Goldstein and E.R. Speer, *J. Stat. Phys.* **34**, 263 (1984).
8. J.-P. Eckmann, C.-A. Pillet and L. Rey-Bellet, *Commun. Math. Phys.* **201**, 657 (1999); *J. Stat. Phys.* **95**, 305 (1999); L. Rey-Bellet and L.E. Thomas, *Commun. Math. Phys.* **215**, 1 (2000) and references therein.
9. T.G. Ho and H. Araki, *Proc. Steklov Math. Institute* **228**, (2000) 191.
10. D. Ruelle, *Comm. Math. Phys.* **224** , 3 (2001).
11. S. Tasaki, T. Matsui, in *Fundamental Aspects of Quantum Physics* eds. L. Accardi, S. Tasaki, World Scientific, (2003) p.100.
12. S. Tasaki, *Chaos, Solitons and Fractals* **12** 2657 (2001); in *Statistical Physics*, eds. M. Tokuyama and H. E. Stanley, 356 (AIP Press, New York, 2000); *Quantum Information III*, eds. T. Hida and K. Saito, 157 (World Scientific, Singapore,2001).
13. V. Jakšić, C.-A. Pillet, *Commun. Math. Phys.* **226**, 131 (2002).
14. J. Fröhlich, M. Merkli, D. Ueltschi, *Ann. Henri Poincaré* **4**, 897 (2003).
15. I. Ojima, H. Hasegawa and M. Ichiyanagi, *J. Stat. Phys.* **50**, 633 (1988).
16. I. Ojima, *J. Stat. Phys.* **56**, 203 (1989); in *Quantum Aspects of Optical Communications*, (LNP **378**,Springer,1991).
17. V. Jakšić and C.-A. Pillet, *Commun. Math. Phys.* **217**, 285 (2001).
18. V. Jakšić and C.-A. Pillet, *J. Stat. Phys.* **108**, 269 (2002).
19. D. Ruelle, *J. Stat. Phys.* **98**, 57 (2000).
20. E. B. Davies, *Quantum Theory of Open Systems*, Academic Press, London, (1976) and references therein.
21. L. Accardi, A. Frigerio, and Y. G. Lu, Commun. Math. Phys. **131**, 537 (1990); L. Accardi, J. Gough, and Y. G. Lu, Rep. Math. Phys. **36**, 155 (1995); L. Accardi, S. V. Kozyrev, and I. V. Volovich, Phys. Lett. A **260**, 31 (1999); L.Accardi, S.V.Kozyrev, *Quantum interacting particle systems*, Lecture Note of Levico school, September 2000, Volterra Preprint N.431
22. L. Accardi, Y. G. Lu, and I. V. Volovich, *Quantum Theory and Its Stochastic*

Limit Springer-Verlag, Berlin, (2002).
23. L. Accardi, K. Imafuku, and Y. G. Lu, Volterra Preprint N.50x (2002)
24. L. Accardi, K. Imafuku, Y.G. Lu, in *Fundamental Aspects of Quantum Physics* eds. L. Accardi, S. Tasaki, World Scientific, (2003) p.1.
25. E. C. Titchmarsh, *Introduction to the Theory of Fourier Integrals*, Clarendon Press, Oxford, (1948).

CONTROL OF DECOHERENCE WITH MULTIPULSE APPLICATION

CHIKAKO UCHIYAMA
Interdisciplinary Graduate School of Medicine and Engineering,
University of Yamanashi,
4-3-11, Takeda, Kofu, Yamanashi 400-8511, JAPAN
E-mail: chikako$_u$chiyama@hotmail.com

We present a strategy to control decoherence with multipulse application. We show that the degradation of a two-level system which linearly interacts with a reservoir characterized by a specific frequency is effectively suppressed by synchronizing the pulse-train application with the dynamical motion of the reservoir. We find that the non-Markovian nature of dynamical motion of the reservoir makes this strategy effective.

1. Introduction

Control of decoherence is a major issue to stabilize the process of quantum computation and to realize quantum information technology. While the "bang-bang" method [1,2,3,4,5,6] gives a possibility to control decoherence, it requires infinite number of pulses with infinitely short pulse interval. At least, the pulse interval is required to be shorter than the characteristic time of reservoir.

In this paper, we propose a strategy to suppress the decoherence effectively with a finite pulse interval. When the two-level system linearly interacts with a boson reservoir that has a characteristic frequency, we find the degradation of the system can be suppressed by synchronizing a π pulse train with the oscillation of the reservoir. In the following discussion, we name this strategy as synchronized pulse control(SPC). To make clear the applicability of SPC, we discuss how the effectiveness of SPC depends on the type of coupling(bath) spectral density such as non-Lorentzian and Lorentzian coupling spectral density.

This paper is organized as follows: In Sec.2, we derive the basic formula for multipulse control on the linear spin-boson model. Next, we discuss the synchronized pulse control in Sec.3 for non-Lorentzian and Lorentzian coupling spectral density. We give discussion and concluding remarks in Sec.4.

2. Formulation

Let us consider a two-level system which is composed of an excited state $|e\rangle$ and a ground state $|g\rangle$ with energy E_e. We consider decoherence of the two-level system when the excited state linearly interacts with a boson reservoir:

$$\mathcal{H}_R = \mathcal{H}_0 + \mathcal{H}_{SB} = (\mathcal{H}_S + \mathcal{H}_B) + \mathcal{H}_{SB}, \tag{1}$$

$$\mathcal{H}_S \equiv E_e|e\rangle\langle e|, \tag{2}$$

$$\mathcal{H}_B \equiv \hbar \sum_k \epsilon_k b_k^\dagger b_k, \tag{3}$$

$$\mathcal{H}_{SB} \equiv \hbar|e\rangle\langle e| \sum_k h_k \epsilon_k (b_k + b_k^\dagger). \tag{4}$$

Supposing that we can apply sufficiently short and strong pulses to suppress the decoherence, we neglect the interaction with the reservoir during pulse application:

$$\mathcal{H}_{SP}(t) = \mathcal{H}_S + \sum_{j=0}^{N} \mathcal{H}_{P,j}(t), \tag{5}$$

$$\mathcal{H}_{P,j}(t) = -\frac{1}{2}\vec{E}_j(t) \cdot \vec{\mu}\, (|e\rangle\langle g|e^{-i\omega t} + |g\rangle\langle e|e^{i\omega t}), \tag{6}$$

where $\vec{E}_j(t)$ is the j-th applied pulse of external field. Here we set the pulse to be on resonance with the two-level system, which means $E_e = \hbar\omega$.

When we apply N pulses with a pulse interval τ_s and pulse duration Δt, the time evolution of the density operator $\rho(t)$ of the total system is written as

$$\rho(t) = e^{-iL_R(t-(N\tau_s+\Delta t))} T_+[e^{-i\int_{N\tau_s}^{N\tau_s+\Delta t} dt' L_{SP}(t')}] \\ \times \{ \prod_{j=0}^{N-1} e^{-iL_R(\tau_s-\Delta t)} T_+[e^{-i\int_{j\tau_s}^{j\tau_s+\Delta t} dt' L_{SP}(t')}] \} \rho(0), \tag{7}$$

where T_+ is the time ordering symbol from right to left. L_{SP} (L_R) indicates the Liouville operator during the pulse application(the interaction with the reservoir) which is defined as

$$iL_\nu \ \cdots \ \equiv \frac{i}{\hbar}[\mathcal{H}_\nu \ , \cdots \ ], \ \ (\nu = \{SP\} \ or \ \{R\}). \tag{8}$$

Let us consider the case where we apply π pulses, after a $\frac{\pi}{2}$ pulse at an initial time ($t = 0$) to generate a superposed two-level state. The degradation of the two-level system is shown in the intensity of off diagonal element of the density operator $\rho(t)$ as

$$I(t) = |Tr_R \langle e|\rho(t)|g\rangle|^2, \tag{9}$$

where Tr_R denotes the operation to trace over the reservoir variable.

Assuming that the boson reservoir is in the vacuum state and the two-level system is in the ground state at the initial time:

$$\rho(0) = |g\rangle\langle g| \otimes |0\rangle\langle 0|, \tag{10}$$

we eliminate the variables of the two-level system that periodically changes its state between $|e\rangle$ and $|g\rangle$ by the π pulse train to obtain for even N,

$$I(t) = |Tr_R \langle e|[(|A_N(t)\rangle|e\rangle)(\langle B_N(t)|\langle g|)]|g\rangle|^2 = |\langle B_N(t)|A_N(t)\rangle|^2, \tag{11}$$

where

$$\begin{aligned} |A_N(t)\rangle &= D(\{\alpha_{N,k}(t)\})|0\rangle = |\{\alpha_{N,k}(t)\}\rangle \ , \\ \langle B_N(t)| &= \langle 0|D(\{\beta_{N,k}(t)\})| = \langle\{\beta_{N,k}(t)\}|. \end{aligned} \tag{12}$$

Here we denoted the displacement operator as

$$D(\{\alpha_k\}) \equiv \exp[\sum_k (\alpha_k b_k^\dagger - \alpha_k^* b_k)], \tag{13}$$

where $\{\cdots\}$ means a set of bosons in the reservoir. In Eq.(12), we defined $\alpha_{N,k}(t)$ and $\beta_{N,k}(t)$ as

$$\begin{aligned} \alpha_{N,k}(t) &\equiv -h_k + \sum_{j=0}^{N} (-1)^j \{h_k e^{-i\epsilon_k (t - j\tau_s)}\} \ , \\ \beta_{N,k}(t) &\equiv \sum_{j=1}^{N} (-1)^{j-1} \{h_k e^{-i\epsilon_k (t - j\tau_s)}\}. \end{aligned} \tag{14}$$

For odd N, we obtain,

$$I(t) = |Tr_R \langle e|[(|B_N(t)\rangle|e\rangle)(\langle A_N(t)|\langle g|)]|g\rangle|^2 = |\langle\{\alpha_{N,k}(t)\}|\{\beta_{N,k}(t)\}\rangle|^2, \tag{15}$$

with

$$\begin{aligned} \alpha_{N,k}(t) &\equiv \sum_{j=0}^{N} (-1)^j \{h_k e^{-i\epsilon_k (t - j\tau_s)}\} \ , \\ \beta_{N,k}(t) &\equiv \sum_{j=1}^{N} (-1)^{j-1} \{h_k e^{-i\epsilon_k (t - j\tau_s)} - h_k\}. \end{aligned} \tag{16}$$

In the evaluation of Eqs.(11) $\sim$ (16), we need to rewrite the summation over k into the energy integral,

$$\sum_k |h_k|^2 f(\omega_k) = \sum_k |h_k|^2 f(\omega_k) \int_0^\infty de\delta(e-\omega_k) = \int_0^\infty deh(e)f(e), \quad (17)$$

where we have defined coupling spectral density $h(e)$ as,

$$h(e) \equiv \sum_k |h_k|^2 \delta(e-\omega_k). \quad (18)$$

In the next section, we evaluate the intensity $I(t)$ for non-Lorentzian and Lorentzian coupling spectral density.

3. Numerical evaluation

Taking as an example of non-Lorentzian coupling spectral density, we consider a semi-elliptic distribution with center frequency ω_p,

$$h_S(e) \equiv s\frac{1}{p}\sqrt{-(e-\omega_p)^2 + p}, \quad (19)$$

where p is defined with half width γ_p as

$$p \equiv \frac{4\gamma_p^2}{3}. \quad (20)$$

The semi-elliptic coupling function has been used to describe the coupling strength between phonons and a localized electron in a solid[7].

We show the time dependence of the intensity $I(t)$ for semi-elliptic coupling function in Fig.1. We have scaled time variable with the center frequency ω_p of the distribution as $\tilde{t} \equiv \omega_p t$. The parameters are set as $\tilde{\gamma}_p \equiv \gamma_p/\omega_p = 0.15$, $s = 3$. This corresponds to the situation where the decay time of the interaction mode is relatively long, the average number of boson which interact with the spin is 3. In Fig.1(a), we show time evolution of $I(t)$ after a single $\frac{\pi}{2}$ pulse at $t = 0$. We see a damped oscillation whose period is 2π, which corresponds to the oscillation period of the center frequency. Here we define the oscillation period as $\tau_p (= \frac{2\pi}{\omega_p})$. When we apply the pulse much faster the oscillation period ($\tau_s = \frac{\tau_p}{20\pi}$), the decay is suppressed (Fig.1(b)), which corresponds to the fact that the "bang-bang" method [1,2,3,4,5,6] tells us. When the pulse interval corresponds to just the half of τ_p, the decay becomes faster than the case without pulse (Fig.1(c)). But, when we synchronize the pulse application with τ_p, we find that the phase coherence recovers at the pulse application time(Fig.1(d)).

We call this strategy for suppression of decoherence as synchronized pulse control(SPC).

While it is necessary to take into account the nonlinear interaction, which causes the pure dephasing phenomena (irreversible processes in the long time region) in many systems[8], we consider only the linear interaction between the spin and the original boson reservoir in this paper. Since the nonlinear interaction is often much weaker than the linear one, the effect of the pure dephasing is not significant in the time region shown in Fig.1.

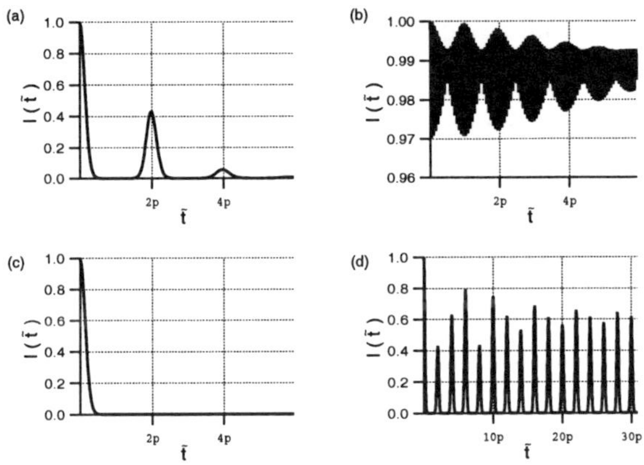

Figure 1. Time evolution of $I(\tilde{t})$ for semi-elliptic coupling spectral density with $r = 1$, $s = 3$, and $\tilde{\gamma}_p = 0.15$; (a) without pulse application, (b) pulse interval $\tau_s = \frac{\tau_p}{20\pi}$, (c) pulse interval $\tau_s = \frac{\tau_p}{2}$, (d) for pulse interval $\tau_s = \tau_p$.[9]

Next we consider the case of the Lorentzian coupling spectral density,

$$h_L(e) \equiv \frac{s}{\pi} \frac{\gamma_p}{(e - \omega_p)^2 + \gamma_p^2}. \tag{21}$$

The Lorentzian coupling function has been often used to describe a relaxation process of an atomic system or quantum dots in a high-Q cavities[10,11,12].

In Fig.2, we show the time evolution of $I(t)$ for the same parameters as in Fig.1. Fig.2(a) shows the time evolution without π pulse application. Fig.2(b) shows that we cannot obtain the sufficient decoherence suppression for short pulse interval $\tau_s = \frac{\tau_p}{20\pi}$ contrary to the semi-elliptic distribution. When we increase the pulse interval to $\tau_s = \frac{\tau_p}{2}$, we find that the degree of suppression becomes worse(Fig.2(c)). We show the time dependence under

SPC in Fig.2(d) to find almost the same time evolution as the one without pulse control.

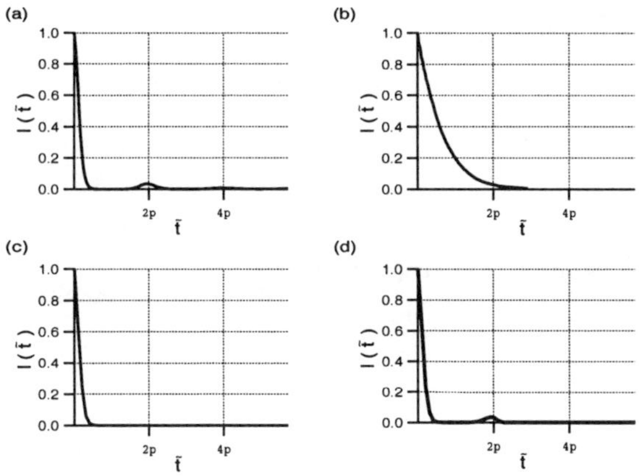

Figure 2. Time evolution of $I(\tilde{t})$ for Lorentzian coupling spectral density with $\tilde{\gamma}_p = 0.15$; (a) without pulse application, (b) pulse interval $\tau_s = \frac{\tau_p}{20\pi}$, (c) pulse interval $\tau_s = \frac{\tau_p}{2}$, (d) pulse interval $\tau_s = \tau_p$.[9]

4. Discussion and Concluding Remarks

We have proposed a strategy for suppression of decoherence by synchronizing a π pulse train with the dynamical motion of reservoir. While we find a periodic recovery of a quantum superposition at the pulse application times for the non-Lorentzian coupling spectral density, we cannot obtain the recovery for the Lorentzian coupling spectral density.

In order to explain the reason why the SPC is ineffective for the case of Lorentzian coupling spectral density, let us introduce a two-step structured reservoir(Fig.3) which consists of a single harmonic oscillator and a new "reservoir". This kind of model has been used in various systems. For atom-cavity system[13,14,15,16,17], the single harmonic oscillator is called as a quasi mode, and for electron-phonon system[18], it is called as an interaction mode. Here we call the new harmonic oscillator as the interaction mode.

On replacing the original reservoir with the two-step structure, we set the coupling function of the coupling between the interaction mode and the new "reservoir" so that the frequency (decay constant) of the motion of the interaction mode corresponds to the center frequency (width) of

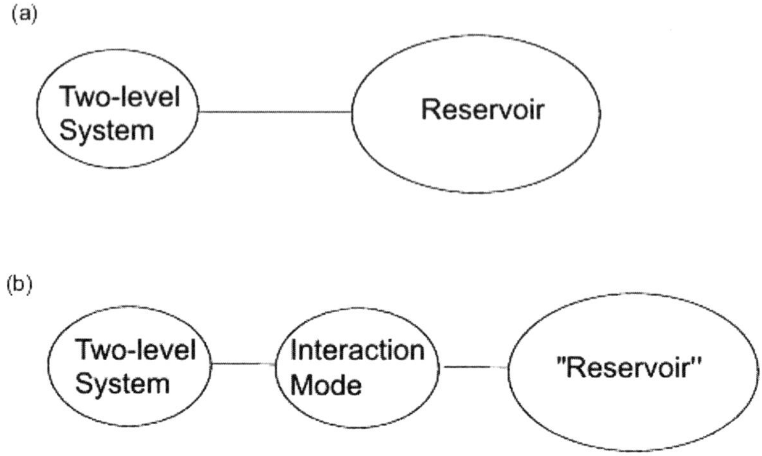

Figure 3. Schematic representation of the two pictures for the boson system: (a) the normal mode picture, (b)the interaction mode picture.[9]

the original coupling spectral density, respectively. In order to represent the case where the coupling function of the original spin-boson interaction is Lorentzian, the coupling function between the interaction mode and the newly introduced "reservoir" is characterized by a flat (white) one[17,7] where the interaction mode shows the Markovian nature. This indicates that the interaction mode shows an irreversible motion. Since the application of a π pulse causes time reversal to the two-level system, the effectiveness of the SPC depends on the degree of reversibility of the interaction mode at the pulse application times. For the case of Lorentzian coupling spectral density, the irreversibility of the motion of the interaction mode makes the SPC ineffective.

Besides, for the case of the non-Lorentzian coupling spectral density, too, we can also use the two-step structured reservoir. In this case, a non-white coupling function between the interaction mode and the "reservoir" represents the original spin-boson interaction, which implies that the time evolution of the interaction mode is non-Markovian. Since partial reversibility remains at the pulse application times in this case, the SPC is effective.

In this paper, we have shown an another kind of method to suppress the decoherence by paying attention to the dynamical motion of the reservoir. We hope that the synchronized pulse control might extend the possibility of the pulse control of decoherence.

Acknowledgements This study is supported by the Grant in Aid for Scientific Research from the Ministry of Education, Science, Sports and Culture of Japan.

References

1. L.Viola and S.Lloyd, Phys. Rev. A58, 2733 (1998).
2. M.Ban, J. Mod. Opt. 45, 2315 (1998).
3. Lu-Ming Duan and Guang-Can Guo, Phys. Lett. A 261, 139(1999).
4. L.Viola, E.Knill, and S.Lloyd, Phys. Rev. Lett. 82, 2417 (1999).
5. L.Viola, S.Lloyd and E.Knill, ibid. 83, 4888 (1999).
6. L.Viola and E.Knill, Phys. Rev. Lett. 90, 037901(2003); quant-ph/0208056.
7. Y.Toyozawa , *Optical Processes in Solids*, (Cambridge University Press,2003).
8. C.Uchiyama and M.Aihara, Phys. Rev. A66,032313(2002).
9. C.Uchiyama and M.Aihara, Phys. Rev. A68,052302(2003).
10. H.M.Lai, P.T.Leung, and K.Young, Phys. Rev. A37,1597(1988).
11. H.J.Kimble, Adv. At. Mol., Opt. Phys. S2, 203(1994).
12. P.Lambropoulos, G.M.Nikolopoulos, T.R.Nielsen and S.Bay, Rep.Prog.Phys.63,455(2000).
13. R.Lang, M.O.Scully, and W.E.Lamb,Jr., Phys. Rev. A7,1788(1973).
14. S.M.Barnett and P.M.Radmore, Opt. Commun. 68,364(1988).
15. B.J.Dalton, S.M.Barnett, and P.L.Knight, J. Mod. Opt. 46,1315(1999);ibid,46,1495(1999).
16. B.J.Dalton and P.L.Knight, J. Mod. Opt. 46,1817(1999);ibid,46,1839(1999)
17. B.J.Dalton, S.M.Barnett, and B.M.Garraway, Phys. Rev. A64,053813(2001).
18. Y.Toyozawa and M.Inoue, J. Phys. Soc. Jpn. 21, 1663(1966).

QUANTUM ENTANGLEMENT, PURIFICATION, AND LINEAR-OPTICS QUANTUM GATES WITH PHOTONIC QUBITS

PHILIP WALTHER AND ANTON ZEILINGER

Institut für Experimentalphysik, Universität Wien, Vienna, Austria
Institut für Quantenoptik und Quanteninformation, Österreichische Akademie der Wissenschaften, Vienna, Austria
E-mail: wspc@wspc.com

1. Entanglement

Strikingly, quantum information processing has its origins in the purely philosophically motivated questions concerning the nonlocality and completeness of quantum mechanics sparked by the work of Einstein, Podolsky and Rosen in 1935 [1]. In experiments using entanglement, the system of spin-$\frac{1}{2}$ particles is realized by the usage of single photons, whose properties are defined by their polarization. Considering the H/V bases, a logical $|0\rangle$ corresponds to a horizontally polarized photon $|H\rangle$, respectively a logical $|1\rangle$ corresponds to a vertically polarized photon $|V\rangle$. A single qubit can be written as a coherent superposition of the form $|\psi\rangle = \alpha|H\rangle + \beta|V\rangle$, where the the probabilities α^2 and β^2 sum up to $\alpha^2 + \beta^2 = 1$. For the two qubit case the four different maximally entangled Bell-states are defined as:

$$|\Phi^{\pm}\rangle_{12} = \frac{1}{\sqrt{2}}(|H\rangle_1|H\rangle_2 \pm |V\rangle_1|V\rangle_2)$$
$$|\Psi^{\pm}\rangle_{12} = \frac{1}{\sqrt{2}}(|H\rangle_1|V\rangle_2 \pm |V\rangle_1|H\rangle_2)$$

The Bell states have the unique feature that all information on polarization properties is completely contained in the (polarization-)correlations between the separate photons, while the individual particle does not have any polarization prior to measurement. In other words, all of the information is distributed among two particles, and none of the individual systems carries any information. This is the essence of entanglement. At the same time, these (polarization-)correlations are stronger than classically allowed since they violate bounds imposed by local realistic theories via the Bell-inequality [2] or they lead to a maximal contradiction between such theories and quantum mechanics as signified by the Greenberger-Horne-Zeilinger theorem [3,4]. Distributed entanglement thus allows to establish non-classical correlations between distant parties and can therefore be considered the quantum analogue to a classical communication channel, a quantum communication channel.

The most widely used source for polarization-entangled photons today utilizes the process of spontaneous parametric down-conversion in nonlinear optical crystals [5]. Occasionally the nonlinear interaction inside the crystal leads to the annihilation of a high frequency pump photon and the simultaneous creation of two lower frequency photons, signal and idler, which satisfy the phase matching condition:

$$\omega_p = \omega_s + \omega_i \qquad \text{and} \qquad \vec{k}_p = \vec{k}_s + \vec{k}_i$$

where ω is the frequency and $\vec{k}$ the wavevector of the pump p, signal s and idler i photon. A typical picture of the emerging radiation is shown in Figure 1.

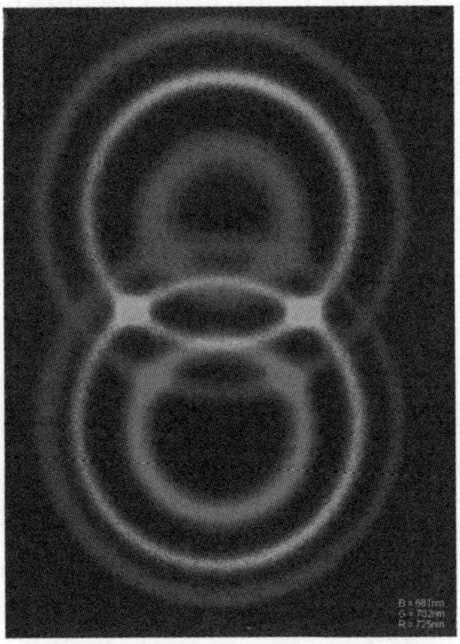

Figure 1. Photograph of the light emitted in type-II parametric down-conversion (false colours). The polarization-entangled photons emerge along the directions of the intersection between the white rings and are selected by placing small holes there

The possibility to establish such quantum communication channels over large distances offers the fascinating perspective to eventually take advantage of these novel communication capabilities in networks of increasing size. Naturally, non-trivial problems emerge in scenarios involving long distances or multiple parties. Experiments based on present fiber technology have demonstrated that entangled photon pairs can be separated by distances ranging from several hundreds of meters up to about 10 km [6,7,8], but no improvements by orders of magnitude are to be expected. Optical free-space links could provide a solution to this problem since they allow in principle for much larger propagation distances of photons because of the low absorption of the atmosphere in certain wavelength ranges. Single optical free-space links have been studied and successfully implemented already for several years for their application in quantum cryptography based on faint classical laser pulses [9,?]. We have recently demonstrated a next crucial step, namely the distribution of quantum entanglement via two simultaneous optical free-space links in an outdoor environment [11]. Polarization-entangled photon pairs have been transmitted across the Danube River in the city of Vienna via optical free-space links to independent receivers separated by 600m and without a line of sight between them (see Figure 2). A Bell inequality between those receivers was violated by more than 4 standard deviations confirming the quality of the entanglement:

$$S = |E(\phi_A, \phi_B) - E(\phi_A, \hat{\phi}_B) + E(\hat{\phi}_A, \phi_B) + E(\hat{\phi}_A, \hat{\phi}_B) \leq 2$$

where S is the "Bell parameter" and E the two photon visibility when polarizers are set to ϕ or $\hat{\phi}$ at receiver A or B. In this experiment, the setup for the source generating the entangled photon pairs has been miniaturized to fit on a small optical breadboard and it could easily be operated completely independent from an ideal laboratory environment.

Obviously, terrestrial free-space links are limited to rather short distances because they suffer from possible obstruction of objects in the line of sight, from atmospheric attenuation and, eventually, from the Earth's curvature. To fully exploit the advantages of free-space links, it will eventually be necessary to use space and satellite technology. By transmitting and/or receiving either photons or entangled photon pairs to and/or from a satellite, entanglement can be distributed over truly large distances. This would allow quantum communication applications on a global scale. From a fundamental point of view, satellite-based distribution of quantum entanglement is also the first step towards exploiting quantum correlations on a scale larger by orders of magnitude than achievable in laboratory and even ground-based experimental environments. State of the art photon sources and detectors would already suffice to achieve a satellite-based quantum communication link over some thousands of kilometers [12,13,14].

2. Quantum Key Distribution using Polarization Entangled Photons

The appeal of quantum cryptography is that its security is based on the laws of nature. In contrast to existing classical schemes of Key Distribution, Quantum (Cryptographic) Key Distribution does not invoke the transport of the key, since it is created at the sender and receiver site immediately. Furthermore, the key is created from a completely random sequence, which is in general an extremely difficult task in classical schemes. Finally, eavesdropping is easily detected due to the fragile nature of the qubits invoked for the quantum key distribution. Those features show that quantum cryptog-

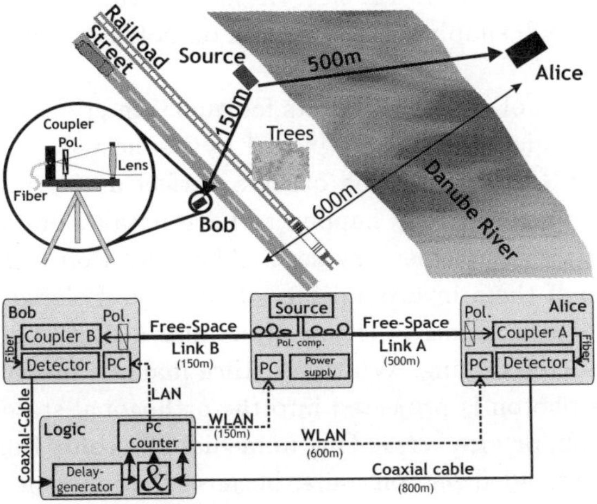

Figure 2. Free-space distribution of polarization-entangled photons [11]. The entangled-photon source was positioned on the bank of the Danube River. The two receivers, Alice and Bob, were located on rooftops and separated by approx. 600m, without a direct line of sight between each other. The inset shows the schematics of the telescopes consisting of a single-mode fibre coupler and a 5cm diameter lens. At the receiver telescopes, polarizers (Pol.) were attached to determine the polarization correlations and eventually violate a Bell-inequality. The lower figure shows a functional block diagram of the experiment. Detection signals from Alice were relayed to Bob using a long BNC cable. Singles and coincidence counting was performed locally at Bob and the results were shared between all three stations using LAN and Wave-LAN connections.

raphy is a superior technology which overcomes limitations and drawbacks of classical cryptographic schemes by utilizing the fascinating properties of quantum physics.

Cryptography (Quantum Key Distribution) allows two physically-seperated parties to create a random secret key without resorting to the services of a courier, and to verify that the key has not been intercepted. This is due to the fact that any measurements of incompatible quantities on a quantum system will inevitably modify the state of this system. This means that an eavesdropper (Eve) might get information out of a quantum channel by performing measurements, but the legitimate users will detect her and hence not use the key.

To ensure privacy of the key in advance, Alice and Bob do not use the quantum channel to transmit information, but only to establish a random sequence of bits, i.e.a key. The security of the key is determined by estimating the error rate after transmission and measuring the qubits. Quantum physics guarantees that any eavesdropping of the quantum channel will necessarily lead to errors in the key. If the key turns out to be insecure, then Alice and Bob simply discard it, and do not use it for encoding their message.

The utilization of entangled qubits for quantum cryptography has been proposed independently by Ekert [15] and by Bennett et al. [16]. As is indicated in Figure 3, Alice and Bob observe perfect anticorrelations of their measurements whenever they happen to have parallel oriented polarizers, leading to bitwise complementary keys. Alice and Bob will obtain identical keys if one of them inverts all bits of the key. Polarization entangled photon pairs offer a means to effectively realize a single photon situation, necessary for secret-sharing. Whenever Alice makes a measurement on her photon, Bob's photon is projected into the orthogonal state which is then analyzed by Bob, or vice versa. One immediately profits from the peculiar properties of entangled photon pairs, because the inherent randomness of quantum mechanical observations renders any analysis of the randomness of the keys or the encoded messages void.

After collecting the keys, Alice and Bob authenticate their keys by openly comparing (via classical communication) a small subset of their keys and evaluating the bit error rate. One advantage of using entangled photons is that the individual results of the measurements on entangled photons are purely random and therefore the randomness of the final key is ascertained . A further advantage is that the entangled photons represent a conditional single photon source, and the probability of having two photon pairs within the coincidence window can be very low.

Most recently an entangled state quantum cryptography prototype system in a typical application scenario was presented in Vienna [17]. It was possible to distribute secure quantum keys on demand between the headquarters of an Austrian bank and the Vienna City Hall using polarization-entangled photon pairs. The produced key was directly handed over to an application that was used to send a quantum secured online wire transfer from the City Hall to the headquarters of the bank.

The quantum cryptography system used (see Figure 4) consists of the source for polarization-entangled photons located at the bank, two combined polarization analysis and detection modules and two electronic units for key generation. These two quantum cryptography units which handled the five steps of secure key generation - real-time acquisition, key sifting , error estimation, error correction and privacy amplification - are based on an embedded electronic design and are compatible with classical telecommunication equipment. The quantum channel between Alice and Bob consists of an optical fiber that has been installed between the two experimental sites in the Vienna sewage system. The exposure of the fibers to realistic environmental conditions such as stress and strain during installation, as well as temperature changes were an important feature of this experiment, as it shows that our system not only works under laboratory conditions, but also in a realistic quantum cryptography scenario.

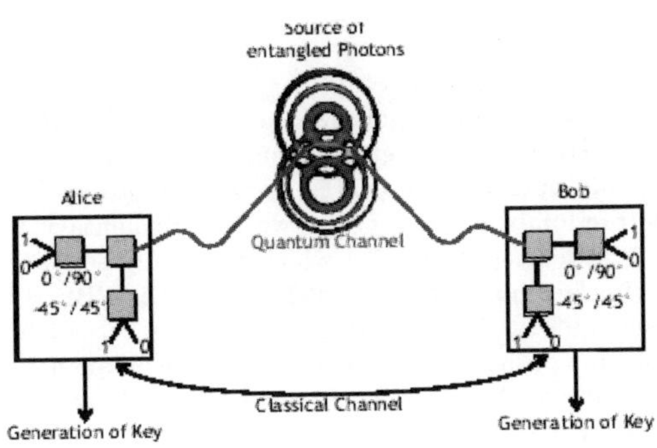

Figure 3. Quantum Cryptography using entangled photons

3. Purification of Entanglement

Owing to unavoidable decoherence in the quantum communication channel, the quality of entangled states generally decreases with the channel length. Entanglement purification is a way to extract a subset of states of high entanglement and high purity from a larger set of less entangled states - and is thus needed to overcome the decoherence of noisy quantum channels. We were able for the first time to experimentally demonstrate a general quantum purification scheme for mixed polarization-entangled two-particle states [18]. The crucial operation for a successful purification step is a bilateral conditional NOT (CNOT) gate, which effectively detects single bit-flip errors in the channel by performing local CNOT operations at Alice's and Bob's side between particles of shared entangled states. The outcome of these measurements can be used to correct for such errors and eventually end up in a less noisy quantum channel [19]. For the case of polarization entanglement, such a "parity-check" on the correlations can be performed in a straight forward way by using polarizing beamsplitters (PBS) [20] that transmit horizontally polarized photons and reflect vertically polarized ones, as seen in Figure 5.

Consider the situation in which Alice and Bob have established a noisy

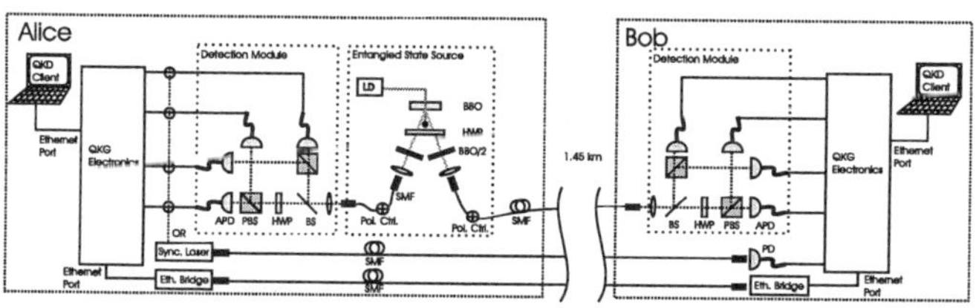

Figure 4. Sketch of the experimental setup. An entangled state source pumped by a violet laser diode at 405 nm produces polarization entangled photon pairs. One of the photons is locally analyzed in Alice's detection module, while the other is sent over a 1.45 km long single-mode optical fiber (SMF) to the remote site (Bob). Polarization measurement is done randomly in one of the two complementary bases ($|H\rangle/|V\rangle$ and $|45\rangle/|-45\rangle$), by using a beamsplitter (BS) which randomly sends incident photons to one of the two polarizing beamplsplitters (PBS). One of the PBS is defined to measure in the $|H\rangle/|V\rangle$ basesm the other in the $|45\rangle/|-45\rangle$ bases turned by a half-wave plate (HWP). The final detection of the photons is done in passively quenched silicon avalanche photodiodes (APD). When a photon is detected in one of Alice's four photodiodes an optical trigger pulse is created (Sync. Laser) and sent over a second fiber to establish a common time bases. at both sides, the trigger pulses and the detection events from the APDs are fed into an dedicated quantum key generator (QKG) device for further processing. This QKG electronic device is an embedded system, which is capable of autonomously doing all necessary calculations for key generation.

quantum channel, i.e. they share a set of equally mixed, entangled states ρ_{AB}. At both sides the two particles of two shared pairs are directed into the input ports a_1, a_2 and b_1, b_2 of a PBS (see Figure 6). Only if the entangled input states have the same correlations, i.e. they have the same parity with respect to their polarization correlations, the four photons will exit in four different outputs (four-mode case) and a projection of one of the photons at each side will result in a shared two-photon state with a higher degree of entanglement. All single bit-flip errors are effectively suppressed.

For example, they might start off with the mixed state

$$\rho_{AB} = F \cdot |\Phi^+\rangle\langle\Phi^+|_{AB} + (1-F) \cdot |\Psi^-\rangle\langle\Psi^-|_{AB}$$

where $|\Phi^+\rangle = (|HH\rangle + |VV\rangle)$ is another Bell state. Then, only the combinations $|\Phi^+\rangle_{a_1,a_2} \otimes |\Phi^+\rangle_{b_1,b_2}$ and $|\Psi^-\rangle_{a_1,a_2} \otimes |\Psi^-\rangle_{b_1,b_2}$ will lead to a four-mode case, while $|\Phi^+\rangle_{a_1,a_2} \otimes |\Psi^-\rangle_{b_1,b_2}$ and $|\Psi^-\rangle_{a_1,a_2} \otimes |\Phi^+\rangle_{b_1,b_2}$ will be rejected. Finally, a projection of the output modes a_4, b_4 into the basis $|\pm\rangle = \frac{1}{\sqrt{2}}(|H\rangle \pm |V\rangle)$ is needed to create the new mixed state

$$\rho'_{AB} = F' \cdot |\Phi^+\rangle\langle\Phi^+|_{AB} + (1-F') \cdot |\Psi^-\rangle\langle\Psi^-|_{AB}$$

with probability $F' = F^2/[F^2 + (1-F)^2]$ for the pure states $|\Phi^+\rangle_{a_3,b_3}$ and

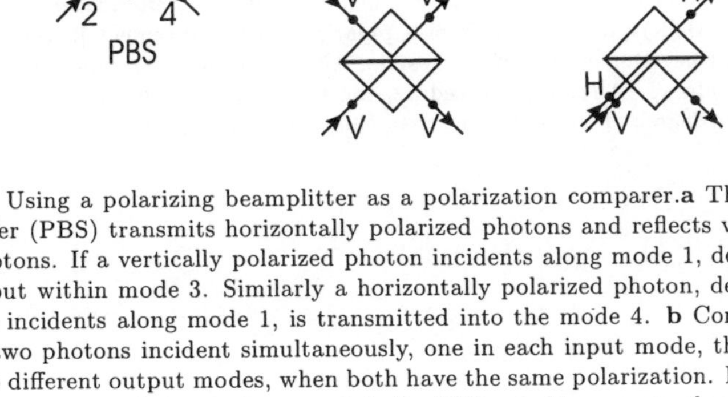

Figure 5. Using a polarizing beamplitter as a polarization comparer.**a** The polarizing beamsplitter (PBS) transmits horizontally polarized photons and reflects vertically polarized photons. If a vertically polarized photon incidents along mode 1, denoted by V, it will go out within mode 3. Similarly a horizontally polarized photon, denoted by H, which also incidents along mode 1, is transmitted into the mode 4. **b** Considering the case that two photons incident simultaneously, one in each input mode, then they will go out into different output modes, when both have the same polarization. For this case, where each output mode has to be occupied, the PBS acts like a party-checker **c** On the other hand, if the two incident photons have opposite polarization, then they will always go out along the same direction.

probability $1 - F'$ for $|\Psi^+\rangle_{a_3,b_3}$, respectively. The fraction F' of the desired state $|\Phi^+\rangle$ becomes larger for $F > \frac{1}{2}$. In other words, the new state ρ'_{AB} shared by Alice and Bob after the bilateral parity operation demonstrates an increased fidelity with respect to a pure, maximally entangled state. This is the purification of entanglement.

Typically, in the experiment, one photon pair of fidelity 92% could be obtained from two pairs, each of fidelity 75%. Also, although only bit-flip errors in the channel have been discussed, the scheme works for any general mixed state, since any phase-flip error can be transformed to a bit-flip by a rotation in a complementary basis.

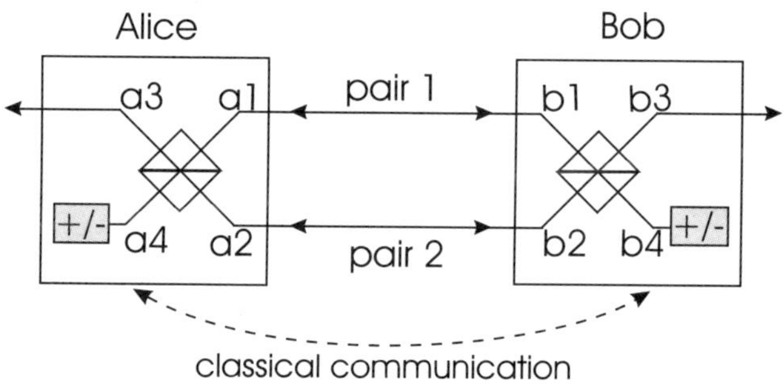

Figure 6. Scheme for entanglement purification of polarization-entangled qubits (from [18]). Two shared pairs of an ensemble of equally mixed, entangled states ρ_{AB} are fed in to the input ports of polarizing beamsplitters that substitute the bilateral CNOT operation necessary for a successful purification step. Alice and Bob keep only those cases where there is exactly one photon in each output mode. This can only happen if no bit-flip error occurs over the channel. Finally, to obtain a larger fraction of the desired pure (Bell-)state they perform a polarization measurement in the $|\pm\rangle$ basis in modes a4 and b4. Depending on the results, Alice performs a specific operation on the photon in mode a3. After this procedure, the remaining pair in modes a3 and b3 will have a higher degree of entanglement than the two original pairs.

References

1. A. Einstein, B. Podolsky, N. Rosen *Phys. Rev.* **75**, 777 (1935).
2. J. Bell *Physics* **1**, 195 (1964).
3. D. M. Greenberger, M. A. Horne, A. Zeilinger Bell's Theorem, Quantum Theory, and Conceptions of the Universe *Kluwer* (Dordrecht), (1989).
4. D. M. Greenberger, M. A. Horne, A. Shimony, A. Zeilinger *Am. J. Phys.* **58**, 1131 (1990).
5. P. G. Kwiat *et al. Phys.Rev.Lett.* **75**, 4337 (1995).
6. P. R. Tapster, J. G. Rarity, P. C. M. Owens *Phys. Rev. Lett.* **73**, 1923 (1994).
7. W. Tittel, J. Brendel, H. Zbinden, N. Gisin *Phys. Rev. Lett.* **81**, 3563 (1998).
8. G. Weihs, T. Jennewein, C. Simon, H. Weinfurter, A. Zeilinger *Phys. Rev. Lett.* **81**, 5039 (1998).
9. W. T. Buttler *et al. Phys. Rev. Lett.* **81**, 3283 (1998).
10. C. Kurtsiefer *et al. Nature* **419**, 450 (2002).
11. M. Aspelmeyer *et al. Science* **301**, 621 (2003).
12. M. Aspelmeyer *et al. European Space Agency (ESA)* 16358/02 (2003).
13. M. Aspelmeyer, T. Jennewein, M. Pfennigbauer, W. Leeb, A. Zeilinger *quant-ph/0305105* (2003).
14. R. Kaltenbaek *et al. quant-ph/0308174* (2003).
15. A. K. Ekert *Phys. Rev. Lett.* **67**, 661 (1991).
16. C. H. Bennett, G. Brassard, N. D. Mermin *Phys. Rev. Lett.* **68**, 557 (1992).
17. A. Poppe *et al. quant-ph/0404115 v2* (2004).
18. J.-W. Pan, S. Gasparoni, R. Ursin, G. Weihs, A. Zeilinger, *Nature* **423**, 417 (2003).
19. C. H. Bennett *et al. Phys. Rev. Lett.* **76**, 722 (1996).
20. J.-W. Pan, C. Simon, C. Brukner, A. Zeilinger *Nature* **410**, 1067 (2001).

ON QUANTUM MUTUAL TYPE MEASURES AND CAPACITY

NOBORU WATANABE
Department of Information Sciences
Tokyo University of Science
Yamazaki 2641, Noda City, Chiba 278-8510, Japan
E-mail: watanabe@is.noda.tus.ac.jp

The study of classical mutual entropy has been extensively done by several researchers like Kolmogorov and Gelfand. In quantum system, there are several definitions of the mutual entropy from classical inputs to quantum outputs, which are called semi-classical mutual entropy. In 1983, Ohya introduced a fully quantum mutual entropy by means of the relative entropy of Umegaki, and he extended it to general quantum systems by using the relative entropy of Araki and Uhlmann. It is showed that the Ohya mutual entropy contains the definitions of the semi-classical mutual entropy including the classical mutual entropy. Recently, Bennet, Nielsen, Shor et al. took the coherent information and so-called Lindblad - Nielsen's entropy for discuss a sort of coding theorem for communication processes. One can be concluded that Ohya mutual entropy is a most suitable one for studying the information transmission in quantum communication systems.

There exist several different types of quantum mutual entropy. The classical mutual entropy was introduced by Shannon to discuss the information transmission from an input system to an output system. It denotes an amount of information correctly transmitted from the input system to the output system through a channel. Kolmogorov, Gelfand and Yaglom gave a measure theoretic expression of the mutual entropy by means of the relative entropy defined by Kullback and Leibler. The Shannon's expression of the mutual entropy is generalized to one for finite dimensional quantum (matrix) case by Holevo,Ingarden,Levitin[9,8,12]. Ohya took the measure theoretic expression by Kolmogorov-Gelfand-Yaglom and defined Ohya mutual entropy [18] by means of quantum relative entropy of Umegaki [37,13] in 1983, he extended it [20] to general quantum systems by using the relative entropy of Araki [2] and Uhlmann [38]. Recently Shor [36] and Bennet, Nielsen et al [6,4] took the entropy exchange and defined the mutual type entropies, that is coherent information and so-called Lindblad-Nielsen's entropy, to discuss a sort of coding theorem for quantum communication systems.

In this paper, we briefly review quantum channels. We briefly explain three type of quantum mutual type entopies and compare with these mutual types entropies in order to obtain most suitable measure to discuss the information transmission for quantum communication processes. Finaly, we show some results [21,23,24,25,28,29,26] of quantum capacity defined by taking the supremum of Ohya mutual entropy according to the subsets of the input state space. The capacity means the ability of the information transmission of the channel, which is used as a measure for construction of channels.

1. Quantum Channels

In development of quantum information theory, the concept of channel has been played an important role. In particular, an attenuation channel introduced in [18] has been paid much attention in optical communication. A quantum channel is a map describing the state change from an initial system to a final system, mathematically. Let us consider the construction of the quantum channels.

Let $\mathcal{H}_1, \mathcal{H}_2$ be the complex separable Hilbert spaces of an input and an output systems, respectively, and let $\mathbf{B}(\mathcal{H}_k)$ be the set of all bounded linear operators on $\mathcal{H}_k$. $\mathfrak{S}(\mathcal{H}_k)$ is the set of all density operators on $\mathcal{H}_k$ $(k = 1, 2) : \mathfrak{S}(\mathcal{H}_k) \equiv \{\rho \in \mathbf{B}(\mathcal{H}_k); \rho \geq 0,\ \rho = \rho^*,\ tr\rho = 1\}$

(1) A map Λ^* from the input system to the output system is called a **(purely) quantum channel**.
(2) The quantum channel Λ^* satisfying the affine property (i.e., $\sum_k \lambda_k = 1\ (\forall \lambda_k \geq 0)\ \Rightarrow\ \Lambda^*\left(\sum_k \lambda_k \rho_k\right) = \sum_k \lambda_k\ \Lambda^*(\rho_k),\ \forall \rho_k \in \mathfrak{S}(\mathcal{H}_1)$) is called a **linear channel**.

A map Λ from $\mathbf{B}(\mathcal{H}_2)$ to $\mathbf{B}(\mathcal{H}_1)$ is called the dual map of $\Lambda^* : \mathfrak{S}(\mathcal{H}_1) \to \mathfrak{S}(\mathcal{H}_2)$ if Λ satisfies

$$tr\rho\Lambda(A) = tr\Lambda^*(\rho)A \tag{1}$$

for any $\rho \in \mathfrak{S}(\mathcal{H}_1)$ and any $A \in \mathbf{B}(\mathcal{H}_2)$.

(3) Λ^* from $\mathfrak{S}(\mathcal{H}_1)$ to $\mathfrak{S}(\mathcal{H}_2)$ is called a **completely positive (CP) channel** if its dual map Λ satisfies

$$\sum_{j,k=1}^{n} B_j^* \Lambda\left(A_j^* A_k\right) B_k \geq 0 \tag{2}$$

for any $n \in \mathbf{N}$, any $B_j \in \mathbf{B}(\mathcal{H}_1)$ and any $A_k \in \mathbf{B}(\mathcal{H}_2)$.

One can consider the quantum channel for more general systems. The input system denoted by $(\mathcal{A}, \mathfrak{S}(\mathcal{A}))$ and an output system by $(\overline{\mathcal{A}}, \overline{\mathfrak{S}}(\overline{\mathcal{A}}))$, where $\mathcal{A}$ (resp. $\overline{\mathcal{A}}$) is a C^*-algebra, $\mathfrak{S}(\mathcal{A})$ (resp. $\overline{\mathfrak{S}}(\overline{\mathcal{A}})$) is the set of all states on $\mathcal{A}$ (resp. $\overline{\mathcal{A}}$).

When some outside effects should be considered in a certain physical process such as noise and an effect of reservoir, it is convenient to extend the system $\mathcal{A}$ to $\mathcal{A} \otimes \mathcal{B}$, where $\mathcal{B}$ describes the outside system. In such cases, the concept of lifting introduced in [1] is useful.

(4) A **lifting** from $\mathcal{A}$ to $\mathcal{A} \otimes \mathcal{B}$ is a continuous map

$$\mathcal{E}^* : \mathfrak{S}(\mathcal{A}) \to \mathfrak{S}(\mathcal{A} \otimes \mathcal{B}). \tag{3}$$

(5) A lifting $\mathcal{E}^*$ is **linear** if it is affine.

(6) A lifting $\mathcal{E}^*$ is **nondemolition for a state** $\varphi \in \mathfrak{S}(\mathcal{A})$ if $\mathcal{E}^*\varphi(A \otimes I) = \varphi(A)$ for any $A \in \mathcal{A}$.

This compound state (lifting) can be extensively used in the sequel sections.

The concept of lifting came from the above cmpound state (nonlinear lifting) [18] and the dual of a transition expectation (linear lifting) [1], hence it is a natural generalization of these concepts.

Let $(\Omega, \mathfrak{F}_\Omega, P(\Omega))$, $\left(\tilde{\Omega}, \mathfrak{F}_{\tilde{\Omega}}, P\left(\tilde{\Omega}\right)\right)$ be input and output probability spaces, where $\mathfrak{F}_\Omega$ (resp. $\mathfrak{F}_{\tilde{\Omega}}$) is a σ-algebra of Ω (resp. $\tilde{\Omega}$) and $P(\Omega)$, $P\left(\tilde{\Omega}\right)$ are the sets of regular probability measures on Ω and $\tilde{\Omega}$.

A channel Ξ^* transmitted from a probability measure to a quantum state is called a classical-quantum (CQ) channel, and a channel $\tilde{\Xi}^*$ from a quantum state to a probability measure is called a quantum-classical (QC) channel. The capacity of both CQ and QC channels have been discussed in several papers [9,21,26]. A channel from the classical input system to the classical output system through CQ, Q and QC channels is now denoted by

$$P(\Omega) \xrightarrow{\Xi^*} \mathfrak{S}(\mathcal{H}) \xrightarrow{\Lambda^*} \mathfrak{S}\left(\tilde{\mathcal{H}}\right) \xrightarrow{\tilde{\Xi}^*} P\left(\tilde{\Omega}\right)$$

One of the examples of the CQ channel Ξ^* is given in [29] as follows:

(1) **CQ channel:** Let $C(\Omega)$ be the set of all continuous functions on Ω. For each $\omega \in \Omega$, we assume that $\rho_\omega \in \mathfrak{S}(\mathcal{H})$ is $\mathfrak{F}_\Omega$ - measurable.

Then the CQ channel Ξ^* is denoted by

$$tr\Xi^*(\mu)(\cdot) = \int_\Omega (tr\rho_\omega(\cdot))\,\mu(d\omega) \tag{4}$$

for any $\mu \in P(\Omega)$. Namely, one of the QC channel is given in [29].

(2) **QC channel:** Let $\{\Pi_n\}$ be the set of all non negative Hermite operators with $\sum_n \Pi_n = I$, which is called a positive operator valued measure (POV). The QC channel $\tilde{\Xi}^*$ given by the measurement process using $\{\Pi_n\}$ is obtained by

$$\tilde{\Xi}^*(\sigma) = \sum_n \Pi_n^{\frac{1}{2}} \sigma \Pi_n^{\frac{1}{2}} \tag{5}$$

for any $\sigma \in \mathfrak{S}\left(\tilde{\mathcal{H}}\right)$.

1.1. *Noisy optical channel*

To discuss the communication system using the laser signal mathematically, it is necessary to formulate the quantum communication theory being able to treat the quantum effects of signals and channels. In order to discuss influences of noise and loss in communication processes, one needs the following two systems [18]. Let $\mathcal{K}_1$, $\mathcal{K}_2$ be the separable Hilbert spaces for the noise and the loss systems, respectively.

Quantum communication process is described by the following scheme [18].

The quantum channel Λ^* is given by the composition of three CP channels a^*, π^*, γ^* such as

$$\Lambda^* \equiv a^* \circ \pi^* \circ \gamma^*. \tag{6}$$

a^* is a CP channel from $\mathfrak{S}(\mathcal{H}_2 \otimes \mathcal{K}_2)$ to $\mathfrak{S}(\mathcal{H}_2)$ defined by

$$a^*(\sigma) \equiv tr_{\mathcal{K}_2}\sigma \tag{7}$$

for any $\sigma \in \mathfrak{S}(\mathcal{H}_2 \otimes \mathcal{K}_2)$, where $tr_{\mathcal{K}_2}$ is a partial trace with respect to $\mathcal{K}_2$. π^* is the CP channel from $\mathfrak{S}(\mathcal{H}_1 \otimes \mathcal{K}_1)$ to $\mathfrak{S}(\mathcal{H}_2 \otimes \mathcal{K}_2)$ depending on the physical property of the device. γ^* is the CP channel from $\mathfrak{S}(\mathcal{H}_2)$ to $\mathfrak{S}(\mathcal{H}_1 \otimes \mathcal{K}_1)$ with a certain noise state $\xi \in \mathfrak{S}(\mathcal{K}_2)$ defined by

$$\gamma^*(\rho) \equiv \rho \otimes \xi \tag{8}$$

for any $\rho \in \mathfrak{S}(\mathcal{H}_1)$. The quantum channel Λ^* with the noise ξ is written by

$$\Lambda^*(\rho) = tr_{\mathcal{K}_2}\pi^*(\rho \otimes \xi) \tag{9}$$

for any $\rho \in \mathfrak{S}(\mathcal{H}_1)$.

Here we briefly review noisy optical channel [27]. A channel Λ^* is called a **noisy optical channel** if π^* and ξ above are given by

$$\xi \equiv |m\rangle\langle m| \text{ and } \pi^*(\cdot) \equiv V(\cdot)V^*, \tag{10}$$

where $|m\rangle\langle m|$ is m photon number state in $\mathcal{H}_1$ and V is a linear mapping from $\mathcal{H}_1 \otimes \mathcal{K}_1$ to $\mathcal{H}_2 \otimes \mathcal{K}_2$ given by

$$V(|n\rangle \otimes |m\rangle) \equiv \sum_{j=0}^{n+m} C_j^{n,m} |j\rangle \otimes |n+m-j\rangle, \tag{11}$$

$$C_j^{n,m} \equiv \sum_{r=L}^{K} (-1)^{n-r} \frac{\sqrt{n!m!j!(n+m-j)!}}{r!(n-r)!(j-r)!(m-j+r)!} \alpha^{m-j+2r} \left(-\bar{\beta}\right)^{n+j-2r} \tag{12}$$

for any $|n\rangle$ in $\mathcal{H}_1$ and $K \equiv \min\{j,n\}$, $L \equiv \max\{j-m,0\}$, where α and β are complex numbers satisfying $|\alpha|^2 + |\beta|^2 = 1$, and $\eta = |\alpha|^2$ is the transmission rate of the channel. In particular, $\rho \otimes \xi$ is given by the tensor products of two coherent states $|\theta\rangle\langle\theta| \otimes |\kappa\rangle\langle\kappa|$, then $\pi^*(\rho \otimes \xi)$ is obtained by

$$\pi^*(\rho \otimes \xi) = |\alpha\theta + \beta\kappa\rangle\langle\alpha\theta + \beta\kappa| \otimes \left|-\bar{\beta}\theta + \bar{\alpha}\kappa\right\rangle\left\langle-\bar{\beta}\theta + \bar{\alpha}\kappa\right|.$$

π^* is called a **generalized beam splittings**. It means that one beam comes and two beams appear after passing through π^*. Here we remark that the attenuation channel Λ_0^* [18] is derived from the noisy optical channel with $m = 0$. That is, the attenuation channel Λ_0^* was formulated in[18] on 1983 such as

$$\Lambda_0^*(\rho) = tr_{\mathcal{K}_2}\pi_0^*(\rho \otimes \xi_0) \tag{13}$$

$$= tr_{\mathcal{K}_2} V_0 (\rho \otimes |0\rangle\langle 0|) V_0^*, \tag{14}$$

where $|0\rangle\langle 0|$ is the vacuum state in $\mathfrak{S}(\mathcal{K}_1)$, V_0 is the mapping from $\mathcal{H}_1 \otimes \mathcal{K}_1$ to $\mathcal{H}_2 \otimes \mathcal{K}_2$ given by

$$V_0(|n_1\rangle \otimes |0\rangle) = \sum_j^{n_1} C_j^{n_1} |j\rangle \otimes |n_1 - j\rangle, \tag{15}$$

$$C_j^{n_1} = \sqrt{\frac{n_1!}{j!(n_1-j)!}\eta^j(1-\eta)^{n_1-j}}. \tag{16}$$

This attenuation channel is most important channel for discussing the optical communication processes. After that, Accardi and Ohya [1] reformulated it by using liftings, which is the dual map of the transition expectation by mean of Accardi.

$$\Lambda_0^*(|\theta\rangle\langle\theta|) = tr_{\mathcal{K}_2}\mathcal{E}_0^*(|\theta\rangle\langle\theta|),$$
$$\mathcal{E}_0^*(|\theta\rangle\langle\theta|) \equiv |\alpha\theta\rangle\langle\alpha\theta| \otimes |-\bar{\beta}\theta\rangle\langle-\bar{\beta}\theta|.$$

It contains the concept of beam splittings, which is extended by Fichtner, Freudenberg and Libsher [7] concerning the mappings on generalized Fock spaces. The generalized beam splittings on symmetric Fock space was formulated by using the compound Hida-Malliavin derivative and the compound Skorohod integral.

2. Ohya Mutual Entropy

The quantum entropy is denoted by

$$S(\rho) \equiv -tr\rho\log\rho$$

for any density operators ρ in $\mathfrak{S}(\mathcal{H}_1)$, which was introduced by von Neumann around 1932 [16]. It denotes the amount of information contained in the quantum state given by the density operator.

The mutual entropy was first discussed by Shannon to study the information transmission in classical systems and its fully general quantum version was formulated in [20]. The classical mutual entropy is determied by an input state and a channel, so that we denote the quantum mutual entropy with respect to the input state ρ and the quantum channel Λ^* by $I(\rho;\Lambda^*)$. This quantum mutual entropy $I(\rho;\Lambda^*)$ should satisfy the following three conditions:

(1) If the channel Λ^* is identity map, then the quantum mutual entropy equals to the von Neumann entropy of the input state, that is, $I(\rho; id) = S(\rho)$.
(2) If the system is classical, then the quantum mutual entropy equals to the classical mutual entropy.
(3) The following fundamental inequalities are satisfied:

$$0 \le I(\rho;\Lambda^*) \le S(\rho).$$

In order to define such a quantum mutual entropy, we need the quantum relative entropy and the joint state, which is called **Ohya compound state**, describing the correlation between an input state ρ and the output state $\Lambda^*\rho$ through a channel Λ^*. Ohya compound state σ_E (corresponding to joint state in CS) of ρ and $\Lambda^* \rho$ was introduced in [18,19], which is given by

$$\sigma_E = \sum_n \lambda_n E_n \otimes \Lambda^* E_n, \tag{17}$$

where E derives from a Schatten decomposition [32] (i.e., one dimension orthogonal decomposition of ρ)$\{\rho = \sum_n \lambda_n E_n\}$ of ρ. **Ohya mutual entropy** with respect to the input state ρ and the quantum channel Λ^* was defined in [18] such as

$$I(\rho; \Lambda^*) \equiv \sup_E S(\sigma_E, \sigma_0), \tag{18}$$

where $\sigma_0 = \rho \otimes \Lambda^* \rho$ and $S(\sigma_E, \sigma_0)$ is the quantum relative entropy defined by Umegaki [37],[13] as follows:

$$S(\rho, \sigma) \equiv \begin{cases} tr\rho(\log \rho - \log \sigma) & (\text{when } \overline{ran\rho} \subset \overline{ran\sigma}) \\ \infty & (\text{otherwise}) \end{cases}$$

There were several trials to extend the relative entropy to more general quantum systems and apply it to some other fields [2,38,22]. This mutual entropy $I(\rho; \Lambda^*)$ satisfies all conditions (1)$\sim$(2) mentioned above, and it satisfies also condition (3) the Shannon's type inequality as follows: $0 \leq I(\rho, \Lambda^*) \leq \min\{S(\rho), S(\Lambda^* \rho)\}$. It represents the amount of information correctly transmitted from the input quantum state ρ to the output quantum system through the quantum channel Λ^*. It is easily shown that we can take orthogonal decomposition instead of the Schatten-von Neumann decomposition [32]. Ohya mutual entropy is completely quantum, namely, they describe the information transmission from a quantum input to a quantum output. When the input system is classical, the state ρ is a probability distribution and the Schatten-von Neumann decomposition is unique with delta measures δ_n such that $\rho = \sum_n \lambda_n \delta_n$. In this case we need to code the classical state ρ by a quantum state, whose process is a quantum coding described by a channel Γ^* such that $\Gamma^* \delta_n = \sigma_n$ (quantum state) and $\sigma \equiv \Gamma^* \rho = \sum_n \lambda_n \sigma_n$. Then Ohya mutual entropy $I(\rho; \Lambda^*)$ becomes Holevo's one, that is,

$$I(\rho; \Lambda^* \circ \Gamma^*) = S(\Lambda^* \sigma) - \sum_n \lambda_n S(\Lambda^* \sigma_n)$$

when $\sum_n \lambda_n S(\Lambda^* \sigma_n)$ is finite.

Holevo proved the following inequality in 1973 [9].

Theorem 2.1. *When* $\mathcal{A} = \mathbb{C}^k$ *and* $\mathcal{B} = \mathbb{C}^m$ *of the above notation*

$$I_{cl} = \sum_{i,j} p_{ji} \log \frac{p_{ji}}{\lambda_i q_j} \leq S(\gamma^* \varphi) - \sum_k \lambda_i S(\gamma^* \varphi_k)$$

holds.

Holevo's upper bound can now be expressed by

$$S(\gamma^* \varphi) - \sum_i \lambda_i S(\gamma^* \varphi_i) = \sum_i \lambda_i S(\gamma^* \varphi_i, \gamma^* \varphi) . \tag{19}$$

The theorem bounds the performance of the detecting scheme. For general quantum case, we have the following inequality according to the lemma of [18].

Theorem 2.2. [23] *When the Schatten decomposition (i.e., one dimensional spectral decomposition)* $\varphi = \sum_i \lambda_i \varphi_i$ *is unique,*

$$I_{cl} \leq I = \sum_i \lambda_i S(\Lambda^* \varphi_i, \Lambda^* \varphi).$$

for any quantum channel Λ^*.

We see that in most cases the bound can not be achieved. Namely, the bound may be achieved in the only case when the output states $\Lambda^* \varphi_i$ *have commuting densities.*

Proposition 2.1. [23] *If the states* $\Lambda^* \varphi_i$, $1 \leq i \leq m$, *do not commute, then*

$$I_{cl} < S(\Lambda^* \varphi) - \sum_i \lambda_i S(\Lambda^* \varphi_i)$$

is a strict inequality.

3. Comparison of various quantum mutual type entropies

3.0.1. *Coherent information and Lindblad-Nielsen's entropy*

Let us discuss the entropy exchange [34,14]. For a state ρ, a quantum channel Λ^* is defined by an operator valued measure $\{A_j\}$ such as

$$\Lambda^* (\cdot) \equiv \sum_j A_j^* (\cdot) A_j$$

Then define a matrix $W = (W_{ij})_{i,j}$ with

$$W_{ij} \equiv trA_i^* \rho A_j,$$

by which the **entropy exchange** is defined by

$$S_e(\rho, \Lambda^*) = -trW \log W.$$

Using the above entropy exchange, two mutual type entropies are defined as below and they are applied to the study of quantum version of Shannon's coding theorem [33,10,11,6,4,31,36]. The first one is called the **coherent information** $I_C(\rho; \Lambda^*)$ [35] and the second one is called the **Lindblad-Nielsen's entropy** $I_L(\rho; \Lambda^*)$ [6], which are defined by

$$I_C(\rho; \Lambda^*) \equiv S(\Lambda^* \rho) - S_e(\rho, \Lambda^*),$$
$$I_L(\rho; \Lambda^*) \equiv S(\rho) + S(\Lambda^* \rho) - S_e(\rho, \Lambda^*).$$

By comparing these mutual entropies for quantum information communication processes, we have the following theorem [30]:

Theorem 3.1. *When* $\{A_j\}$ *is a projection valued measure and* $\dim A_j = 1$ *for arbitrary state* ρ *we have*
(1) $0 \le I(\rho, \Lambda^*) \le \min\{S(\rho), S(\Lambda^* \rho)\}$ ***(Ohya mutual entropy)****,*
(2) $I_C(\rho, \Lambda^*) = 0$ ***(coherent information)****,*
(3) $I_L(\rho, \Lambda^*) = S(\rho)$ ***(Lindblad-Nielsen's entropy)****.*

Theorem 3.2. *For the attenuation channel*

$$\Lambda_0^*(\rho) = tr_{\mathcal{K}_2} V_0^* (\rho \otimes |0\rangle \langle 0|) V_0^* = \sum_{n=0} A_n \rho A_n^*,$$

$$A_n = \left\{ \sum_j |z_j\rangle (\langle z_j| \otimes \langle 0|) \right\} V_0 \left\{ \sum_i (|z_i\rangle \otimes |0\rangle) \langle z_i| \right\},$$

one can obtain the following results for any input states $\rho = \sum_n \lambda_n E_n, \quad (E_n = |n\rangle \langle n|)$.
(1) $0 \le I(\rho, \Lambda^*) \le \min\{S(\rho), S(\Lambda^* \rho)\}$ ***(Ohya mutual entropy)****,*
(2) $I_C(\rho, \Lambda^*) = 0$ ***(coherent information)****,*
(3) $I_L(\rho, \Lambda^*) = S(\rho)$ ***(Lindblad-Nielsen's entropy)****.*

From this theorem, Ohya mutual entropy $I(\rho, \Lambda^*)$ only satisfies the inequality held in classical systems, so that only Ohya mutual entropy can be a candidate as quantum extension of the classical mutual entropy. Other two entropies can describe a sort of entanglement between input and output, such a correlation can be also described by quasi-mutual entropy, a slight generalization of $I(\rho, \Lambda^*)$, discussed in [20,5].

4. Quantum Capacity

The capacity of purely quantum channel was studied in [23,21,26,24,28,29].

Let $\mathcal{S}$ be the set of all input states satisfying some physical conditions. Let us consider the ability of information transmition for the quantum channel Λ^*. The answer of this question is the **capacity** of quantum channel Λ^* for a certain set $\mathcal{S} \subset \mathfrak{S}(\mathcal{H}_1)$ defined by

$$C_q^{\mathcal{S}}(\Lambda^*) \equiv \sup\{I(\rho;\Lambda^*); \rho \in \mathcal{S}\}. \tag{20}$$

When $\mathcal{S} = \mathfrak{S}(\mathcal{H}_1)$, the capacity of quantum channel Λ^* is denoted by $C_q(\Lambda^*)$. Then the following theorem for the attenuation channel was proved in [23].

Theorem 4.1. [23] *For a subset* $\mathcal{S}_n \equiv \{\rho \in \mathfrak{S}(\mathcal{H}_1); \dim s(\rho) = n\}$, *the capacity of the noisy optical channel* Λ^* *satisfies*

$$C_q^{\mathcal{S}_n}(\Lambda^*) = \log n,$$

where $s(\rho)$ *is the support projection of* ρ.

When the mean energy of the input state vectors $\{|\tau\theta_k\rangle\}$ can be taken infinite, i.e.,

$$\lim_{\tau\to\infty} |\tau\theta_k|^2 = \left|x^{(\infty)}\right|^2 = \infty$$

the above theorem tells that the quantum capacity for the noisy optical channel Λ^* with respect to $\mathcal{S}_n$ becomes $\log n$. It is a natural result, however it is impossible to take the mean energy of input state vector infinite. Therefore we have to compute the quantum capacity

$$C_q^{\mathcal{S}_e}(\Lambda^*) = \sup\{I(\rho;\Lambda^*); \rho \in \mathcal{S}_e\} \tag{21}$$

under some constraint $\mathcal{S}_e \equiv \{\rho \in \mathcal{S}; E(\rho) < e\}$ on the mean energy $E(\rho)$ of the input state ρ. In [20,23], we also considered the **pseudo-quantum capacity** $C_{pq}^{\mathcal{S}_e}(\Lambda^*)$ defined by

$$C_{pq}^{\mathcal{S}_e}(\Lambda^*) = \sup\{I_p(\rho;\Lambda^*); \rho \in \mathcal{S}_e\} \tag{22}$$

with the **pseudo-mutual entropy** $I_{pq}(\rho;\Lambda^*)$

$$I_{pq}(\rho;\Lambda^*) = \sup\left\{\sum_k \lambda_k S(\Lambda^*\rho_k, \Lambda^*\rho); \rho = \sum_k \lambda_k \rho_k, \text{finite decomposition}\right\}, \tag{23}$$

where the supremum is taken over all finite decompositions instead of all orthogonal pure decompositions for purely quantum mutual entropy. A **pseudo-quantum code** is a probability distribution on $\mathfrak{S}(\mathcal{H})$ with finite support in the set of product states. So $\{(\lambda_k),(\rho_k)\}$ is a pseudo-quantum code if (λ_k) is a probability vector and ρ_k are product states of $B(\mathcal{H})$. The quantum states ρ_k are sent over the quantum mechanical media, for example, optical fiber, and yield the output quantum states $\Lambda^*\rho_k$. The performance of coding and transmission is measured by the **pseudo-mutual entropy (information)**

$$I_{pq}((\lambda_k),(\rho_k);\Lambda^*) = I_{pq}(\rho;\Lambda^*) \tag{24}$$

with $\rho = \sum_k \lambda_k \rho_k$. Taking the supremum over certain classes of pseudo-quantum codes, we obtain various capacities of the channel. The supremum is over product states because we have mainly product (that is, memoryless) channels in our mind. Here we consider a subclass of pseudo-quantum codes. A **quantum code** is defined by the additional requirement that $\{\rho_k\}$ is a set of pairwise orthogonal pure states [18]. However the pseudo-mutual entropy is not well-matched to the conditions explained in Sec.3, and it is difficult to compute numerically [24]. From the monotonicity of the mutual entropy [22], we have

$$0 \le C_q^{\mathcal{S}_0}(\Lambda^*) \le C_{pq}^{\mathcal{S}_0}(\Lambda^*) \le \sup\{S(\rho);\rho\in\mathcal{S}_0\}.$$

Let us consider the quantum capacity for a general system [23] given by N-fold tensor products. Let $\mathcal{H}_j^{(N)}$, $B(\mathcal{H}_j^{(N)})$ and $\mathfrak{S}(\mathcal{H}_j^{(N)})$ be the N-fold tensor products of $\mathcal{H}_j$, $B(\mathcal{H}_j)$ and $\mathfrak{S}(\mathcal{H}_j)$ given by

$$\mathcal{H}_j^{(N)} = \bigotimes_{i=1}^{N}\mathcal{H}_j,\ B(\mathcal{H}_j^{(N)}) = \bigotimes_{i=1}^{N} B(\mathcal{H}_j),\ \mathfrak{S}(\mathcal{H}_j^{(N)}) = \bigotimes_{i=1}^{N}\mathfrak{S}(\mathcal{H}_j) \quad (j=1,2).$$

The quantum channel Λ_N^* from $\mathfrak{S}(\mathcal{H}_1^{(N)})$ to $\mathfrak{S}(\mathcal{H}_2^{(N)})$ defined by

$$\Lambda_N^* = \Lambda^*\otimes\cdots\otimes\Lambda^* \tag{25}$$

is called a memoryless channel, where Λ^* is the quantum channel from $\mathfrak{S}(\mathcal{H}_1)$ to $\mathfrak{S}(\mathcal{H}_2)$. The quantum mutual entropy with respect to the input state $\rho^{(N)}\in\mathfrak{S}(\mathcal{H}_1^{(N)})$ and the channel Λ_N^* is decribed by

$$I(\rho^{(N)};\ \Lambda_N^*) = \sup\left\{\sum_j p_j^{(N)} S(\Lambda_N^*\rho_j^{(N)},\Lambda_N^*\rho^{(N)});\rho^{(N)} = \sum_j p_j^{(N)}\rho_j^{(N)}\right\}, \tag{26}$$

where the supremum is taken over all von Neumann - Schatten decompositions $\sum_j p_j^{(N)} \rho_j^{(N)}$ of $\rho^{(N)}$. The pair $((p_j^{(N)}),\ (\rho_j^{(N)}))$ of $(p_j^{(N)})$ and $(\rho_j^{(N)})$ is called a quantum code of order N if $(p_j^{(N)})$ is a priori probability vector and $(\rho_j^{(N)})$ is the orthogonal pure states in $\mathfrak{S}(\mathcal{H}_1^{(N)})$. Moreover $((p_j^{(N)}),\ (\rho_j^{(N)}))$ is called a pseudo-quantum code of order N if $(p_j^{(N)})$ is a priori probability vector and $(\rho_j^{(N)})$ is pure states in $\mathfrak{S}(\mathcal{H}_1^{(N)})$. We denote the set of all quantum codes (resp. pseudo-quantum codes) of order N by $\mathcal{C}_q^{(N)}$ (resp. $\mathcal{C}_{pq}^{(N)}$). The quantum mutual entropy with respect to the quantum codes (resp. pseudo-quantum codes) and the quantum channel Λ_N^* is defined by

$$I(((p_j^{(N)}),(\rho_j^{(N)})),\Lambda_N^*) = I(\rho^{(N)};\ \Lambda_N^*) \tag{27}$$

where $\rho^{(N)} = \sum_j p_j^{(N)} \rho_j^{(N)}$. Based on the quantum mutual entropy, the quantum capacity and the pseudo quantum capacity for the quantum channel Λ_N^* are formulated as

$$C_q(\Lambda_N^*) = \sup\left\{I(((p_j^{(N)}),(\rho_j^{(N)})),\Lambda_N^*);\ ((p_j^{(N)}),(\rho_j^{(N)})) \in \mathcal{C}_q^{(N)}\right\}, \tag{28}$$

$$C_{pq}(\Lambda_N^*) = \sup\left\{I(((p_j^{(N)}),(\rho_j^{(N)})),\Lambda_N^*);\ ((p_j^{(N)}),(\rho_j^{(N)})) \in \mathcal{C}_{pq}^{(N)}\right\}. \tag{29}$$

From the above definitions, the inequality

$$C_q(\Lambda_N^*) \le C_{pq}(\Lambda_N^*) \tag{30}$$

is held. Moreover the mean quantum capacity and the mean pseudo-quantum capacity per single use are given by

$$C_q^{(\infty)}(\Lambda^*) = \lim_{N\to\infty} \frac{1}{N} C_q(\Lambda_N^*) \tag{31}$$

$$C_{pq}^{(\infty)}(\Lambda^*) = \lim_{N\to\infty} \frac{1}{N} C_{pq}(\Lambda_N^*) \tag{32}$$

In order to estimate the quantum mutual entropy , we introduce the concept of **divergence center**. Let $\{\omega_i : i \in I\}$ be a family of states and $R > 0$. We say that the state ω is a divergence center for $\{\omega_i :\ i \in I\}$ with radius $\le R$ if

$$S(\omega_i, \omega) \le R \qquad \text{for every } i \in I.$$

In the following discussion about the geometry of relative entropy (or divergence as it is called in information theory) the ideas of [3] can be recognized very well.

Lemma 4.1. [23] *Let* $((\lambda_k), (\rho_k))$ *be a quantum code for the channel* Λ^* *and* ω *a divergence center with radius* $\leq R$ *for* $\{\Lambda^* \rho_k\}$. *Then*

$$I_{pq}((\lambda_k), (\rho_k); \Lambda^*) \leq R.$$

Lemma 4.2. [23] *Let* ψ_0, ψ_1 *and* ω *be states of* $B(\mathcal{K})$ *such that the Hilbert space* $\mathcal{K}$ *is finite dimensional and set* $\psi_\lambda = (1-\lambda)\psi_0 + \lambda\psi_1$ $(0 \leq \lambda \leq 1)$. *If* $S(\psi_0, \omega)$, $S(\psi_1, \omega)$ *are finite and*

$$S(\psi_\lambda, \omega) \geq S(\psi_1, \omega) \quad (0 \leq \lambda \leq 1)$$

then

$$S(\psi_1, \omega) + S(\psi_0, \psi_1) \leq S(\psi_0, \omega).$$

Lemma 4.3. [23] *Let* $\{\omega_i : i \in I\}$ *be a finite set of states of* $B(\mathcal{K})$ *such that the Hilbert space* $\mathcal{K}$ *is finite dimensional. Then the exact divergence center is unique and it is in the convex hull on the states* ω_i.

Theorem 4.2. [23] *Let* $\Lambda^* : \mathfrak{S}(\mathcal{H}) \to \mathfrak{S}(\mathcal{K})$ *be a channel with finite dimensional* $\mathcal{K}$. *Then the capacity* $C_p(\Lambda^*)$ *is the divergence radius of the range of* Λ^*.

References

1. L. Accardi and M. Ohya, *Compond channnels, transition expectation and liftings*, Appl. Math, Optim. **39**, 33–59, 1999.
2. H. Araki, *Relative entropy for states of von Neumann algebras*, Publ. RIMS Kyoto Univ. **11**, 809–833, 1976.
3. I. Csiszar, *I - divergence geometry of probability distributions and minimization problems*, Ann. Probab. **3**, 146–158, 1975.
4. H. Barnum, M.A. Nielsen and B.W. Schumacher, *Information transmission through a noisy quantum channel*, Physical Review A, **57**, No.6, 4153-4175, 1998.
5. V.P. Belavkin and M. Ohya, *Quantum entanglements and entangled mutual entropy*, Proc. R. Soc. Lond. A. **458**, 209-231, 2002.
6. C.H. Bennett, P.W. Shor, J.A. Smolin and A.V. Thapliyalz, *Entanglement-Assisted Capacity of a Quantum Channel and the Reverse Shannon Theorem*, quant-ph/0106052.
7. K.H. Fichtner, W. Freudenberg and V. Liebscher, *Beam splittings and time evolutions of Boson systems*, Fakultät für Mathematik und Informatik, **Math/ Inf/96/ 39**, Jena, 105, 1996.
8. R.S.Ingarden, *Quantum information theory*, Rep. Math. Phys., **10**, 43-73, 1976.
9. A.S.Holevo, *Some estimates for the amount of information transmittable by a quantum communication channel (in Russian)*, Problemy Peredachi Informacii, **9**, 3-11, 1973.

10. R. Jozsa and B.W. Schumacher, *A new proof of the quantum noiseless coding theorem*, Journal of Modern Optics, **41**, 2343-2349, 1994.
11. C. King and A. Lesniewski, *Quantum source and quantum coding theorem*, Journal of Mathematical Physics, **39** (1), 88-101, 1998.
12. L.B.Levitin, *Physical information theory for 30 years: basic concepts and results*, Springer Lect. Note in Phys., **378**, 101-110, 1991.
13. G.Lindblad, *Entropy, information and quantum measurements*, Commun. Math. Phys., **33**, 111-119, 1973.
14. S. Lloyd, *Capacity of the noisy quantum channel*, Physical Review A, **55**, 1613, 1997.
15. N.Muraki, M.Ohya and D.Petz, *Note on entropy of general quantum systems*, Open Systems and Information Dynamics, **1**, No.1, 43–56, 1992.
16. von Neumann, *Die Mathematischen Grundlagen der Quantenmechanik*, Springer-Berlin, 1932.
17. M.Ohya, *Quantum ergodic channels in operator algebras*, J. Math. Anal. Appl. **84**, 318-327, 1981.
18. M. Ohya, *On compound state and mutual information in quantum information theory*, IEEE Trans. Information Theory, **29**, 770–777, 1983.
19. M.Ohya, *Note on quantum probability*, L. Nuovo Cimento, **38**, 402–406, 1983.
20. M. Ohya, *Some aspects of quantum information theory and their applications to irreversible processes*, Rep. Math. Phys. **27**, 19–47, 1989.
21. M. Ohya, *Fundamentals of quantum mutual entropy and capacity*, Open Systems and Information Dynamics, 6. No.1, 69-78, 1999.
22. M. Ohya and D. Petz, *Quantum Entropy and Its Use*, Springer, 1993.
23. M. Ohya, D. Petz and N. Watanabe, *On capacity of quantum channels*, Probability and Mathematical Statistics, **17**, 179-196, 1997.
24. M. Ohya, D. Petz and N. Watanabe, *Numerical computation of quantum capacity*, International J. Theor. Phys., **38** , No.1, 507-510, 1998.
25. M. Ohya, D. Petz and N. Watanabe, *Quantum capacity of a noisy optical channel*, TUS preprint.
26. M. Ohya, I. Volovich, *On quantum capacity and its bound,* Infinite Dimensional Analysis, Quantum Probability and Related Topics, **6**, No.2, 301-310, 2003.
27. M. Ohya and N. Watanabe, *Construction and analysis of a mathematical model in quantum communication processes*, Electronics and Communications in Japan, Part 1, **68**, No.2, 29-34, 1985.
28. M. Ohya and N. Watanabe, *Quantum capacity of noisy quantum channel*, Quantum Communication and Measurement, **3**, 213-220, 1997.
29. M. Ohya and N. Watanabe, *Foundation of Quantum Communication Theory (in Japanese)*, Makino Pub. Co., 1998.
30. M. Ohya and N. Watanabe, *Comparison of mutual entropy - type measures*, TUS preprint.
31. D. Petz and M Mosonyi, *Stationary quantum source coding*, Journal of Mathematical Physics, **42**, 4857-4864, 2001.
32. R. Schatten; *Norm Ideals of Completely Continuous Operators*, Springer-Verlag, 1970.

33. B.W. Schumacher, *Quantum coding*, Physical Review A, **51**, 2738-2747, 1995.
34. B.W. Schumacher, *Sending entanglement through noisy quantum channels*, Physical Review A, **54**, 2614, 1996.
35. B.W. Schumacher and M.A. Nielsen, *Quantum data processing and error correction*, Physical Review A, **54**, 2629, 1996.
36. P. Sho, *The quantum channel capacity and coherent information*, Lecture Notes, MSRI Workshop on Quantum Computation, 2002.
37. H. Umegaki, *Conditional expectations in an operator algebra IV (entropy and information)*, Kodai Math. Sem. Rep. **14**, 59–85, 1962.
38. A.Uhlmann, *Relative entropy and the Wigner-Yanase-Dyson-Lieb concavity in interpolation theory*, Commun. Math. Phys., **54**, 21–32, 1977.